Introduction to
# Tsallis Entropy Theory
# in Water Engineering

# Introduction to
# Tsallis Entropy Theory in Water Engineering

Vijay P. Singh

**CRC Press**
Taylor & Francis Group
Boca Raton London New York

CRC Press is an imprint of the
Taylor & Francis Group, an **informa** business

CRC Press
Taylor & Francis Group
6000 Broken Sound Parkway NW, Suite 300
Boca Raton, FL 33487-2742

First issued in paperback 2017

© 2016 by Taylor & Francis Group, LLC
CRC Press is an imprint of Taylor & Francis Group, an Informa business

No claim to original U.S. Government works

ISBN-13: 978-1-4987-3660-2 (hbk)
ISBN-13: 978-1-138-74794-4 (pbk)

**Visit the Taylor & Francis Web site at**
**http://www.taylorandfrancis.com**

**and the CRC Press Web site at**
**http://www.crcpress.com**

*Dedicated to*

*my wife, Anita,*
*daughter, Arti,*
*son, Vinay,*
*daughter-in-law, Sonali,*
*and*
*grandsons, Ronin and Kayden*

# Contents

# SECTION II   Hydraulic Engineering

# SECTION III    Hydrology

# SECTION IV   Water Resources Engineering

# Preface

In 1988, Tsallis began to study a new form of entropy, called the Tsallis entropy, and in subsequent years, he developed the whole theory, which can be rightly referred to as the Tsallis entropy theory. This theory has since been applied to a wide spectrum of areas in physics and chemistry, and new topics applying this entropy are emerging each year. In the area of water engineering, the past few years have witnessed a range of applications of the Tsallis entropy. The literature shows the theory has enormous potential.

Currently, there seems to be no book on the Tsallis entropy for water engineering readership. Therefore, there exists a need for a book that deals with basic concepts of the Tsallis entropy theory and applications of these concepts to a range of water engineering problems. This book is an attempt to cater to this need.

The subject matter of the book is divided into 14 chapters organized in 4 sections. Section I, comprising two chapters, deals with preliminaries. Chapter 1 discusses the Tsallis entropy theory for both discrete and continuous variables. It then goes on to discuss the properties of the Tsallis entropy, partial Tsallis entropy, and constrained Tsallis entropy. The chapter is concluded with a discussion of generalized entropies. Frequency analysis constitutes the subject matter of Chapter 2. Beginning with a discussion of the procedure for deriving probability distributions, it goes on to present maximum entropy–based distributions with regular moments as constraints, the use of *m*-expectation, and choosing expectation value.

Section II consists of six chapters dealing with some aspects of hydraulics. One-dimensional velocity distributions are discussed in Chapter 3, which presents velocity distributions based on different constraints or the specification of information. It also discusses the relation between mean velocity and maximum velocity, simplification of the velocity distribution, and estimation of mean velocity. Chapter 4 presents two-dimensional velocity distributions using the Chiu coordinate system and the generalized framework. It deals with different characteristics of the velocity distribution.

Chapter 5 discusses sediment concentration. Starting with a discussion of the methods for determining sediment concentration, it presents a step-by-step procedure for the derivation of entropy-based suspended sediment concentration and the characteristics of the derived distribution. Chapter 6 treats the subject of sediment discharge in three ways. First, it considers velocity as entropy based but not sediment concentration. The second considers sediment concentration as entropy-based but not entropy-based velocity. The third considers both velocity and sediment concentration as entropy-based. The sediment concentration in debris flow is presented in Chapter 7. It presents a step-by-step methodology for determining the debris flow concentration and concludes with the treatment of reparameterization and equilibrium debris flow concentration. Chapter 8 deals with the stage–discharge rating curve. It first discusses errors and randomness in rating curves and forms thereof. It then discusses the derivation of rating curves, reparameterization, relation between maximum discharge and drainage area, relation between mean discharge and drainage area, relation between entropy parameter and drainage area, and extension of rating curves.

Hydrology is the subject of Section III, which comprises four chapters. Chapter 9 discusses precipitation variability and deals with intensity entropy, apportionment entropy, entropy scaling, hydrological zoning, and the assessment of water resources availability. Infiltration is discussed in Chapter 10, which presents the derivation of six infiltration equations, including the equations of Horton, Kostiakov, Philip, Green and Ampt, Overton, and Holtan. Chapter 11 is on soil moisture. Providing a short introduction to soil moisture profiles and their estimation, it presents the derivation of soil moisture profiles for wetting, drying, and mixed phases and the variation of soil moisture in time. Chapter 12 deals with flow duration curves. Discussing first the use and construction of flow duration curves, it presents a step-by-step procedure for deriving flow duration curves, reparameterization, mean flow and ratio of mean to maximum flow, prediction of flow duration curves for ungagged sites, forecasting of flow duration curve, and variation of entropy with time scale.

The concluding Section IV is on water resources engineering; it contains two chapters. Eco-index constitutes the subject matter of Chapter 13, containing indicators of hydrologic alteration (IHA), probability distributions of IHA parameters, and computation of nonsatisfaction eco-level and eco-index. Chapter 14 discusses measures of redundancy for water distribution networks. Presenting the optimization of water distribution networks, it deals with reliability, the Tsallis entropy, redundancy measures, the development of redundancy measures under different conditions, and the relation between redundancy and reliability.

**Vijay P. Singh**
*College Station, Texas*


MATLAB® is a registered trademark of The MathWorks, Inc. For product information, please contact:

The MathWorks, Inc.
3 Apple Hill Drive
Natick, MA 01760-2098 USA
Tel: 508-647-7000
Fax: 508-647-7001
E-mail: info@mathworks.com
Web: www.mathworks.com

# Acknowledgments

The book draws from the works of tens of scientists and engineers that have been inspiring. I have tried to acknowledge these works specifically. Any omission on my part has been entirely inadvertent and I offer my apologies in advance.

I have had a number of graduate students and visiting scholars over the years who have helped me in myriad ways, and I am grateful for their help. I would particularly like to acknowledge Dr. Z. Hao from Beijing Normal University, China; Dr. Mrs. H. Cui and Dr. Clement Sohoulande from Texas A&M University; and Dr. Deepthi Rajasekhar from Stanford University. Without their support, this book would not have been completed.

Finally, I would like to take this opportunity to acknowledge the support of my brothers and sisters in India and my family here in the United States that they have given me over the years. My wife, Anita, daughter, Arti, son, Vinay, and daughter-in-law, Sonali are always there to lend me a helping hand. My grandsons, Ronin and Kayden, are my future: they make my life complete. Therefore, I dedicate this book to them, for without their support and affection, this book would not have come to fruition.

# Author

**Vijay P. Singh, PhD, DSc, PE, PH, Hon. DWRE**, is a university distinguished professor and holds the Caroline and William N. Lehrer Distinguished Chair in Water Engineering at Texas A&M University, College Station, Texas. He currently serves as editor-in-chief of Springer's *Water Science and Technology Library Book Series*, the *Journal of Groundwater Research*, and *De Gruyter Open Journal of Agriculture*. He is also an associate editor of more than 15 other journals. He has won more than 72 national and international awards—including the Chow Award, Arid Lands Hydraulic Engineering Award, Torrens Award, Normal Medal, and Lifetime Achievement Award of ASCE; Linsley Award and Founders' Award of American Institute of Hydrology (AIH); and three honorary doctorates—for his technical contributions and professional service. Professor Singh has been president and senior vice-president of AIH and president of the Louisiana Section of AIH. He is a distinguished member of American Society of Civil Engineers (ASCE) and an honorary member of American Water Resources Association (AWRA), and a fellow of Environmental and Water Resources Institute (EWRI), Institution of Engineers (IE), Indian Society of Agricultural Engineers (ISAE), Indian Water Resources Society (IWRS), and Indian Association of Soil and Water Conservationists (IASWC), as well as a member or fellow of 10 international science/engineering academies. He is a member of numerous committees of ASCE and AWRA and is currently serving as chair of Watershed Council of ASCE. He has extensively published in the areas of surface water hydrology, groundwater hydrology, hydraulic engineering, irrigation engineering, environmental engineering, water resources, and stochastic and mathematical modeling.

# Section I

## Preliminaries

# 1 Introduction to Tsallis Entropy Theory

The concept of entropy originated in thermodynamics and has a history of over a century and half dating back to Clausius in 1850. In 1870, Boltzmann developed a statistical definition of entropy and hence connected it to statistical mechanics. The concept of entropy was further advanced by Gibbs in thermodynamics and by von Neumann in quantum mechanics. Outside of the world of physics, it is Shannon who developed, in the late 1940s, the mathematical foundation of entropy and connected it to information. The informational entropy is now frequently called Shannon entropy or sometimes called Boltzmann–Gibbs–Shannon entropy. Kullback and Leibler (1951) developed the principle of minimum cross entropy (POMCE) and in the late 1950s Jaynes (1957a,b) developed the principle of maximum entropy (POME). Koutsoyiannis (2013, 2014) has given an excellent historical perspective on entropy. The Shannon entropy, POME, and POMCE constitute the entropy theory that has witnessed a wide spectrum of applications in virtually every field of science and engineering and social and economic sciences, and each year new applications continue to be reported (Singh, 2013, 2014, 2015). A review of entropy applications in hydrological and earth sciences is given in Singh (1997, 2010, 2011).

In 1988, Tsallis postulated a generalization of the Boltzmann–Gibbs–Shannon entropy, now popularly called the Tsallis entropy, and discussed its mathematical properties. The definition and properties of the Tsallis entropy constitute the Tsallis entropy theory. In physics, the Tsallis entropy has received tremendous attention (Tsallis, 2001). Recently, this entropy has been applied to a number of geophysical, hydrological, and hydraulic processes. Because of its interesting properties, it is expected to receive increasing attention in water engineering in the years ahead. This chapter introduces the Tsallis entropy and presents its properties that are of particular interest in environmental and water engineering.

## 1.1 DEFINITION OF TSALLIS ENTROPY

First, it is useful to define the Boltzmann–Gibbs–Shannon entropy (henceforth, simply Shannon entropy). For a discrete random variable $X = \{x_i, i = 1, 2, …, N\}$ that has a probability distribution $P = \{p_i, i = 1, 2, …, N\}$ [$p_i$ is the probability of $X = x_i$], the Shannon entropy $H_s$ can be defined as

$$H_s = -k \sum_{i=1}^{N} p_i \log p_i \tag{1.1}$$

where $k$ is a conventional positive constant and is often taken as unity and log is taken to the base of 2, $e$ or 10, and accordingly, the unit of entropy becomes bit, nat, or docit.

Scaling $p_i$ to $p_i^m$, where $m$ is any real number, Tsallis (1988) postulated

$$H_m = k \frac{1 - \sum_{i=1}^{N} p_i^m}{m-1} = \frac{k}{m-1} \sum_{i=1}^{N} \left[ p_i - p_i^m \right] \tag{1.2}$$

where
$H_m$ is the Tsallis entropy
$k$ is often taken as unity

For $m \to 1$, the Tsallis entropy reduces to the Shannon entropy. Quantity $m$ is often referred to as nonextensivity index or Tsallis entropy index or simply entropy index. Entropy index $m$ characterizes the degree of nonlinearity and is related to the microscopic dynamics of the system. The value of $m$ can be positive or negative. The Tsallis entropy is often referred to as nonextensive statistic, $m$-statistic, or Tsallis statistic. Tsallis (2002) noted that superextensivity, extensivity, and subextensivity occur when $m < 1$, $m = 1$, or $m > 1$, respectively. For $m \geq 0$, $m < 1$ corresponds to the rare events and $m > 1$ corresponds to frequent events (Tsallis, 1998; Niven, 2004) pointing to the stretching or compressing of the entropy curve to lower or higher maximum entropy positions.

From an informational perspective, the information gain from the occurrence of any event $i$ is a power function and can be expressed as

$$\Delta I_i = \frac{1}{m-1} \left( 1 - p_i^{m-1} \right), \quad \sum_{i=1}^{N} p_i = 1 \tag{1.3}$$

where
$\Delta I_i$ is the gain in information from an event $i$ that occurs with probability $p_i$
$m$ is any real number
$N$ is the number of events

Equation 1.3 is a generalization of the Shannon gain function describing the information from an event expressed in logarithmic terms. For $N$ events, the average or expected gain function is the weighted average of Equation 1.3

$$H_m = E[\Delta I_i] = \sum_{i=1}^{N} p_i \left[ \frac{1}{m-1} \left( 1 - p_i^{m-1} \right) \right] = \frac{1}{m-1} \sum_{i=1}^{N} p_i \left( 1 - p_i^{m-1} \right) \tag{1.4}$$

where $H_m$ is designated as the Tsallis entropy or $m$-entropy.

In a similar manner, the information gain for the Shannon entropy, $\Delta H_{si}$, can be written as

$$\Delta H_{si} = -\log p_i \tag{1.5}$$

Therefore,

$$H_s = \sum_{i=1}^{N} H_{si} = -\sum_{i=1}^{N} p_i \log p_i \qquad (1.6)$$

If random variable $X$ is nonnegative continuous with a probability density function (PDF), $f(x)$, then the Shannon entropy can be written as

$$H_s(X) = H_s(f) = -k \int_0^\infty f(x) \log f(x) dx \qquad (1.7)$$

Likewise, the Tsallis entropy can be expressed (Koutsoyiannis, 2005a,b,c) as

$$H_m(X) = H_m(f) = \frac{k}{m-1} \int_0^\infty \{f(x) - [f(x)]^m\} dx = \frac{k}{m-1} \left\{ 1 - \int_0^\infty [f(x)]^m \right\} dx \qquad (1.8)$$

Frequently, $k$ is taken as 1. From now onward, subscript $m$ will be deleted and $H_m$ will be simply denoted by $H$.

A plot of $H/k$ versus $p$ for $m = -1, -0.5, 0, 0.5, 1,$ and 2 is given in Figure 1.1. For $m < 0$, the Tsallis entropy is concave and for $m > 0$ it becomes convex. For $m = 0$, $H = k(N-1)$ for all $p_i$. For $m = 1$, it converges to the Shannon entropy. For all cases, the Tsallis entropy decreases as $m$ increases.

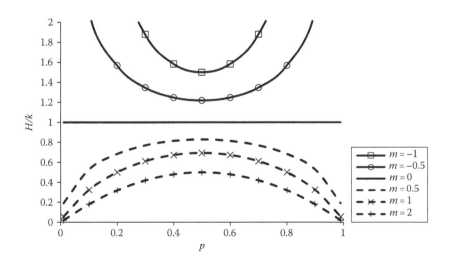

**FIGURE 1.1** Plot of $H/k$ for $N = 2$ for $m = -1, -0.5, 0, 0.5, 1,$ and 2.

### Example 1.1

Plot the gain function defined by the Tsallis entropy for different values of probability: 0.1, 0.2, 0.3, 0.4, 0.5, 0.6, 0.7, 0.8, 0.9, and 1.0. Take $k$ as 1, and $m$ as −1, 0, 1, and 2. What do you conclude from this plot?

### Solution

Using Equation 1.3 the gain function is computed, as shown in Table 1.1. Figure 1.2 shows the gain function for $m = -1$, 0, $m = 1$, and 2. It is seen from the figure that the gain in information decreases with the increase in the probability value regardless of the value of $m$. For increasing value of $m$, the gain diminishes for the same

**TABLE 1.1**

**Computation of Gain Function**

| | $\Delta I$ | | | |
|---|---|---|---|---|
| $p$ | $m = -1$ | $m = 0$ | $m = 1$ | $m = 2$ |
| 0.1 | 49.50 | 9.00 | 2.30 | 0.90 |
| 0.2 | 12.00 | 4.00 | 1.61 | 0.80 |
| 0.3 | 5.06 | 2.33 | 1.20 | 0.70 |
| 0.4 | 2.63 | 1.50 | 0.92 | 0.60 |
| 0.5 | 1.50 | 1.00 | 0.69 | 0.50 |
| 0.6 | 0.89 | 0.67 | 0.51 | 0.40 |
| 0.7 | 0.52 | 0.43 | 0.36 | 0.30 |
| 0.8 | 0.28 | 0.25 | 0.22 | 0.20 |
| 0.9 | 0.12 | 0.11 | 0.11 | 0.10 |
| 1 | 0.00 | 0.00 | 0.00 | 0.00 |

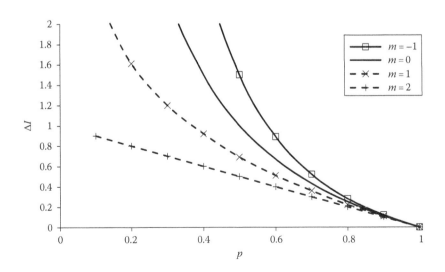

**FIGURE 1.2**    Gain function for $m = -1$, 0, 1, and 2.

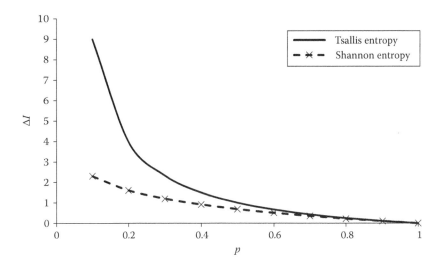

**FIGURE 1.3**   Comparison of the Shannon and Tsallis gain functions.

probability value. For $m = 1$, the Tsallis entropy converges to the Shannon entropy. The two gain functions are shown in Figure 1.3. The Tsallis gain function has a much longer tail showing very low values of gain as the probability increases.

### Example 1.2

Consider a two-state variable taking on values $x_1$ and $x_2$. Assume that $p(x_1) = 0.0$, 0.1, 0.2, 0.3, 0.4, 0.5, 0.6, 0.7, 0.8, 0.9, and 1.0. Note that $p(x_2) = 1 - p(x_1)$. Compute and plot the Tsallis entropy. Take $m$ as 1.5 and 2.0. What do you conclude from the plot?

**Solution**

The Tsallis entropy is given by Equation 1.2. Let $a = p(x_1)$. For any given value of $p$, one can write the Tsallis entropies $H_1$ and $H_2$, respectively, for $x_1$ and $x_2$ as

$$H_1 = \frac{k}{m-1} a(1 - a^{m-1})$$

$$H_2 = \frac{k}{m-1} (1-a)[1 - (1-a)^{m-1}]$$

Then, the Tsallis entropy is

$$H = H_1 + H_2$$

where each component is a weighted gain function. Thus, the Tsallis entropy is computed as shown in Table 1.2. The computed Tsallis entropy for $k = 1$ and $m = 1.5$ and 2 is shown in Figure 1.4. The Tsallis entropy plot shows a little skewness from the Shannon entropy and also predicts the maximum entropy at $p(x) = 0.5$. It also can be observed that the Tsallis entropy value decreases with an increase in the value of $m$.

**TABLE 1.2**

**Computation of the Tsallis Entropy for $k = 1$, $m = 1.5$, and $m = 2$**

| $p(x)$ | $1 - p(x)$ | $m = 1.5$ | | | $m = 2.0$ | | |
|---|---|---|---|---|---|---|---|
| | | $H_1$ | $H_2$ | $H = H_1 + H_2$ | $H_1$ | $H_2$ | $H = H_1 + H_2$ |
| 0 | 1 | 0 | 0 | 0 | 0 | 0 | 0 |
| 0.1 | 0.9 | 0.137 | 0.092 | 0.229 | 0.090 | 0.090 | 0.180 |
| 0.2 | 0.8 | 0.221 | 0.169 | 0.390 | 0.160 | 0.160 | 0.320 |
| 0.3 | 0.7 | 0.271 | 0.229 | 0.500 | 0.210 | 0.210 | 0.420 |
| 0.4 | 0.6 | 0.294 | 0.270 | 0.565 | 0.240 | 0.240 | 0.480 |
| 0.5 | 0.5 | 0.293 | 0.293 | 0.586 | 0.250 | 0.250 | 0.500 |
| 0.6 | 0.4 | 0.270 | 0.294 | 0.565 | 0.240 | 0.240 | 0.480 |
| 0.7 | 0.3 | 0.229 | 0.271 | 0.500 | 0.210 | 0.210 | 0.420 |
| 0.8 | 0.2 | 0.169 | 0.221 | 0.390 | 0.160 | 0.160 | 0.320 |
| 0.9 | 0.1 | 0.092 | 0.137 | 0.229 | 0.090 | 0.090 | 0.180 |
| 1 | 0 | 0 | 0 | 0 | 0 | 0 | 0 |

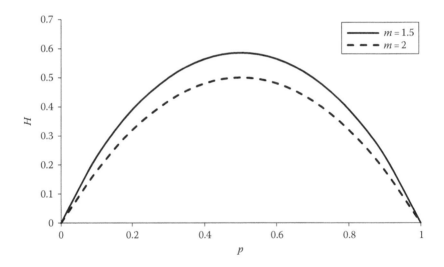

**FIGURE 1.4**    Tsallis entropy for $k = 1$ and $m = 1.5, 2.0$.

## 1.2 DERIVATION OF SHANNON ENTROPY FROM TSALLIS ENTROPY

It may be useful to show that the Tsallis entropy is a generalization of the Shannon entropy. One can express

$$p_i^m = p_i \exp[(m-1)\ln p_i] \tag{1.9}$$

The Tsallis entropy given by Equation 1.2 can be written as

$$H = \frac{k}{m-1}\left\{1 - \sum_{i=1}^{N} p_i \exp[(m-1)\ln p_i]\right\} \qquad (1.10)$$

It must now be shown that when $m$ tends to unity

$$H = k \lim_{m \to 1} \frac{1 - \sum_{i=1}^{N} p_i \exp[(m-1)\ln p_i]}{m-1} \qquad (1.11)$$

leads to the Shannon entropy given by Equation 1.1.

Now, consider L'Hospital's rule for the division of two arbitrary functions $f(a)$ and $g(a)$:

$$\lim_{a \to b} \frac{f(a)}{g(a)}, \quad \text{if } \lim_{a \to b} g(a) = 0 \text{ or } \infty, \quad \lim_{a \to b} g(a) = 0 \text{ or } \infty \qquad (1.12)$$

where $b$ is some value and may even approach infinity. For example,

$$\lim_{m \to 1} f(m) = \lim_{m \to 1}\left(1 - \sum_{i=1}^{N} p_i^m\right) = 1 - \sum_{i=1}^{N} p_i = 0 \qquad (1.13)$$

$$\lim_{m \to 1} g(m) = \lim_{m \to 1}(m-1) = 0 \qquad (1.14)$$

Now

$$f(m) = 1 - \sum_{i=1}^{N} p_i \exp[(m-1)\ln p_i] \qquad (1.15)$$

or

$$f'(m) = -\sum_{i=1}^{N} p_i \ln p_i \exp[(m-1)\ln p_i] = -\sum_{i=1}^{N} p_i^m \ln p_i \qquad (1.16)$$

$$g(m) = m - 1 \qquad (1.17)$$

$$g'(m) = 1 \qquad (1.18)$$

Therefore, taking the limit on Equation 1.11,

$$\lim_{m \to 1} H = \lim_{m \to 1} \frac{f(m)}{g(m)} = \lim_{m \to 1} \frac{f'(m)}{g'(m)} = \lim_{m \to 1} \sum_{i=1}^{N} (-1) p_i^m \ln p_i$$

$$= -\lim_{m \to 1} \sum_{i=1}^{N} p_i^m \ln p_i = -\sum_{i=1}^{N} p_i \ln p_i \qquad (1.19)$$

which is the Shannon entropy.

## 1.3  PROPERTIES OF TSALLIS ENTROPY

Following Tsallis (1988, 2004), some interesting and useful properties of the Tsallis entropy are briefly summarized here.

### 1.3.1  *m*-ENTROPY

Analogous to surprise or unexpectedness defined in the Shannon entropy, the $m$-surprise or $m$-unexpectedness is defined as $\log_m(1/p_i)$. Hence, the $m$-entropy can be defined as

$$H = E \left[ \log_m \frac{1}{p_i} \right] \qquad (1.20)$$

which coincides with the Tsallis entropy:

$$H = E \left[ \frac{1 - p_i^{m-1}}{m - 1} \right] \qquad (1.21)$$

in which $E$ is the expectation. Recalling the definition

$$\lim_{n \to 0} \left[ \frac{w^n - 1}{n} \right] = \log w \qquad (1.22)$$

where
   $n$ is any number
   $w$ is any variable

Then, Equation 1.21 is the same as Equation 1.20. For small values of $n$, $w^n$ will behave as $\log w$. A plot of function $(w^n - 1)/n$ is shown in Figure 1.5 that shows its approximation by the logarithmic function.

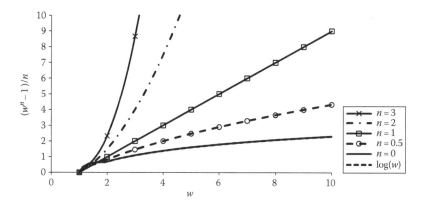

**FIGURE 1.5**   Plot of function $(w^n - 1)/n$ versus $w$ for various values of $n$.

### 1.3.2   MAXIMUM VALUE

Equation 1.2 attains an extreme value for all values of $m$ when all $p_i$ are equal, that is, $p_i = 1/N$. For $m > 0$, it attains a maximum value and for $m < 0$, it attains a minimum value. The extremum of $H$ becomes

$$H = k \frac{N^{m-1} - 1}{1 - m} \tag{1.23}$$

If $m = 1$, applying L'Hopsital's rule to Equation 1.23 or 1.22, one gets

$$H = k \ln N \tag{1.24}$$

which is the Boltzmann entropy, $H_B$. Plotting $H/k$ versus $N$ using Equation 1.23, as shown in Figure 1.6, it is seen that $H$ diverges for $m < 1$. The Tsallis entropy, given

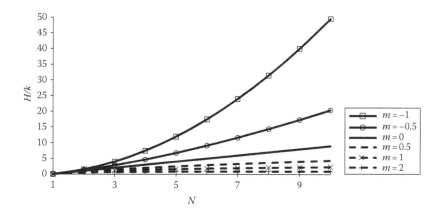

**FIGURE 1.6**   Plot of $H/k$ versus $N$ for $m = -1$, $-0.5$, $0$, $0.5$, $1$, $2$ when all $p_i$ are equal (from Equation 1.18).

by Equation 1.23, diverges if $m < 1$, is maximum for $m > 1$ and is minimum for $m < 1$, and is $k(N - 1)$ for all equal $p_i$. Interestingly, for any value of $m$, the entropy extreme can be expressed in terms of the entropy for $m = 1$ as follows. For $m = 1$, Equation 1.24 can be written as $N = \exp(H/k)$. Substituting it into Equation 1.23, the result is

$$\frac{H_m}{k} = \frac{\exp[(1-m)H_B/k]-1}{1-m} \qquad (1.25)$$

### 1.3.3  CONCAVITY

Consider two probability distributions $P = \{p_i, i = 1, 2, ..., N\}$ and $Q = \{q_i, i = 1, 2, ..., N\}$ corresponding to a unique set of $N$ possibilities. Then, an intermediate probability distribution $G = \{g_i, i = 1, 2, ..., N\}$ can be defined for a real $a$ such that $0 < a < 1$ as

$$g_i = ap_i + (1-a)q_i \qquad (1.26)$$

for all $i$. It can be shown that for $m > 0$,

$$H[G] \geq aH[P] + (1-a)H[Q] \qquad (1.27)$$

and for $m < 0$,

$$H[G] \leq aH[P] + (1-a)H[Q] \qquad (1.28)$$

Functional $H(G) \geq 0$ if $m > 0$ and is hence concave; $H(G) = 0$ if $m = 0$; and $H(G) \leq 0$ if $m < 0$ and is, therefore, convex. These inequalities, given by Equations 1.27 and 1.28, are true for $m \neq 0$ and $p_i = q_i, \forall i$.

**Example 1.3**

Consider $N = 3$, $m = 3$, and $P = \{0.2, 0.4, 0.4\}$ and $G = \{0.1, 0.3, 0.6\}$ and $a = 0.3$. Compute $H(P)$ and $H(G)$, and then show if Equation 1.27 holds. If $m = -0.5$, then show if Equation 1.28 holds.

**Solution**

$$H_m = E[\Delta l_i] = \sum_{i=1}^{N} p_i \left[ \frac{1}{m-1}\left(1-p_i^{m-1}\right)\right] = \frac{1}{m-1}\sum_{i=1}^{N} p_i \left[1-p_i^{m-1}\right]$$

Given $a = 0.3$, from Equation 1.26, $Q$ can be computed as

$$q_i = \frac{g_i - ap_i}{(1-a)}$$

$$q_1 = \frac{0.1 - 0.3 \times 0.2}{1 - 0.3} = 0.06$$

$$q_2 = \frac{0.3 - 0.3 \times 0.4}{1 - 0.3} = 0.26$$

$$q_3 = \frac{0.6 - 0.3 \times 0.4}{1 - 0.3} = 0.68$$

When $m = 3$,

$$H(P) = \frac{1}{3-1}[(0.2 - 0.2^3) + (0.4 - 0.4^3) + (0.4 - 0.4^3)] = 0.432$$

$$H(Q) = \frac{1}{3-1}[(0.06 - 0.06^3) + (0.26 - 0.26^3) + (0.68 - 0.68)] = 0.330$$

$$H(G) = \frac{1}{3-1}[(0.1 - 0.1^3) + (0.3 - 0.3^3) + (0.6 - 0.6^3)] = 0.378$$

$$H[G] \geq aH[P] + (1-a)H[Q] = 0.3 \times 0.432 + (1-0.3) \times 0.330 = 0.361$$

Equation 1.27 holds.

When $m = -0.5$

$$H(P) = \frac{1}{-0.5-1}[(0.2 - 0.2^{-0.5}) + (0.4 - 0.4^{-0.5}) + (0.4 - 0.4^{-0.5})] = 2.932$$

$$H(Q) = \frac{1}{-0.5-1}[(0.06 - 0.06^{-0.5}) + (0.26 - 0.26^{-0.5}) + (0.68 - 0.68^{-0.5})] = 4.242$$

$$H(G) = \frac{1}{-0.5-1}[(0.1 - 0.1^{-0.5}) + (0.3 - 0.3^{-0.5}) + (0.6 - 0.6^{-0.5})] = 3.519$$

$$H[G] \leq aH[P] + (1-a)H[Q] = 0.3 \times 2.932 + (1-0.3)4.242 = 3.849$$

Equation 1.28 holds.

### 1.3.4  ADDITIVITY

Let there be two independent systems $A$ and $B$ with ensembles of configurational possibilities $E^A = \{1, 2, \ldots, N\}$ with probability distribution $P^A = \{p_i^A, i = 1, 2, \ldots, N\}$ and configurational possibilities $E^B = \{1, 2, \ldots, M\}$ with probability distribution

$P^B = \{p_j^B, j = 1, 2, \ldots, M\}$. Then, one needs to deal with the union of two systems $A \cup B$ and their corresponding ensembles of possibilities $E^{A \cup B} = \{(1,1), (1,2), \ldots, (i, j), \ldots, (N, M)\}$. If $p_{ij}^{A \cup B}$ represents the corresponding probabilities then by virtue of independence the joint probability will be equal to the product of individual probabilities, that is $p_{ij}^{A \cup B} = p_i^A p_j^B$ or $p_{ij}(A + B) = p_i(A)p_j(B)$ for all $i$ and $j$. Hence,

$$\sum_{i,j}^{N,M} \left( p_{ij}^{A \cup B} \right)^m = \left[ \sum_{i=1}^{N} \left( p_i^A \right)^m \right]\left[ \sum_{j=1}^{M} \left( p_j^B \right)^m \right] \tag{1.29}$$

Taking the logarithms of Equation 1.29, one obtains

$$\log\left[ \sum_{i,j}^{N,M} \left( p_{ij}^{A \cup B} \right)^m \right] = \log\left[ \sum_{i=1}^{N} \left( p_i^A \right)^m \right] + \log\left[ \sum_{j=1}^{M} \left( p_j^B \right)^m \right] \tag{1.30}$$

Each term of Equation 1.30 is now considered. The left side of Equation 1.30 can be written in terms of the Tsallis entropy as

$$\log\left[ \sum_{i,j}^{N,M} \left( p_{ij}^{A \cup B} \right)^m \right] = \log\left\{ 1 - \frac{(m-1)\left[ 1 - \sum_{i=1,j=1}^{N,M} \left( p_{ij}^{A \cup B} \right)^m \right]}{(m-1)} \right\}$$

$$= \log[1 - (m-1)H^{A \cup B}] \tag{1.31}$$

Similarly, terms on the right side of Equation 1.31 can be written as

$$\log\left[ \sum_{i}^{N} \left( p_i^A \right)^m \right] = \log\left\{ 1 - \frac{(m-1)\left[ 1 - \sum_{i=1}^{N} \left( p_i^A \right)^m \right]}{(m-1)} \right\} = \log[1 - (m-1)H^A] \tag{1.32}$$

$$\log\left[ \sum_{j}^{M} \left( p_j^B \right)^m \right] = \log\left\{ 1 - \frac{(m-1)\left[ 1 - \sum_{i=1}^{N} \left( p_j^B \right)^m \right]}{(m-1)} \right\} = \log[1 - (m-1)H^B] \tag{1.33}$$

Equation 1.31 is equal to the sum of Equations 1.32 and 1.33:

$$\log[1 - (m-1)H^{A \cup B}] = \log[1 - (m-1)H^A] + \log[1 - (m-1)H^B] \tag{1.34}$$

Equation 1.34 can be recast as

$$1-(m-1)H^{A\cup B} = [1-(m-1)H^A][1-(m-1)H^B] \tag{1.35}$$

Equation 1.35 can be simplified as

$$1-(m-1)H^{A\cup B} = [1-(m-1)H^A -(m-1)H^B +(m-1)^2 H^A H^B] \tag{1.36}$$

Equation 1.36 reduces to

$$H^{A\cup B} = H^A + H^B -[(m-1)H^A H^B] \tag{1.37}$$

Equation 1.37 is often expressed as

$$H(A+B) = H(A)+ H(B)+(1-m)H(A)H(B) \tag{1.38}$$

Equation 1.38 can also be expressed as

$$\frac{\log[1+(1-m)H(A+B)]}{1-m} = \frac{\log[1+(1-m)H(A)]}{1-m} + \frac{\log[1+(1-m)H(B)]}{1-m} \tag{1.39}$$

In the limit as $m \to 1$, Equation 1.38 can be written as the sum of marginal entropies

$$H^{A\cup B} = H^A + H^B \quad \text{or} \quad H(A,B) = H(A)+H(B) \tag{1.40}$$

Equations 1.37 through 1.39 describe the additivity property. This property can be extended to any number of systems. In all cases, $H \geq 0$ (nonnegativity property). If systems $A$ and $B$ are correlated, then

$$p_{ij}^{A\cup B} \neq \left[\sum_{i=1}^{N} p_{ij}^{A\cup B}\right]\left[\sum_{j=1}^{M} p_{ij}^{A\cup B}\right] \tag{1.41}$$

for all $(i, j)$. One may define mutual information or transinformation $S$ as

$$T\left[\left(p_{ij}^{A\cup B}\right)\right] = H^{A\cup B}\left[\left(p_{ij}^{A\cup B}\right)\right] - H^A\left[\left(\sum_{i=1}^{N} p_{ij}\right)\right] - H^B\left[\left(\sum_{j=1}^{M} p_{ij}\right)\right] \tag{1.42}$$

Considering Equation 1.42, $T(p_{ij}) = 0$ for all $m$, if $X$ and $Y$ are independent, and Equation 1.42 will reduce to Equation 1.38. For correlated $X$ and $Y$, $T(p_{ij}) < 0$ for $m = 1$, and $T(p_{ij}) = 0$ for $m = 0$. For arbitrary values of $m$, it will be sensitive to $p_{ij}$; it can take on negative or positive values for both $m < 1$ and $m > 1$ with no particular regularity and can exhibit more than one extremum.

## Example 1.4

Consider a system $A$ that has two states with probabilities $p_1^A = 0.4$ and $p_2^A = 0.6$. Consider another system designated as $B$ with two states having probabilities $p_1^B = 0.3$ and $p_1^B = 0.7$. Both systems are independent. Compute the joint Tsallis entropy of the two systems. Take $m = 3$. Also compute the Shannon entropy.

### Solution

For system A, $p_1^A, p_2^A$ and $p_1^A + p_2^A = 1.0$. Therefore,

$$H^A = \frac{1}{3-1}[(0.4 - 0.4^3) + (0.6 - 0.6^3)] = 0.36$$

$$H^B = \frac{1}{3-1}[(0.3 - 0.3^3) + (0.7 - 0.7^3)] = 0.315$$

$$H(A + B) = 0.36 + 0.315 - (3 - 1) \times 0.36 \times 0.315 = 0.448$$

The joint Shannon entropy can be computed as follows:

$$H^A = -[0.4 \log_2 0.4 + 0.6 \log_2 0.6] = 0.971$$

$$H^B = -[0.3 \log_2 0.3 + 0.7 \log_2 0.7] = 0.881$$

$$H(A + B) = 0.971 + 0.881 = 1.852$$

In this case, the Shannon entropy is much larger than the Tsallis entropy because $m$ is much greater than unity.

### 1.3.5 COMPOSIBILITY

The entropy $H(A + B)$ of a system comprising two subsystems $A$ and $B$ can be computed from the entropies of subsystems, $H(A)$ and $H(B)$, and the entropy index $m$.

### 1.3.6 INTERACTING SUBSYSTEMS

Consider a set of $N$ possibilities arbitrarily separated into two subsystems with $N_1$ and $N_2$ possibilities, where $N = N_1 + N_2$. Defining $P_{N_1} = \sum_{i=1}^{N_1} p_i$ and $P_{N_2} = \sum_{j=1}^{N_2} p_j$, $P_{N_1} + P_{N_2} = 1 = \sum_{k=1}^{N} p_k$. It can be shown that

$$H(P_N) = H(P_{N_1}, P_{N_2}) + P_{N_1}^m H(\{p_i | P_{N_1}\}) + P_{N_2}^m H(\{p_j | P_{N_2}\}) \tag{1.43}$$

where $\{p_i|P_{N_1}\}$ and $\{p_j|P_{N_2}\}$ are the conditional probabilities. Note that $p_i^m > p_i$ for $m < 1$ and $p_i^m < p_i$ for $m > 1$. Hence, $m < 1$ corresponds to rare events and $m > 1$ frequent events (Tsallis, 2001). This property can be extended to any number $R$ of interacting subsystems: $N = \sum_{j=1}^{R} N_j$. Then, defining $w_j = \sum_{i=1}^{N_j} p_i, j = 1, 2, \ldots, N_j$, $\sum_{j=1}^{N} w_j = 1$, Equation 1.43 can be generalized as

$$H(\{p_i\}) = H(\{w_j\}) + \sum_{j=1}^{R} w_j^m H(\{p_i|w_j\}) \tag{1.44}$$

Here, $p_j = w_j$.

## Example 1.5

Consider a set of five possibilities, $p_i = \{0.1, 0.15, 0.2, 0.25, 0.3\}$, separated into two subsets $N_1 = 3$, $p_{N_1} = \{0.1, 0.15, 0.2\}$, and $N_2 = 2$, $p_{N_2} = \{0.25, 0.3\}$. Compute the Tsallis entropy for this system. Then, use Equation 1.43 to compute the Tsallis entropy and show that both ways the entropy is the same.

### Solution

First, the Tsallis entropy can be computed as

$$H(P_N) = \frac{1}{3-1}[(0.1 - 0.1^3) + (0.15 - 0.15^3) + (0.2 - 0.2^3) + (0.25 - 0.25^3) + (0.4 - 0.4^3)]$$

$$= 0.473$$

or consider as two subsystems as

$$P_{N_1} = \sum_{i=1}^{N_1} p_i = 0.1 + 0.15 + 0.2 = 0.45$$

$$P_{N_2} = \sum_{j=1}^{N_2} p_j = 0.25 + 0.3 = 0.55$$

$$H(\{p_i|P_{N_1}\}) = \frac{1}{3-1}\left\{\left[\frac{0.1}{0.45} - \left(\frac{0.1}{0.45}\right)^3\right] + \left[\frac{0.1}{0.45} - \left(\frac{0.1}{0.45}\right)^3\right] + \left[\frac{0.1}{0.45} - \left(\frac{0.1}{0.45}\right)^3\right]\right\}$$

$$= 0.432$$

$$H(\{p_i|P_{N_2}\}) = \frac{1}{3-1}\left\{\left[\frac{0.25}{0.55}-\left(\frac{0.25}{0.55}\right)^3\right]+\left[\frac{0.3}{0.55}-\left(\frac{0.3}{0.55}\right)^3\right]\right\} = 0.372$$

Thus, using Equation 1.43,

$$H(P_N) = H(P_{N_1},P_{N_2}) + P_{N_1}^m H(\{p_i|P_{N_1}\}) + P_{N_2}^m H(\{p_j|P_{N_2}\})$$

$$= 0.371 + 0.45^3 \times 0.432 + 0.55^3 \times 0.372 = 0.473$$

It shows that the entropy computed from Equation 1.43 is the same as given by the definition.

### 1.3.7 OTHER FEATURES

Many complex systems exhibit a power like behavior and they may be in stationary but nonequilibrium states. This may often be the case for geomorphological systems. The Tsallis statistics (Tsallis, 2004) is particularly useful for describing such systems. This statistics exhibits three interesting features (Ferri et al., 2010). First, the PDFs, based on the Tsallis entropy, that describe metastable or stationary systems are proportional to what is called $m$-exponential defined as

$$\exp_m(-\alpha x) = [1-(1-m)\alpha x]^{1/(1-m)} \tag{1.45}$$

in which $m$ and $\alpha$ are constants. Figure 1.7 shows a plot of Equation 1.45 for different values of $\alpha$ and $m$. In the limit $m \to 1$, $m$-exponential becomes the ordinary exponential, that is $\exp_1(x) = \exp(x)$. Further, if $m \to 1$ and $x = y^2$ then $\exp_m(-\alpha x)$ becomes an $m$-Gaussian.

The inverse of $m$-exponential is referred to as $m$-logarithm defined as

$$\ln_m(x) = \frac{x^{1-m}-1}{1-m}, \quad \ln_1(x) = \ln(x), \quad \ln_m[\exp_m(x)] = \exp[\ln_m(x)] = 1 \tag{1.46}$$

Stationary systems are characterized by nonextensivity index $m = m_{stat}$. Figure 1.8 shows a plot of Equation 1.46 for different values of $m$.

Second, stationary states show $m$-exponential sensitivity to initial conditions or weak chaos with a parameter $m = m_{sens}$. This means that small differences between adjacent states grow in an $m$-exponential fashion. Third, microscopic variables decrease $m$-exponentially with a parameter $m = m_{rel}$.

In this manner, a stationary or metastable system can be characterized by a triplet of $m$ values, often referred to as the Tsallis $m$-triplet, that is $(m_{stat}, m_{sens}, m_{rel}) \neq (1, 1, 1)$, in which $m_{stat} > 1$, $m_{sens} < 1$, and $m_{rel} < 1$ (Ferri et al., 2010). Ausloos and Petroni (2007) and Petroni and Ausloos (2007) reported the values of $m_{stat}$ for daily variation of the El Nino Southern Oscillation (ENSO) index.

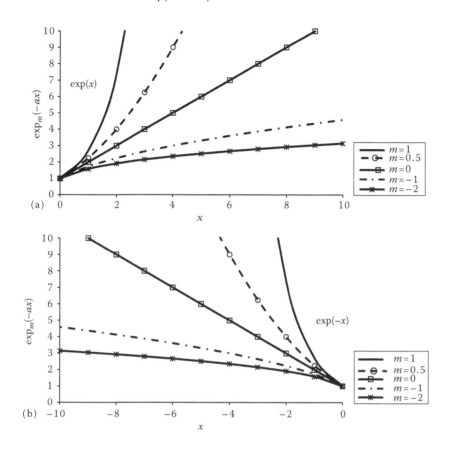

**FIGURE 1.7**    *m*-Exponential for various *m* values with (a) *a* = −1 and (b) *a* = 1.

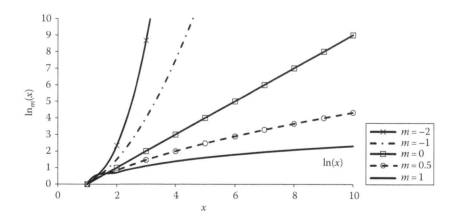

**FIGURE 1.8**    *m*-Logarithm for various *m* values.

## 1.4   MODIFICATION OF TSALLIS ENTROPY

Yamano (2001a) provided a modification of the Tsallis entropy. It may be worth recalling that the Shannon entropy function is uniquely determined not because of the definition of the mean value of information but because of the additivity of the uncertainty of information that the source contains. Considering the amount of information as the $m$-logarithmic function of probability

$$I_m(p) = -\ln_m p(x) \tag{1.47}$$

where

$$\ln_m p(x) = \frac{(p(x^{1-m})-1)}{1-m} = \frac{1-p(x^{1-m})}{m-1} \tag{1.48}$$

Function $I_m(p)$ is a monotonically decreasing function and so is $-\ln p$. The unit of measurement in this case is nat, not bit. In the limit, as $m$ tends to 1, the information content becomes $-\ln p$.

Taking the normalized $m$-average (or escort average) of the information content or entropy, one obtains

$$\frac{-\sum_{i=1}^{N} p^m(x_i)\ln_m p(x_i)}{\sum_{i=1}^{N} p^m(x_i)} = \frac{1-\sum_{i=1}^{N} p^m(x_i)}{(m-1)\sum_{i=1}^{N} p^m(x_i)} = H_m(X) \tag{1.49}$$

which is the modified form of the Tsallis entropy and is obtained by dividing the Tsallis entropy by factor $\sum_{i=1}^{N} p^m(x_i)$.

### Example 1.6

Let $N = 3$ and $p(x_i) = \{0.2, 0.3, 0.5\}$. Compute the $m$-average entropy and ordinary entropy.

### Solution

Let $m = 3$, the $m$-average entropy is computed as

$$H_m(X) = \frac{1-(0.2^3 + 0.3^3 + 0.5^3)}{(3-1)(0.2^3 + 0.3^3 + 0.5^3)} = 2.625$$

and the ordinary entropy is

$$H(X) = \frac{1}{(3-1)}[(0.2-0.2^3)+(0.3-0.3^3)+(0.5-0.5^3)] = 0.42$$

where the factor

$$\sum_{i=1}^{N} p^m(x_i) = (0.2^3 + 0.3^3 + 0.5^3) = 0.16 = \frac{H(X)}{H_m(X)}$$

Yamano (2001b) discussed the properties of the modified Tsallis entropy, which are briefly presented in the following. For two random variables $X$ and $Y$, their joint entropy can be expressed as

$$H_m(X,Y) = \frac{1 - \sum_{x,y} p^m(x,y)}{(m-1)\sum_{x,y} p^m(x,y)} \tag{1.50}$$

and a nonadditive conditional entropy $H_m(Y|x)$ can be written as

$$\frac{\sum_{i=1}^{N}(p^m(x_i)/[1-(m-1)H_m(Y|x)])}{\sum_{i=1}^{N} p^m(x_i)} = [1+(m-1)H_m(Y|X)]^{-1} \tag{1.51}$$

The mutual information $T_m(Y; X)$ can now be defined in the usual way as common information between $X$ and $Y$, which is equal to the reduction in uncertainty in one variable due to the knowledge of another variable:

$$T_m(Y;X) = H_m(Y) - H_m(Y|X) = \frac{H_m(X) + H_m(Y) - H_m(X,Y) + (m-1)H_m(X)H_m(Y)}{1+(m-1)H_m(X)} \tag{1.52}$$

This will converge to the usual mutual information or transinformation in the additive limit $m$ tending to 1. Following Yamano (2001b), the following relations hold for $X$, $Y$, and $Z$ random variables:

1.      $$H_m(X;Y) = H_m(X) + H_m(Y|X) + (m-1)H_m(X)H_m(Y|X) \tag{1.53}$$

2.   $$H_m(X_1, X_2, \ldots, X_n) = \sum_{i=1}^{n}[1+(m-1)H_m(X_{i-1},\ldots,X_1)]H_m(X_i|X_{i-1},\ldots,X_1) \tag{1.54}$$

3.      $$H_m(X_1, X_2, \ldots, X_n) \le \sum_{i=1}^{n}[1+(m-1)H_m(X_{i-1},\ldots,X_1)]H_m(X_i) \tag{1.55}$$

4.      $$T_m(X;Y) = H_m(X) - [1+(m-1)H_m(Y)]H_m(X|Y) \tag{1.56}$$

The mutual information becomes symmetric in $X$ and $Y$:

$$T_m(X,Y) = T_m(Y,X) = H_m(X) + H_m(Y) - H_m(X,Y) \tag{1.57}$$

5.  $[1+(m-1)H_m(X)]H_m(Y|X)$

$$= H_m(Y,Z|X) - H_m(Z|Y,X) + (m-1)\{H_m(X)H_m(Y,Z|X) - H_m(X,Y)H_m(Z|Y,X)Z\} \tag{1.58}$$

It is seen that mutual information becomes symmetric in $X$ and $Y$. In the limit $m$ tending to 1, these relations reduce to the ones satisfied by the Shannon entropy.

6.  The Kullback–Leibler (KL) cross entropy between two distributions $p(x)$ and $q(x)$ can be written in a Tsallis entropy sense as

$$D_m[p(x),q(x)] = \frac{\sum_{i=1}^{N} q^m(x_i)\ln_m q(x_i)}{\sum_{i=1}^{N} q^m(x_i)} - \frac{\sum_{i=1}^{N} p(x_i)\ln_m p(x_i)}{\sum_{i=1}^{N} p^m(x_i)} \tag{1.59}$$

The KL cross entropy satisfies

$$D_m[p(x),q(x)]\begin{cases} \geq 0 \ (m>0) \\ < (m<0) \end{cases} \tag{1.60}$$

and equals 0 if $p(x) = q(x)$.

7.  The generalized mutual information can be defined in terms of the generalized KL cross entropy as

$$T_m(X,Y) = D_m[P(x,y)|P(x)P(y)]$$

$$= \frac{1/(1-m)\left[1 - \sum_{x,y} p(x,y)(p(x)p(y)/p(x,y))^{1-m}\right]}{\sum_{x,y} p^m(x,y)} \tag{1.61}$$

## 1.5   MAXIMIZATION

Consider a case where $H$ given by Equation 1.2 is to be maximized subject to the following constraints:

$$\sum_{i=1}^{N} P_i = 1 \tag{1.62}$$

and

$$\sum_{i=1}^{N} p_i x_i = \overline{x} \tag{1.63}$$

where $\{x_i\}$ and $\bar{x}$ are real numbers. Following the method of the Lagrange multipliers, the Lagrange function can be defined as

$$L = H + \lambda_0 \left[ \sum_{i=1}^{N} p_i - 1 \right] + \lambda_1 \left[ \sum_{i=1}^{N} p_i x_i - \bar{x} \right] \tag{1.64}$$

where $\lambda_0$ and $\lambda_1$ are the Lagrange multipliers. Following Tsallis (1988), Equation 1.64 can be recast as

$$L = H + \lambda_0 \sum_{i=1}^{N} p_i + \lambda_0 \lambda_1 (m-1) \sum_{i=1}^{N} p_i x_i - [\lambda_0 + \lambda_0 \lambda_1 (m-1)\bar{x}] \tag{1.65}$$

It may be noted that the term within brackets on the right side of Equation 1.65 does not influence the maximization of entropy. Therefore, for entropy maximizing Equation 1.65 can simply be written as

$$L = H + \lambda_0 \sum_{i=1}^{N} p_i + \lambda_0 \lambda_1 (m-1) \sum_{i=1}^{N} p_i x_i \tag{1.66}$$

Differentiating $L$ in Equation 1.66 with respect to $p_i$ and equating to zero for all $i$, one obtains

$$p_i = \frac{[1 - \lambda_1 (m-1) x_i]^{1/(m-1)}}{Z} \tag{1.67}$$

where $Z$ is the partition function defined as

$$Z = \sum_{i=1}^{N} [1 - \lambda_1 (m-1) x_i]^{1/(m-1)} \tag{1.68}$$

If $m$ tends to one, Equation 1.67 reduces

$$p_i = \frac{1}{Z} \exp(-\lambda_1 x_i) \tag{1.69}$$

in which

$$Z = \sum_{i=1}^{N} \exp(-\lambda_1 x_i) \tag{1.70}$$

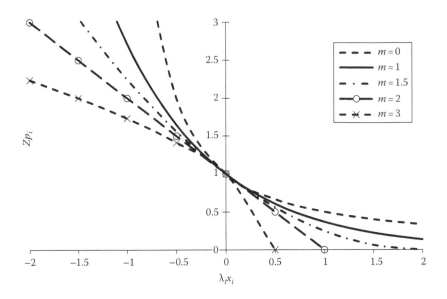

**FIGURE 1.9**   Plot of distribution given by Equation 1.67 parameterized by $m$.

Equation 1.67 expresses a power law distribution (Tsallis et al., 1998; Evans et al., 2000). This suggests that one way to obtain a power distribution is to extremize the Tsallis entropy with the constraint: $\sum_{i=1}^{N} p_i x_i^m = \overline{x^m}$, instead of $\overline{x}$. This distribution is plotted in Figure 1.9 for $m = 0$, 1, 1.5, 2, 3; the $x$-axis is taken as $\lambda_1 x_i$ and the $y$-axis is taken as $Z p_i$. For $m = 1$, this leads to an exponential distribution. For $m > 1$, it shows a cutoff at $\lambda_1 x_i = 1/(m - 1)$, where the slope is 0 for $m < 2$, $-1$ for $m = 2$, and $-\infty$ for $m > 2$ and diverges for $\lambda_1 x_i$ tending to $-\infty$. For $m < 1$, the distribution diverges at $\lambda_1 x_i = -1/(1 - m)$ and vanishes when $\lambda_1 x_i$ tends to $+\infty$.

## 1.6  PARTIAL TSALLIS ENTROPY

Let $H_i$ denote the Tsallis entropy for the $i$th system state whose probability is $p_i$. Then, the Tsallis entropy for the system can be expressed as

$$H = \sum_{i=1}^{N} H_i = \sum_{i=1}^{N} H(p_i), \quad H_i = H(p_i) \tag{1.71}$$

The partial Tsallis entropy can be defined as (Niven, 2004)

$$H_i = -p_i^m \ln_m p_i = \frac{p_i - p_i^m}{m - 1} \tag{1.72}$$

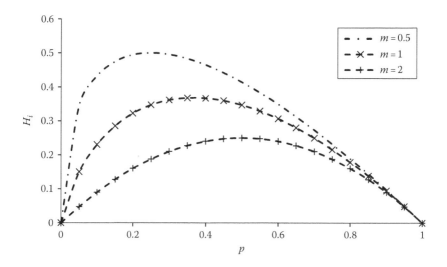

**FIGURE 1.10**   Partial Tsallis entropy for $m = 0.5$, 1, and 2. The curve of $m = 1$ is equivalent to the partial Shannon entropy.

Recall the $m$-logarithmic function:

$$\ln_m p = \frac{p^{1-m} - 1}{1 - m}, \quad p > 0 \tag{1.73}$$

In the limit, as $m \to 1$,

$$\ln_m p \to \ln p \tag{1.74}$$

The partial Tsallis entropy is plotted in Figure 1.10 for various values of $m$. Also plotted is the partial Shannon entropy. It is interesting to note that $H_i$ is bounded by two endpoint minima at $H_i(0) = 0$ and $H_i(1) = 0$ and has the maximum at

$$\max H_i = H_i(m^{1/(1-m)}) = \frac{1}{m-1}[m^{1/(1-m)} - m^{m/(1-m)}] \tag{1.75}$$

Equation 1.75 shows that the maximum Tsallis entropy depends on the entropy index $m$ and is independent of state, but its position can vary. In the case of the Shannon partial entropy, the position is fixed. Further, it implies that it does not accommodate the local effect of constraints. It may be noted that $H_i(m = 0) = 1 - p_i$, which is a linear relation; $H_i(m \to \infty) = 0$ in the limit $m \to 1$, $H_i$ reduces to the Shannon entropy.

For $m \leq 0$, Figure 1.11 shows that the Tsallis partial entropy does not have a real-valued extremum and tends to infinity as $p_i \to 0$ for $-1 < m < 0$ and/or over some finite range of $p_i$ for $-\infty < m < 1$ The Tsallis partial entropy is nonnegative but the constrained partial Shannon entropy may not be.

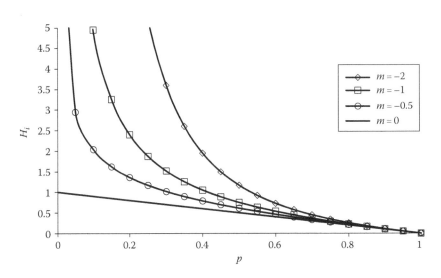

**FIGURE 1.11**  Partial Tsallis entropy for $m = 0$, $-0.5$, $-1$, and $-2$.

## 1.7  CONSTRAINED TSALLIS ENTROPY

For defining the constrained Tsallis entropy, the Lagrange function $L$ is constructed for entropy maximizing. To that end, constraints are defined as the normalizing constraint given by Equation 1.62 and the general constraint given by

$$\sum_{i=1}^{N} p_i g_i(x) = \overline{g_i(x)} \tag{1.76}$$

Now the Lagrangian function $L$ can be written as

$$L = \frac{1}{m-1}\sum_{i=1}^{N}\left(p_i - p_i^m\right) - \lambda_0\left(\sum_{i=1}^{N} p_i - 1\right) - \sum_{j=1}^{N}\lambda_j\left(\sum_{i=1}^{N} p_i g_{ji} - \overline{g_j}\right) \tag{1.77}$$

For entropy maximization, the Lagrangian function can simply be written as

$$L = \frac{1}{m-1}\sum_{i=1}^{N}\left(p_i - p_i^m\right) - \lambda_0\sum_{i=1}^{N} p_i - \sum_{j=1}^{N}\lambda_j\sum_{i=1}^{N} p_i g_{ji} \tag{1.78}$$

Differentiation of Equation 1.78 with respect to $p_i$ yields

$$\frac{dL}{dp_i} = 0 = \frac{1}{m-1}\left(1 - mp_i^{m-1}\right) - \lambda_0 - \sum_{j=1}^{n}\lambda_j g_{ji} \tag{1.79}$$

Equation 1.79 yields the maximum entropy-based probability distribution denoted by $p_i^*$ :

$$p_i^* = \left\{ \frac{1}{m}\left[1-(m-1)\lambda_0-(m-1)\sum_{j=1}^{n}\lambda_j g_{ji}\right]\right\}^{1/(m-1)} \tag{1.80}$$

From Equation 1.80, the zeroth Lagrange multiplier $\lambda_0$ can be expressed as

$$\lambda_0 = \frac{1}{m-1}\left[1-mp_i^{*(m-1)}-(m-1)\sum_{i=1}^{n}\lambda_j g_{ji}\right] \tag{1.81}$$

Substituting Equation 1.81 in Equation 1.78, one obtains

$$L = \frac{1}{m-1}\sum_{i=1}^{N}\left(p_i-p_i^m\right)-\left\{\frac{1}{m-1}\left[1-mp_i^{*(m-1)}-(m-1)\sum_{j=1}^{n}\lambda_j g_{ji}\right]\right\}$$

$$\times \sum_{i=1}^{N}p_i-\sum_{j=1}^{n}\lambda_j\sum_{i=1}^{N}p_i g_{ji} \tag{1.82}$$

Equation 1.82 simplifies to

$$L = \frac{1}{m-1}\left[mp_i^{*(m-1)}\sum_{i=1}^{N}p_i-\sum_{i=1}^{N}p_i^m\right] \tag{1.83}$$

Equation 1.83 is the constrained Lagrange function. Sometimes, this is also referred to as constrained Tsallis entropy:

$$H^C = \frac{1}{m-1}\left[mp_i^{*(m-1)}\sum_{i=1}^{N}p_i-\sum_{i=1}^{N}p_i^m\right] \tag{1.84}$$

Therefore, the partial constrained Tsallis entropy is given as

$$H_i^C(p,p^*) = \frac{1}{m-1}\left[mp_i^{*(m-1)}p_i-p_i^m\right] \tag{1.85}$$

Figures 1.12 and 1.13 plot the constrained Tsallis entropy for various values of $m$ and $p_i$.

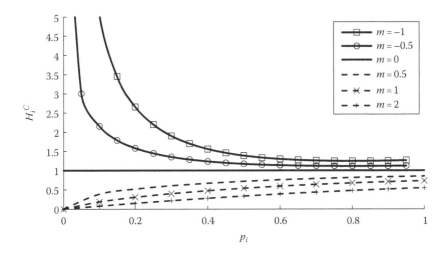

**FIGURE 1.12**   Constrained Tsallis entropy for various values of $p_i$ ($m = 1$).

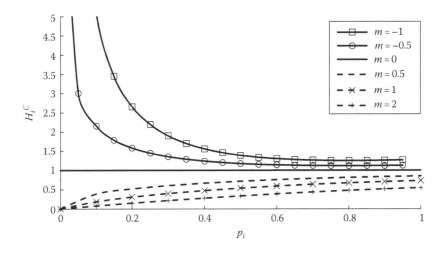

**FIGURE 1.13**   Constrained Tsallis entropy for various values of $m$ ($p_i = 0.8$).

## 1.8   GENERALIZED RELATIVE ENTROPIES

Relative entropy plays an important role in comparing two distributions. Let the two distributions be $P$: $\{p_i, i = 1, 2, \ldots, N\}$ and $R$: $\{r_i, i = 1, 2, \ldots, N\}$. Two types of generalized relative entropies have been defined in nonextensive statistical mechanics. One is of Bregman (1967) type described by Naudts (2004):

$$J_m[P|R] = \frac{1}{m-1} \sum_{i=1}^{N} p_i \left[ p_i^m - r_i^{m-1} \right] - \sum_{i=1}^{N} (p_i - r_i) r_i^{m-1} \tag{1.86}$$

and the other of Csiszar (1972) type described by Tsallis (1998):

$$K_m[P|R] = \frac{1}{1-m}\left[1 - \sum_{i=1}^{N} p_i^m r_i^{1-m}\right] \tag{1.87}$$

Here, $R = \{r_i, i = 1, 2, ..., N\}$ is the prior or reference probability distribution. It may be noted that the first type of relative entropy is associated with ordinary expectation and the second type with the normalized expectation value. Both entropies are nonnegative.

In the limit $m \to 1$, both relative entropies $J_m[P|R]$ and $K_m[P|R]$ reduce to the KL cross entropy defined as

$$H[P|R] = \sum_{i=1}^{N} p_i \ln\left(\frac{p_i}{r_i}\right) \tag{1.88}$$

Equation 1.88 is often written in the differential form as

$$H(P|R) = \frac{d}{dx} \sum_{i=1}^{N} p_i^x r_i^{1-x}\bigg|_{x \to 1} \tag{1.89}$$

Abe and Bagci (2005) have shown that physically $J_m[P|R]$ and $K_m[P|R]$ are essentially free energy differences. It can be shown that $J_m[P|R]$ is convex in $P$ but not in $R$, whereas $K_m[P|R]$ is convex in both $P$ and $R$. Furthermore, like the KL relative entropy, $K_m[P|R]$ is composable but $J_m[P|R]$ is not. To illustrate it, consider a composite system $(A, B)$ where the joint probability distribution can be factorized as $p_{ij}(A, B) = p_{(1)i}(A)p_{(2)j}(B)$ and $r_{ij}(A, B) = r_{(1)i}(A) r_{(2)j}(B)$. Then one can write

$$K_m[p_{(1)}p_{(2)}|r_{(1)}r_{(2)}] = K_m[p_{(1)}|r_{(1)}] + K_m[p_{(2)}|r_{(2)}]$$

$$+ (m-1)K_m[p_{(1)}|r_{(1)}]K_m[p_{(2)}|r_{(2)}] \tag{1.90}$$

However, such a relation does not exist for $J_m[P|R]$.

The POMCE (Shore and Johnson, 1980) supports $K_m[P|R]$ but not $J_m[P|R]$. The relative entropy $H[P|R]$ satisfying POMCE will have the following form:

$$H[P|R] = \sum_{i=1}^{N} p_i h\left(\frac{p_i}{r_i}\right) \tag{1.91}$$

where $h(x)$ is some function $[x = p_i/r_i]$. Function $h(x)$ exists for $K_m[P|R]$ and can be written as

$$h(x) = \frac{1}{1-m}(1 - x^{m-1}) \tag{1.92}$$

However, it does not seem feasible to cast $J_m[P|R]$ in the form of Equation 1.91. This suggests that POMCE supports the normalized $m$-expectation value but not the ordinary expectation value.

## REFERENCES

Abe, S. and Bagci, G.B. (2005). Necessity of q-expectation value in the nonextensive statistical mechanics. *Physical Review*, E71, 016139.

Ausloos, M. and Petroni, F. (2007). Tsallis nonextensive statistical mechanics of El Nino Southern Oscillation Index. *Physica A: Statistical Mechanics and Applications*, 373, 721–736.

Bregman, L.M. (1967). The relaxation method of finding the common point of convex sets and its application to the solution of problems in convex programming. *USSR Computational Mathematics and Mathematical Physics*, 7, 200–217.

Csiszar, I. (1972). A class of measures of informativity of observation channels. *Periodica Mathematica Hungarica*, 2, 191–213.

Evans, M., Hastings, N., and Peacock, B. (2000). *Statistical Distributions*. John Wiley & Sons, New York.

Ferri, G.L., Reynoso Savio, M.F., and Plastino, A. (2010). Tsallis's q-triplet and the ozone layer. *Physica A*, 389, 1829–1833.

Jaynes, E.T. (1957a). Information theory and statistical mechanics, I. *Physical Review*, 106, 620–630.

Jaynes, E.T. (1957b). Information theory and statistical mechanics, II. *Physical Review*, 108, 171–190.

Koutsoyiannis, D. (2005a). Uncertainty, entropy, scaling and hydrological stochastics: Additional information for the study of marginal distributional properties of hydrological processes. Internal Report, Department of Water Resources, National Technical University of Athens, Athens, Greece.

Koutsoyiannis, D. (2005b). Uncertainty, entropy, scaling and hydrological stochastics: 1. Marginal distributional properties of hydrological processes and state scaling. *Hydrological Sciences Journal*, 50(3), 381–404.

Koutsoyiannis, D. (2005c). Uncertainty, entropy, scaling and hydrological stochastics: 2. Time dependence of hydrological processes and state scaling. *Hydrological Sciences Journal*, 50(3), 405–426.

Koutsoyiannis, D. (2013). Physics of uncertainty, the Gibbs paradox and indistinguishable particles. *Studies in History and Philosophy of Modern Physics*, 44, 480–489.

Koutsoyiannis, D. (2014). Entropy: From thermodynamics to hydrology. *Entropy*, 16, 1287–1314.

Kullback, S. and Leibler, R.A. (1951). On information and sufficiency. *Annals of Mathematical Statistics*, 22, 79–86.

Naudts, J. (2004). The q-exponential family in statistical physics. *Reviews of Mathematical Physics*, 16(6), 809–822.

Niven, R. (2004). The constrained entropy and cross entropy functions. *Physica A*, 334, 444–458.

Petroni, F. and Ausloos, M. (2007). High frequency (daily) data analysis of the Southern Oscillation Index: Tsallis nonextensive statistical mechanics approach. *The European Physical Journal (Special Topics)*, 143, 201–208.

Shore, J.E. and Johnson, R.W. (1980). Properties of cross-entropy minimization. NRL Memorandum Report 4189, Naval Research Laboratory, Washington, DC, 36p.

Singh, V.P. (1997). The use of entropy in hydrology and water resources. *Hydrological Processes*, 11, 587–626.

Singh, V.P. (2010). Entropy theory for earth science modeling. *Indian Geological Congress Journal*, 2(2), 5–40.

Singh, V.P. (2011). Hydrologic synthesis using entropy theory: Review. *Journal of Hydrologic Engineering*, 16(5), 421–433.

Singh, V.P. (2013). *Entropy Theory and Its Application in Environmental and Water Engineering*. John Wiley, Sussex, U.K., 642p.

Singh, V.P. (2014). *Entropy Theory in Hydraulic Engineering: Introduction*. ASCE Press, Reston, VA, 784p.

Singh, V.P. (2015). *Entropy Theory in Hydrologic Science and Engineering*. McGraw-Hill Education, New York, 824p.

Tsallis, C. (1988). Possible generalizations of Boltzmann-Gibbs statistics. *Journal of Statistical Physics*, 52(1/2), 479–487.

Tsallis, C. (1998). On the fractal dimension of orbits compatible with Tsallis statistics. *Physical Review*, E58(2), 1442–1445.

Tsallis, C. (2001). Nonextensive statistical mechanics and thermodynamics: Historical background and present status. In: *Nonextensive Statistical Mechanics and Its Applications*, eds. S. Abe and Y. Okamoto. Springer, Berlin, Germany, pp. 3–98.

Tsallis, C. (2002). Entropic nonextensivity: A possible measure of complexity. *Chaos, Solitons and Fractals*, 12, 371–391.

Tsallis, C. (2004). Nonextensive statistical mechanics: Construction and physical interpretation. In: *Nonextensive Entropy-Interdisciplinary Applications*, eds. M. Gell-Mann and C. Tsallis. Oxford University Press, Oxford, U.K.

Tsallis, C., Mendes, R.S., and Plastino, A.R. (1998). The role of constraints within generalized nonextensive statistics. *Physica A*, 261, 534–554.

Yamano, T. (2001a). A possible extension of Shannon's information theory. *Entropy*, 3, 280–292.

Yamano, T. (2001b). Information theory based nonadditive information content. *Physical Review*, E63, 046105-1–046105-7.

# 2 Probability Distributions

The Tsallis entropy has found only limited application in hydrological frequency analysis. This may partly be due to the complexity in analytically deriving probability distributions by the Tsallis entropy maximizing when more than one constraint is specified. In a seminal work, Koutsoyiannis (2004a,b, 2005a,b) was probably the first to derive different distributions using the Tsallis entropy and showed applications of some of them to extreme rainfall analysis. Since then, some distributions and their applications have been reported in the hydrological literature. Much of the discussion in this chapter is drawn from the work of Koutsoyiannis and his associates (Koutsoyiannis, 2005a,b, 2006; Papalexiou and Koutsoyiannis, 2012, 2013). The objective of this chapter is to discuss the procedure for deriving probability distributions for hydrological frequency analysis using the Tsallis entropy.

## 2.1 PROCEDURE FOR DERIVING A PROBABILITY DISTRIBUTION

Derivation of a probability distribution using the Tsallis entropy entails (1) defining the Tsallis entropy, (2) specifying constraints, (3) maximizing the entropy using the method of Lagrange multipliers, (4) obtaining the probability distribution, (5) determining the Lagrange multipliers in terms of the specified constraints, and (6) determining the maximum entropy as well as the properties of the distribution.

### 2.1.1 DEFINING TSALLIS ENTROPY

In this discussion, the continuous version of the Tsallis entropy is employed, meaning random variables are assumed continuous. Then, the Tsallis entropy $H$ can be written as

$$H = \int_0^\infty f(x) \frac{1 - [f(x)]^{m-1}}{m-1} dx \qquad (2.1)$$

where
  $x$ is a specific value of random variable $X$
  $f(x)$ is the probability density function (PDF) of $X$
  $m$ is the entropy index

### 2.1.2 SPECIFICATION OF CONSTRAINTS

It should be noted that for any PDF, $f(x)$, of a random variable $X$, the total probability must equal one, that is,

$$\int_0^\infty f(x)dx = 1 \tag{2.2}$$

Equation 2.2 is, in a true sense, not a constraint but is often referred to as normalization or natural constraint.

In the Tsallis entropy formalism, constraints are specified in two ways. The first way is to specify the constraints in terms of regular moments. The second way is to specify the constraints in terms of $m$-expectations, where $m$ is the Tsallis entropy index. Both methods are illustrated for deriving a probability distribution using simple constraints in order to keep the algebra easily tractable.

Empirical observations contain the information that we are looking for and we have to find a way to extract and express that information. Constraints encode the information or summarize the knowledge that can be garnered from empirical observations or theoretical considerations.

Tsallis et al. (1998) discussed the role of constraints in the context of the Tsallis entropy formalism. Papalexiou and Koutsoyiannis (2012) provided an excellent discussion on the rationale for choosing different types of constraints in hydrology. Entropy maximizing shows that there is a unique correspondence between a probability distribution and the constraints that lead to it. Therefore, choosing appropriate constraints is extremely important in the entropy formalism.

At the outset, it is important to determine or at least have a good idea as to the type of probability distribution that will best represent the empirical hydrological data, such as rainfall, runoff, temperature, extreme low flows, extreme high flows, sediment yield, and so on. Once an idea about the distribution shape is gathered, the issue of constraints is addressed.

Clearly, there can be a large, if not an infinite, number of constraints that can perhaps summarize the information on the random variable. The question then arises: How should the appropriate constraints be chosen? First, constraints should be simple but simplicity is subjective, and therefore, quantitative criteria are needed. Second, constraints should be as few as absolutely needed.

When empirical data suggest, without any consideration of entropy, a particular shape of probability distribution, it is important to keep in mind that these data represent only a small part of the past. Therefore, any inference on the random variable characteristics may vary in the future, especially because of looming climate change, global warming, and land use change. This suggests that constraints should be defined such that they are more or less preserved in the future. It may be noted that some constraints, especially lower-order moments, such as mean and variance, are less susceptible to change than are others, especially higher-order moments, such as kurtosis.

When defining constraints, it would be desirable to express them in terms of the laws of conservation of mass, momentum, and energy. However, water resources and hydrological processes are complicated and do not often lend themselves to allow for defining constraints in this manner except for very simple cases. For example in case of groundwater flow, it is possible to express constraints in terms of mass conservation and Darcy's law when solving very simple problems. Likewise, for the movement of soil moisture, infiltration, velocity distribution, and suspended sediment concentration, mass conservation can be used as a constraint. However, for a majority of cases, constraints need to be inferred keeping the empirical evidence with respect to the probability distribution shape in mind.

If the observed data show that the random variable can be described by a bell-shaped distribution then all that is needed is the mean and variance. That means these constitute the constraints. Likewise, if the distribution is heavy tailed or light tailed, then appropriate constraints need to be specified accordingly. However, few hydrological processes follow a normal distribution. Of course, most hydrological variables are nonnegative and with the nonnegative condition imposed, the resulting distribution would be truncated normal, if the empirical data suggested a normal distribution. The Tsallis entropy maximizing leads to a symmetric bell-shaped distribution with power-type tails. If the mean is zero, then this becomes the Tsallis distribution. For nonzero mean, this is the Pearson-type VII distribution. A majority of hydrological and environmental processes exhibit a rich variety of asymmetries. Examples are rainfall extreme corresponding to small time intervals, flow maxima, flow minima, extreme temperatures, extreme winds, among others.

Constraints can be defined in terms of moments as

$$C_i = \int_0^\infty g_i(x)f(x)dx = \overline{g_i(x)}, \quad i = 1,2,\ldots,M \tag{2.3}$$

where
$g_i(x)$ is some function of $x$
$C_i$ is the $i$th constraint
$M$ is the number of constraints

If $g_0(x) = 1$, then Equation 2.3 reduces to Equation 2.2. If $g_1(x) = x$, then the constraint becomes mean.

Mean, a measure of central tendency, is one of the most frequently specified constraints

$$E[X] = \mu = \int_0^\infty xf(x)dx \tag{2.4}$$

which is approximated by the sample mean denoted as $\overline{x}$ where over bar denotes sample average. If the random variable takes on only nonnegative values, then geometric mean, denoted as $\mu_G$, which is smaller than the arithmetic mean, is another useful constraint for hydrological processes. For a sample size $N$, an estimate of $\mu_G$ can be defined as

$$\mu_G = \left(\prod_{i=1}^{N} x_i\right)^{1/N} = [\exp][\ln]\left(\prod_{i=1}^{N} x_i\right)^{1/N} = \exp\left[\ln\left(\prod_{i=1}^{N}\right)^{1/N}\right] = \exp\left[\frac{1}{N}\sum_{i=1}^{N}\ln x_i\right]$$

$$= \exp\left[\overline{\ln x}\right] \tag{2.5}$$

Taking the logarithm of both sides of Equation 2.5, a constraint for entropy maximizing can be expressed as

$$E[\ln X] = \ln \mu_G \tag{2.6}$$

This constraint would be useful when empirical data are positively skewed or even heavy tailed. The logarithm of geometric mean, because of its logarithmic character, is likely to be preserved in the future.

The second moment or variance, a measure of dispersion about the central tendency or mean, is also expressed as a constraint

$$\sigma^2 = \int_{0}^{\infty} x^2 f(x)dx \tag{2.7}$$

Here, the expected value, $\mu$, is already subtracted from the $x$ values. Papalexiou and Koutsoyiannis (2012) reasoned that if the second moment is preserved then its square root, the standard deviation, is even more likely to be preserved and more robust to outliers. This reasoning can be extended to lower-order fractional moments.

Recalling for the logarithmic function,

$$\lim_{a\to 0}\frac{x^a - 1}{a} = \ln x \tag{2.8}$$

where $a$ is an arbitrary exponent or power. For small values of $a$, it can be argued that $x^a$ would behave like $\ln(x)$. Therefore, it may be deemed logical to allow the order of the moment to remain unspecified so that even fractional values of order can be accommodated. Thus, a moment of order $r$, $M_r$, as a constraint can be expressed as

$$M_r = E[X^r] = \int_{0}^{\infty} x^r f(x)dx \tag{2.9}$$

Further, recalling the limiting definition of the exponential function:

$$\exp(x^a) = \lim_{b\to 0}(1 + bx^a)^{1/b} \tag{2.10}$$

where $b$ is any arbitrary quantity. One can then define by taking the logarithm of Equation 2.10:

$$x_b^a = \frac{\ln(1+bx^a)}{b}$$

(2.11)

For $b = 0$, Equation 2.11 becomes a power $x^a$ as

$$x_0^a = \lim_{b \to 0} \frac{\ln(1+bx^a)}{b} = x^a$$

(2.12)

Now the regular moments, designated as $p$-moments of order $r$, can be generalized as

$$M_p^r = E\left[X_p^r\right] = \frac{1}{p}\int_0^\infty \ln(1+px^r)f(x)dx$$

(2.13)

This is one generalization among many that can be constructed. Papalexiou and Koutsoyiannis (2012) provided a rationale for this generalization as follows: (1) $p$-moments are simple and for $p = 0$ they reduce to regular moments. (2) Their basis is the $x_p^r$ function that has the desirable properties of $\ln x$ function and therefore are appropriate for positively skewed random variables. (3) The $p$-moments lead to flexible power-type distributions, including the Pareto (for $m = 1$) and Tsallis ($m = 2$) distributions. (4) These moments are no more arbitrary than generalized entropy measures. (5) These moments lead to distributions that represent many hydrological processes as well.

It should be noted that the entropy index $m$ may be different, depending on the way the constraints are introduced: (1) in a regular way and (2) in a non-normalized way:

$$\int f^m(x)x^j dx = E\left[x_m^j\right] = \overline{x_m^j}$$

(2.14)

or (3) in a normalized way:

$$\frac{\int f^m(x)x^j dx}{\int f^m(x)dx}$$

(2.15)

In a similar way, constraints can be defined for the discrete case. The third way is like the escort probability way. The escort probabilities are defined as

$$P_i = \frac{p_i^m}{\sum_{j=1}^N p_j^m}, \quad \sum_{i=1}^N P_i = 1$$

(2.16)

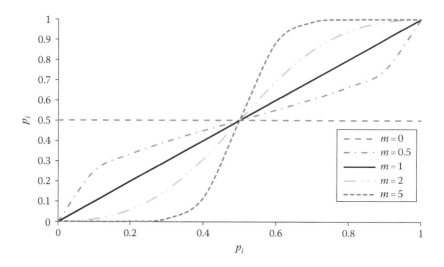

**FIGURE 2.1**    Escort probabilities for $N = 2$.

From Equation 2.16, the following inverse relation stems:

$$p_i = \frac{[P_i]^{1/m}}{\sum_{j=1}^{N}[P_j]^{1/m}} \tag{2.17}$$

Since

$$\frac{[P_i]^{1/m}}{\sum_{j=1}^{N}[P_j]^{1/m}} = \frac{p_i / \left[\sum_{j=1}^{N} p_j^m\right]^{1/m}}{\sum_{j=1}^{N} p_j / \left[\sum_{j=1}^{N} p_j^m\right]^{1/m}} = \frac{p_i}{\sum_{j=1}^{N} p_j} = p_i$$

thus

$$\sum_{i=1}^{N} p_i^m = \frac{\sum_{i=1}^{N}[[P_i]^{1/m}]^m}{\left[\sum_{i=1}^{N}[P_j]^{1/m}\right]^m} = \frac{1}{\left[\sum_{j=1}^{N}[P_j]^{1/m}\right]^m} \tag{2.18}$$

Figure 2.1 shows the escort probabilities for $N = 2$.

### 2.1.3    ENTROPY MAXIMIZING

In order to derive the least-biased PDF $f(x)$, the Tsallis entropy is maximized in accordance with the principle of maximum entropy (Jaynes, 1957a,b), subject to specified constraints. For entropy maximizing, the method of Lagrange multipliers is employed. Therefore, the Lagrangian function $L$, with the use of constraints given by Equations 2.2 and 2.3, can be expressed as

$$L = \int_0^\infty f(x) \frac{1 - [f(x)]^{m-1}}{m-1} dx + \left(\lambda_0 - \frac{1}{m-1}\right) \left[\int_0^\infty f(x)dx - 1\right]$$

$$+ \sum_{i=1}^M \lambda_i \left[\int_0^\infty g_i(x)f(x)dx - \overline{g_i(x)}\right] \tag{2.19}$$

where $\lambda_i$, $i = 0, 1, 2, ..., M$, are the Lagrange multipliers. Note that $-1/(m-1)$ is added to the zeroth Lagrange multiplier for simplifying the algebra a little bit. Differentiating Equation 2.19 with respect to $f(x)$ and equating the derivative to zero, we obtain

$$\frac{dL}{df(x)} = 0 = \frac{1}{m-1} - [f(x)]^{m-1} + \lambda_0 - \frac{1}{m-1} + \sum_{i=1}^M \lambda_i g_i(x) \tag{2.20}$$

### 2.1.4 PROBABILITY DISTRIBUTION

Equation 2.20 yields the least-biased probability distribution of $X$:

$$f(x) = \left[\lambda_0 + \sum_{i=1}^M \lambda_i g_i(x)\right]^{1/(m-1)} \tag{2.21}$$

The cumulative distribution function $F(x)$ is obtained by integrating Equation 2.21 as

$$F(x) = \int_0^x \left[\lambda_0 + \sum_{i=1}^M \lambda_i g_i(x)\right]^{1/(m-1)} dx \tag{2.22}$$

The properties of the probability distribution can be described for a general value of $m$, once functions $g_i(x)$ are known and $M$ is given.

### 2.1.5 DETERMINATION OF THE LAGRANGE MULTIPLIERS

Equation 2.21 has $M$ unknown Lagrange multipliers that can be determined with the use of Equations 2.2 and 2.3. Substituting Equation 2.21 in Equations 2.2 and 2.3, one gets, respectively,

$$\int_0^\infty \left[\lambda_0 + \sum_{i=1}^M \lambda_i g_i(x)\right]^{1/(m-1)} dx = 1 \tag{2.23}$$

$$\int_0^\infty g_i(x) \left[ \lambda_0 + \sum_{i=1}^M \lambda_i g_i(x) \right]^{1/(m-1)} dx = \overline{g_i(x)}, \quad i = 1, 2, \dots, M \qquad (2.24)$$

Equations 2.23 and 2.24 can be solved numerically for the unknown Lagrange multipliers $\lambda_i$, $i = 1, 2, \dots, M$. It may be noted that $\lambda_0$, with the use of Equation 2.23, can be expressed as a function of the other Lagrange multipliers and is, therefore, not an unknown multiplier.

### 2.1.6  MAXIMUM ENTROPY

Substitution of Equation 2.21 in Equation 2.1 leads to the maximum Tsallis entropy:

$$H = \frac{1}{m-1} \left\{ 1 - \int_0^\infty \left[ \lambda_0 + \sum_{i=1}^M \lambda_i g_i(x) \right]^{m/(m-1)} dx \right\} \qquad (2.25)$$

## 2.2  MAXIMUM ENTROPY–BASED DISTRIBUTIONS WITH REGULAR MOMENTS AS CONSTRAINTS

### 2.2.1  MEAN AS A CONSTRAINT

Now the mean constraint is defined as

$$\int_0^\infty x f(x) dx = E(x) = \mu_1 \qquad (2.26)$$

where $\mu_1$ is the mean of $X$ and it is approximated by the sample mean $\bar{x}$.

In order to obtain the least-biased $f(x)$, subject to Equations 2.2 and 2.26, Equation 2.1 can be maximized for $m > 0$ using the method of Lagrange multipliers. To that end, the Lagrangian function $L$ can be written as

$$L = \int_0^\infty f(x) \frac{1 - [f(x)]^{m-1}}{m-1} dx - \lambda_0 \left[ \int_0^\infty f(x) dx - 1 \right] - \lambda_1 \left[ \int_0^\infty x f(x) dx - \mu_1 \right] \qquad (2.27)$$

where $\lambda_0$ and $\lambda_1$ are the Lagrange multipliers. Differentiating $L$ with respect to $f(x)$ and equating the derivative to zero, one obtains

$$f(x) = \left[ \frac{1}{m} + \frac{(1-m)}{m} [\lambda_0 + \lambda_1 x] \right]^{1/(m-1)} \qquad (2.28)$$

Defining $k = (1 - m)/m$ (Koutsoyiannis, 2005a) and $\alpha_i = m\lambda_i, i = 0,1$, Equation 2.28 can be written as

$$f(x) = (1+k)^{-1-1/k}[1+k(\alpha_0 + \alpha_1 x)]^{-1-1/k} \tag{2.29}$$

Equation 2.29 is the Tsallis entropy–based PDF of power type.

The Lagrange multipliers, $\lambda_i$, $i = 0$, 1, and consequently $\alpha_i = m\lambda_i$, $i = 0$, 1 can be determined using Equations 2.2 and 2.26. Substituting Equation 2.29 in Equation 2.2, one obtains

$$\int_0^\infty (1+k)^{-(1+k)/k}[1+k(\alpha_0 + \alpha_1 x)]^{-(1+k)/k} dx = 1 \tag{2.30}$$

whose solution yields

$$\alpha_1 = (1+k)^{-1-1/k}(1+k\alpha_0)^{-1/k} \tag{2.31}$$

or

$$\alpha_0 = -1 + \frac{1}{k}[\alpha_1(1+k)^{1+1/k}]^k \tag{2.32}$$

Now, inserting Equation 2.29 in Equation 2.26, one gets

$$(1+k)^{-(1+m)/m} \int_0^\infty x\{(1+k)^{-(1+k)/k}[1+k(\alpha_0 + \alpha_1 x)^{-(1+k)/k}]dx\} = \bar{x} \tag{2.33}$$

Integration of Equation 2.33 by parts yields

$$-\frac{(1+k)^{-1-1/k}}{k(1-(1/k))}(1+k\alpha_0)^{1-1/k} = \alpha_1^2 \bar{x} \tag{2.34}$$

Equations 2.31 and 2.34 can be utilized to determine $\alpha_0$ and $\alpha_1$. It may be noted that one can also determine the other Lagrange multiplier by inserting Equation 2.31 or 2.32 in Equation 2.29 and the resulting function in Equation 2.26 and then integrating.

Inserting Equation 2.31 in Equation 2.29, the PDF becomes

$$f(x) = (1+k)^{-1-1/k}(1+k\alpha_0)^{-1-1/k}[1+k(1+k)^{-1-1/k}k(1+k\alpha_0)^{-1-1/k}x] \tag{2.35}$$

Equation 2.35 can be simplified by defining

$$\beta = [(1+k)(1+k\alpha_0)]^{-1-1/k} \tag{2.36}$$

Equation 2.35, with the use of Equation 2.36, becomes

$$f(x) = \frac{1}{\beta}\left[1 + \frac{kx}{\beta}\right]^{-1-1/k} \tag{2.37}$$

Equation 2.37 is a two-parameter generalized Pareto distribution.
   Let

$$y = x + \frac{\beta}{k} \Rightarrow x = y - \frac{\beta}{k} \tag{2.38}$$

Equation 2.37 then becomes

$$f(y) = \frac{1}{\beta}\left[\frac{ky}{\beta}\right]^{-1-1/k} \tag{2.39}$$

Equation 2.39 is a two-parameter Pareto distribution.
   If $k$ tends to 0, Equation 2.39 leads to an exponential distribution:

$$f(y) = \frac{1}{\beta}\exp\left[\frac{y}{\beta}\right] \tag{2.40}$$

### 2.2.2 MEAN AND VARIANCE AS CONSTRAINTS

The first and second moments constitute the mean and variance constraints, which can be, respectively, defined by Equations 2.26 and 2.7 as

$$\int_0^\infty x^2 f(x)dx = E(x^2) = \mu_2 \tag{2.41}$$

where $\mu_2$ is the second moment about the mean. In order to obtain the least-biased $f(x)$, subject to Equations 2.2, 2.26, and 2.41, Equation 2.1 is maximized for $m > 0$ using the method of Lagrange multipliers as before. The Lagrangian function $L$ can be written as

$$L = \int_0^\infty f(x)\frac{1-[f(x)]^{m-1}}{m-1}dx - \lambda_0\left[\int_0^\infty f(x)dx - 1\right] - \lambda_1\left[\int_0^\infty xf(x)dx - \mu_1\right]$$

$$- \lambda_2\left[\int_0^\infty x^2 f(x)dx - \mu_2\right] \tag{2.42}$$

Differentiating $L$ with respect to $f(x)$ and equating the derivative to zero, one obtains

$$f(x) = \left\{ \frac{1}{m} + \frac{(1-m)}{m}[\lambda_0 + \lambda_1 x + \lambda_2 x^2] \right\}^{1/(m-1)} \tag{2.43}$$

Using the definition of $k = (1 - m)/m$ and $\alpha_i = m\lambda_i$, $i = 0, 1, 2$, Equation 2.43 is written as

$$f(x) = (1+k)^{-1-1/k}[1 + k(\alpha_0 + \alpha_1 x + \alpha_2 x^2)]^{-1-1/k} \tag{2.44}$$

Equation 2.44 is, again, the PDF of power type. The Lagrange multipliers, $\lambda_i$, $i = 0$, 1, 2, and consequently $\alpha_i = m\lambda_i$, $i = 0, 1, 2$ can be estimated using Equations 2.2, 2.26, and 2.41.

For purposes of comparison, Koutsoyiannis (2005a,b) proposed a form similar to Equation 2.44 as

$$f(x) = [1 + k(\alpha_0 + \alpha_1 x^{c_1})]^{-1-1/k} x^{c_2-1} \tag{2.45}$$

where $c_1$ and $c_2$ are shape parameters. Thus, Equation 2.45 has four parameters: scale parameter $\alpha_1$ and shape parameters $k$ and $c_1$ and $c_2$. Note that $\alpha_0$ is not a parameter, because it is a constant based on the satisfaction of Equation 2.2.

Using appropriate transformations of random variable $X$ and limiting values of shape parameters, Koutsoyiannis (2005a) showed that Equation 2.45 would lead to several exponential and power-type probability distributions as given in the following.

1. Random variable $X^{c_2}$ would have beta prime (also referred to as beta of the second kind) distribution (Evans et al., 2000). Then the distribution of $X$ would be referred to as power-transformed Beta Prime (PBP): Beta prime ($c_1 = 1$) (Figure 2.2),

$$f(x) = (1 + k\alpha_0 + k\alpha_1 x)^{-(1+k)/k} x^{c_2-1} \tag{2.46}$$

2. PBP-L1 ($k \to 0$) (Figure 2.3),

$$f(x) = \exp(-\alpha_0 - \alpha_1 x^{c_1}) x^{c_2-1} \tag{2.47}$$

3. PBP-L2 ($k \to \infty$, $k\alpha_0 \to k_0$, $k\alpha_1 \to k_1$) (Figure 2.4),

$$f(x) = \frac{x^{c_2-1}}{1 + k_0 + k_1 x^{c_1}} \tag{2.48}$$

4. Gamma ($k \to 0$, $c_1 \to 1$) (Figure 2.5),

$$f(x) = \exp(-\alpha_0 - \alpha_1 x) x^{c_2-1} \tag{2.49}$$

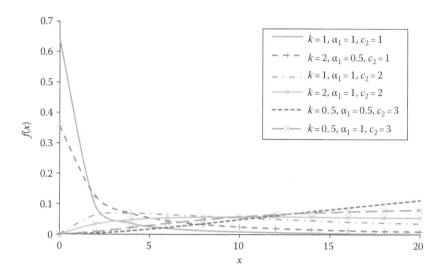

**FIGURE 2.2**  PBP distribution.

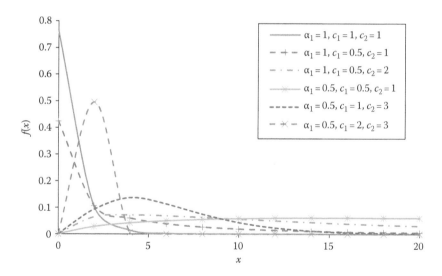

**FIGURE 2.3**  PBP-L1 distribution.

5. Weibull ($k \to 0, c_2 = c_1$) (Figure 2.6),

$$f(x) = \exp(-\alpha_0 - \alpha_1 x^{c_1}) x^{c_1 - 1} \qquad (2.50)$$

6. Pareto ($c_2 = c_1 = 1$) (Figure 2.7),

$$f(x) = (1 + k\alpha_0 + k\alpha_1 x)^{-(1+k)/k} \qquad (2.51)$$

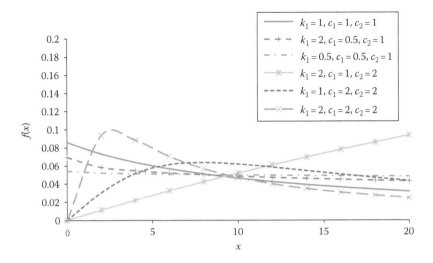

**FIGURE 2.4**   PBP-L2 distribution.

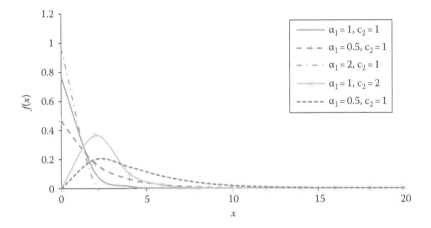

**FIGURE 2.5**   Gamma distribution.

Following Koutsoyiannis (2005a), for the Tsallis entropy, if $\lambda_2 = 0$, the Pareto distribution is the result:

$$f(x) = \frac{1}{\lambda}\left(1 + \frac{k}{\lambda}x\right)^{-(1+k)/k} \tag{2.52}$$

$$F^*(x) = \left(1 + \frac{k}{\lambda}x\right)^{-1/k} ; \quad k = \frac{(1-m)}{m} \tag{2.53}$$

where $F^* = 1 - F(x)$, $F(x)$ is the cumulative probability distribution function.

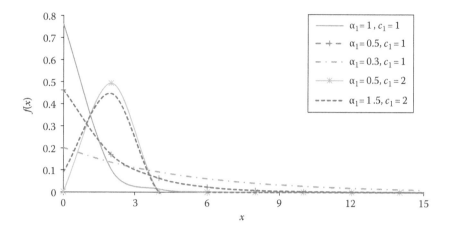

**FIGURE 2.6**    Weibull distribution.

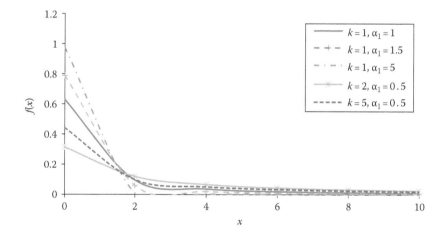

**FIGURE 2.7**    Pareto distribution.

## Example 2.1

Plot the entropy-based PDFs versus the random variable for several values of the coefficient of variation and do the same if the variable is standardized. Take the mean value as 1.

### Solution

For different values of coefficient of variation ($\sigma/\mu = 0.1$, 0.5, 1, 1.5, and 5), the second constraint obtained from Equation 2.41 becomes 0.01, 0.25, 1, 1.25, and 25. By substituting Equation 2.43 into Equations 2.26 and 2.41, the Lagrange multipliers can be computed. Thus, the PDFs can be obtained, as shown in Figure 2.8. For standardized variable $(x - \mu)/\sigma$, the PDF is computed in the same way, as plotted in Figure 2.9.

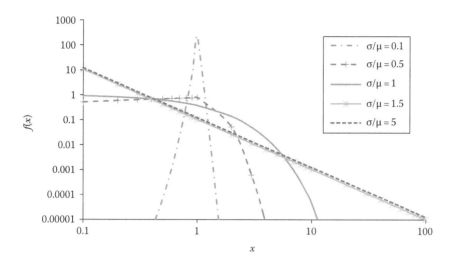

**FIGURE 2.8**  Probability density function.

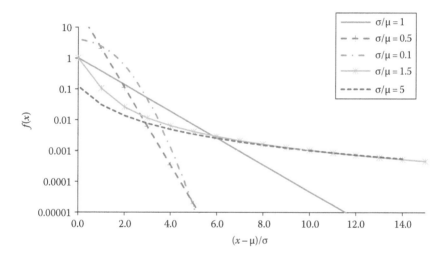

**FIGURE 2.9**  Probability density function.

## Example 2.2

Plot the values of random variable or quantile versus the return period [1–100,000] for various values of the coefficient of variation.

### Solution

Consider random variable as computed in Example 2.1 in which $\mu = 1$ and $\sigma/\mu = 0.1, 0.5, 1, 1.5,$ and 5. The return period $T$ is computed as $1/[1 - F(x)]$ and plotted in Figure 2.10, where $F(x)$ is the cumulative distribution of Example 2.1.

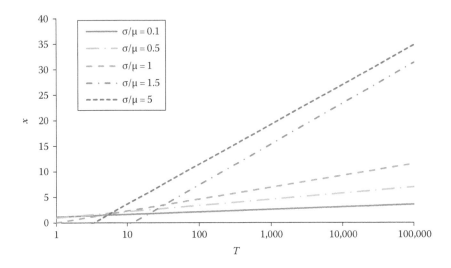

**FIGURE 2.10**  Quantile versus return period.

## Example 2.3

Plot the maximum Tsallis entropy, using Equation 2.45, for various values of $c_1$ for specified values of $k$. Take mean as 1 and the coefficient of variation as 1.5 and $c_2 = 1$. Also plot on the same graph $c_1$ versus $k$. Plot maximum entropy versus the coefficient of variation [0.01–100]; $m = 1$ will lead to truncated normal, $m = 1$ to Pareto and $m < 1$ to Pareto.

### Solution

Let $c_2 = 1$, the Tsallis entropy is computed by inputting Equation 2.45 into Equation 2.1:

$$H = \int_0^\infty f(x)\frac{1-[f(x)]^{m-1}}{m-1}dx = \int_0^\infty f(x)\frac{1-[1+k(\alpha_0+\alpha_1 x^{c_1})]}{m-1}dx$$

$$= \left(-1-\frac{1}{k}\right)\int_0^\infty f(x)[k(\alpha_0+\alpha_1 x^{c_1})]dx$$

$$= -(k+1)\left[k\alpha_0 + k\alpha_1\overline{x^{c_1}}\right]$$

With $\mu = 1$ and $\sigma/\mu = 1.5$, for various values of $c_1 = 1, 1.5, 2$, and 3, the entropy is computed and plotted in Figure 2.11.

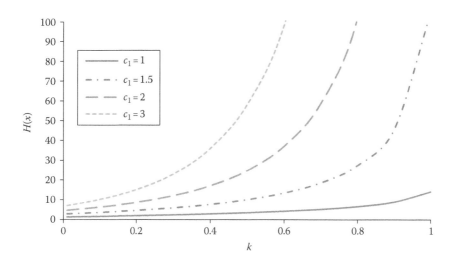

**FIGURE 2.11**  Tsallis entropy for $c_2 = 1$.

### 2.2.3  VARIANCE AS A CONSTRAINT

Let the constraints be defined by Equation 2.2 and

$$\int_{-\infty}^{\infty} \frac{1}{2} x^2 f(x)\,dx = \frac{\overline{x^2}}{2} \tag{2.54}$$

Then, the Lagrangian function $L$ can be expressed as

$$L = \frac{1 - \int_{-\infty}^{\infty} f^m(x)\,dx}{m-1} + \lambda_0 \left[ \int_{-\infty}^{\infty} f(x)\,dx - 1 \right] + \lambda_0\lambda_1(m-1)\left[ \int_{-\infty}^{\infty} \frac{x^2}{2} f(x)\,dx - \frac{\overline{x^2}}{2} \right] \tag{2.55}$$

Differentiating Equation 2.55 with respect to $f(x)$, one gets

$$\frac{dL}{df(x)} \rightarrow -\frac{m}{m-1} f^{m-1}(x) + \lambda_0 + \lambda_0\lambda_1(m-1)\frac{x^2}{2} = 0 \tag{2.56}$$

This yields

$$f(x) = \left\{ \frac{m-1}{m}\left[ \lambda_0 + \lambda_0\lambda_1(m-1)\frac{x^2}{2} \right] \right\}^{1/(m-1)} \tag{2.57}$$

Substituting Equation 2.57 in Equation 2.2, one obtains

$$\left[\frac{m}{m-1}\lambda_0\right] = \frac{1}{\int_{-\infty}^{\infty}[1+\lambda_1(m-1)(x^2/2)]^{1/(m-1)}dx} \tag{2.58}$$

Substituting Equation 2.57 in Equation 2.26, the PDF becomes

$$f(x) = \frac{[1+(m-1)\lambda_1(x^2/2)]^{1/(m-1)}}{\int_{-\infty}^{\infty}[1+(m-1)\lambda_1(x^2/2)]^{1/(m-1)}dx} = \frac{[1+(m-1)\lambda_1(m-1)(x^2/2)]^{1/(m-1)}}{Z} \tag{2.59}$$

where $Z$ is the partition function defined as

$$Z = \int_{-\infty}^{\infty}\left[1+(m-1)\lambda_1\frac{x^2}{2}\right]^{1/(m-1)}dx \tag{2.60}$$

Beck (2000) has shown that the integral in Equation 2.59 exists for $1 \le m < 3$ and it integrates to

$$Z = \sqrt{\frac{2}{(m-1)\lambda_1}}B\left(\frac{1}{2}; \frac{1}{m-1}-\frac{1}{2}\right) \tag{2.61}$$

in which $B(\cdot;\cdot)$ is the beta function defined as

$$B(a,b) = \frac{\Gamma(a)\Gamma(b)}{\Gamma(a+b)} \tag{2.62}$$

where $\Gamma(\cdot)$ is the gamma function of $(\cdot)$. If $k = (1/m - 1) \ge 2$ is an integer, the partition function can be further evaluated as

$$Z = \sqrt{\frac{2k}{\lambda_1}}\frac{\pi(2k-3)!!}{(k-1)!2^{k-1}} \tag{2.63}$$

Figure 2.12 sketches the PDF with unit variance for $k = 1/(m - 1) = 3, 4, \ldots, 20$. As $k$ changes to smaller values, the PDF would exhibit a transition from almost Gaussian to a stretched one.

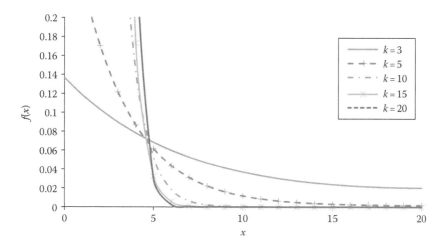

**FIGURE 2.12** Probability density function.

## 2.3 USE OF *m*-EXPECTATION

The $m$-expectation for a random variable $X$ can be defined as

$$\int_0^\infty xf^m(x)dx = \overline{x_m} \tag{2.64}$$

Now the Tsallis entropy is maximized, subject to Equations 2.2 and 2.64, using the method of Lagrange multipliers. The Lagrange multiplier $L$ is constructed as

$$L = \frac{1}{m-1}\left[1-\int_0^\infty f^m(x)dx\right]-\lambda_0\left[\int_0^\infty f(x)dx\right]-\lambda_1\left[\int_0^\infty xf^m(x)dx - \overline{x_m}\right] \tag{2.65}$$

Differentiating Equation 2.65 with respect to $f(x)$ and equating the derivative to 0, one obtains

$$\frac{\partial L}{\partial f} = 0 \Rightarrow -\frac{m}{m-1}f^{m-1}(x)-\lambda_0-\lambda_1 xmf^{m-1}(x) \tag{2.66}$$

This yields

$$f(x) = \left[\frac{1-m}{m}\frac{\lambda_0}{1+\lambda_1(m-1)x}\right]^{1/(m-1)} \tag{2.67}$$

The Lagrange parameters $\lambda_0$ and $\lambda_1$ are determined with the use of Equations 2.2 and 2.64. Substituting Equation 2.67 in Equation 2.2, the result is

$$\int_0^\infty \left[\frac{1}{1+(m-1)x\lambda_1}\right]^{1/(m-1)} dx = \left(\frac{1-m}{m}\lambda_0\right)^{-1/(m-1)} \tag{2.68}$$

Inserting Equation 2.67 in Equation 2.64, one obtains

$$\int_0^\infty x\left[\frac{1-m}{m}\frac{\lambda_0}{1+(m-1)\lambda_1 x}\right]^{m/(m-1)} dx = \overline{x_m} \tag{2.69}$$

Equations 2.68 and 2.69 can be solved to obtain the Lagrange multipliers $\lambda_0$ and $\lambda_1$. Inserting Equation 2.68 in Equation 2.67, the probability distribution becomes

$$f(x) = \frac{[1-(1-m)\lambda_1 x]^{1/(1-m)}}{\displaystyle\int_0^\infty [1-(1-m)\lambda_1 x]^{1/(1-m)} dx} \tag{2.70}$$

Equation 2.70 is the Tsallis distribution that is a power law. Thus, when a phenomenon exhibits a power-law behavior then it is logical to maximize the Tsallis entropy that can accommodate this type of behavior. There are many geophysical phenomena, such as recurrence intervals between floods, braided gravel hydraulic conductivity, wind patterns, and riverbed patterns, to name but a few, that exhibit a power-law type behavior. In the limit $m \to 1$, Equation 2.70 reduces to the conventional exponential distribution.

An $m$-exponential function is defined as

$$\exp_m(x) = e_m^x = [1+(1-m)x]^{1/(1-m)} \tag{2.71}$$

With the use of Equation 2.71, Equation 2.70 can be written as

$$f(x) = \frac{e_m^{-\lambda_1 x}}{\displaystyle\int_0^\infty e_m^{-\lambda_1 x} dx} = \frac{\exp_m(-\lambda_1 x)}{\displaystyle\int_0^\infty \exp_m(-\lambda_1 x) dx} \tag{2.72}$$

## 2.4   CHOOSING EXPECTATION VALUE

Abe and Bagci (2005) dealt with the choice of expectation value by examining the entropies associated with two kinds of expectation: non-normalized and normalized. Following Bashkirov (2004), consider a Lagrange function $L$.

Two operations, namely translation ($T$) and dilatation ($D$), under which function $L$ is invariant, are performed. The generators, corresponding to these operations, can be expressed as

$$T_i = \frac{\delta}{\delta p_i} \tag{2.73}$$

and

$$D_i = \sum_{i=1}^{N} p_i \frac{\delta}{\delta p_i} \tag{2.74}$$

Since $L$ is invariant when subjected to these operations, one can write

$$T_i L = \frac{\delta L}{\delta p_i} = 0 \tag{2.75}$$

$$D_i L = \sum_{i=1}^{N} p_i \frac{\delta L}{\delta p_i} = 0 \tag{2.76}$$

This is now applied to the Tsallis entropy indexed by $m$.

### 2.4.1 ORDINARY EXPECTATION

The Lagrangian function to be maximized, subject to Equations 2.2 and 2.26, can be written as

$$L = \frac{1}{m-1}\left[1 - \sum_{i=1}^{N} p_i^m\right] - \lambda_0\left(\sum_{i=1}^{N} p_i - 1\right) - \lambda_1\left(\sum_{i=1}^{N} p_i x_i - \bar{x}\right) \tag{2.77}$$

where $\lambda_0$ and $\lambda_1$ are the Lagrange multipliers. Applying Equation 2.75 to Equation 2.77, the result is

$$\frac{\delta L}{\delta p_i} = 0 = \frac{m}{1-m} p_i^{m-1} - \lambda_0 - \lambda_1 x_i \tag{2.78}$$

This yields

$$\lambda_0 = \frac{m}{1-m} p_i^{m-1} - \lambda_1 x_i \tag{2.79}$$

Now applying Equation 2.76 to Equation 2.77, one obtains

$$\sum_{i=1}^{N} p_i \frac{\delta L}{\delta p_i} = 0 = \sum_{i=1}^{N} p_i \left[ \frac{m}{1-m} p_i^{m-1} - \lambda_0 - \lambda_1 x_i \right] = \frac{m}{1-m} \sum_{i=1}^{N} p_i^m - \lambda_0 \sum_{i=1}^{N} p_i - \lambda_1 \sum_{i=1}^{N} p_i x_i$$

$$= \frac{m}{m-1} \left[ -1 + \sum_{i=1}^{N} p_i^m + 1 \right] - \lambda_0 - \lambda_1 \bar{x}$$

(2.80)

Equation 2.80 can be written in terms of the Tsallis entropy as

$$\lambda_0 = \frac{m}{1-m} [1 + (1-m)H_m] - \lambda_1 \bar{x}$$ 

(2.81)

Equations 2.79 and 2.81 are solved for $p_i$. Equating Equation 2.79 to Equation 2.81, one gets

$$\frac{m}{1-m} p_i^{m-1} - \lambda_1 x_i = \frac{m}{1-m} [1 + (1-m)H_m] - \lambda_1 \bar{x}$$

(2.82)

Equation 2.82 is simplified sometimes as follows:

$$p_i = \{ [1 + (1-m)H_m] + \frac{1-m}{m} \lambda_1 (x_i - \bar{x}) \}^{1/(m-1)}$$

$$= [1 + (1-m)H_m]^{1/(m-1)} \left[ 1 - \frac{m-1}{m} \frac{\lambda_{1*}(x_i - \bar{x})}{1 + (1-m)H_m} \right]^{1/(m-1)}$$

(2.83)

With the use of the definition of $H$, Equation 2.83 can be cast as

$$p_i = [1 + (1-m)H_m]^{1/(m-1)} \left[ 1 - \frac{m-1}{m} \frac{\lambda_1 (x_i - \bar{x})}{1 + (1-m)\left( 1 - (1/(m-1)) \sum_{j=1}^{N} p_j^m \right)} \right]^{1/(m-1)}$$

(2.84)

That is, Equation 2.84 can now be expressed as

$$p_i = [1 + (1-m)H]^{1/(m-1)} \times \left[ 1 - \frac{m-1}{m} \lambda_{1*}(x_i - \bar{x}) \right]^{1/(m-1)}$$

(2.85)

where

$$\lambda_{1*} = \frac{\lambda_1}{\sum_{j=1}^{N} p_j^m} \tag{2.86}$$

## 2.4.2 Normalized m-Expectation

$$\sum_{i=1}^{N} P_i x_i = \overline{x_m} = \langle x_m \rangle \tag{2.87}$$

where

$$P_i = \frac{p_i^m}{\sum_{j=1}^{N} p_j^m} \tag{2.88}$$

where

$P_i$ is often referred to as the escort distribution associated with the basic distribution $p_i$

$X$ denotes a physical random variable with its specific value $x_i$

Note the ordinary expectation value $\overline{x}$ is denoted as

$$\overline{x} = \sum_{i=1}^{N} p_i x_i \tag{2.89}$$

Following the same procedure as earlier, the Lagrange function becomes

$$L = \frac{1}{m-1} \left[ 1 - \sum_{i=1}^{N} p_i^m \right] - \lambda_0 \left( \sum_{i=1}^{N} p_i - 1 \right) - \lambda_1 \left[ \frac{\sum_{i=1}^{N} x_i p_i^m}{\sum_{i=1}^{N} p_i^m} - \overline{x_m} \right] \tag{2.90}$$

Here, $\overline{x_m}$ denotes the $m$-expectation of $X$ defined by Equation 2.87. Applying Equations 2.75 and 2.76 to Equation 2.90, the result is

$$T_i L = 0 = -\frac{m}{m-1} p_i^{m-1} - \lambda_0 - m\lambda_{1*} \left( x_i - \overline{x_m} \right) p_i^{m-1} \tag{2.91}$$

$$DL = 0 \Rightarrow \lambda_0 = \frac{m}{1-m} [1 + (1-m)H_m] \tag{2.92}$$

Equations 2.91 and 2.92 lead to

$$p_i = \frac{1}{Z}\left[1-(1-m)\lambda_{1*}\left(x_i - \overline{x_m}\right)\right]^{1/(1-m)} \tag{2.93}$$

where

$$Z = [1+(1-m)H_m]^{1/(1-m)} = \sum_{i=1}^{N}\left[1-(1-m)\lambda_{1*}\left(x_i - \overline{x_m}\right)\right]^{1/(1-m)} \tag{2.94}$$

Abe and Bagci (2005) have shown an interesting property of the expectation definitions:

$$\frac{\partial H_m^{(ordinary)}}{\partial x} = \lambda_1 \tag{2.95}$$

$$\frac{\partial H_m^{(normalized)}}{\partial \overline{x_m}} = \lambda_1 \tag{2.96}$$

### 2.4.3 USE OF *m*-EXPECTATION

Let the Tsallis entropy be defined as

$$H_m[p(x)] = \frac{1}{m-1}\left[1 - \int_{-\infty}^{\infty}\frac{1}{\sigma}[\sigma p(x)]^m\, dx\right] \tag{2.97}$$

Let the constraints be defined as Equation 2.2 and

$$\frac{\int_{-\infty}^{\infty} x^2 [p(x)]^m\, dx}{\int_{-\infty}^{\infty} [p(x)]^m\, dx} = \left\langle x\right\rangle_m = \left\langle x_m \right\rangle = \overline{x_m^2} \tag{2.98}$$

Optimization of Equation 2.97 with constraints given by Equations 2.98 and 2.2 yields for $m > 1$

$$p(x) = \frac{1}{\sigma}\left[\frac{m-1}{\pi(3-m)}\right]^{1/2}\frac{\Gamma(1/(m-1))}{\Gamma((3-m)/(2(m-1)))}\frac{1}{[1+((m-1)/(3-m))(x^2/\sigma^2)]^{1/(m-1)}} \tag{2.99}$$

and for $m < 1$

$$p(x) = \frac{1}{\sigma}\left[\frac{1-m}{\pi(3-m)}\right]^{1/2} \frac{\Gamma((5-3m)/(2(1-m)))}{\Gamma((2-m)/(1-m))}\left[1-\frac{1-m}{3-m}\frac{x^2}{\sigma^2}\right]^{1/(1-m)} \quad (2.100)$$

provided $x < \sigma[(3-m)/(1-m)]^{1/2}$ and 0 otherwise. Prato and Tsallis (1999) have derived this result. For $m \to 0$, we obtain

$$p(x) = \frac{1}{Z}\exp(-\beta x^2) \quad (2.101)$$

$$Z = \int_{-\infty}^{\infty} \exp(-\beta x^2)dx = \left(\frac{\pi}{\beta}\right)^{1/2} \quad (2.102)$$

$$\beta = \frac{1}{2\sigma^2} \quad (2.103)$$

Special cases: $m < 5/3 (m \geq 5/3)$ corresponds to a finite (or infinite) second moment $\langle x^2 \rangle_m$. For $m < 1$, there is a cutoff; for $m > 1$, there is a $1/|x|^{2/(m-1)}$ tail at $|x| \geq \sigma$. For $m < 5/3$, the second moment is finite given as

$$\langle x^2 \rangle = \sigma^2\left[\frac{3-m}{5-3m}\right] \quad (2.104)$$

$m \to \infty$: uniform distribution
$m \to 2$: Cauchy–Lorentz distribution
$m \to 3$: completely flat distribution
$m \geq 3$: no distribution, that is Equation 2.67 cannot be satisfied
$m \to 1$: Gaussian distribution

Escort distribution

$$P_m(x) = \frac{[p(x)]^m}{\int_{-\infty}^{\infty}[p(x)]^m dx} \quad (2.105)$$

For $m > 1$, this is given as

$$p(x) = \frac{1}{\sigma}\left[\frac{m-1}{\pi(3-m)}\right]^{1/2} \frac{\Gamma(m/(m-1))}{\Gamma((m+1)/(2(m-1)))}\frac{1}{[1+((m-1)/(3-m))x^2/\sigma^2]^{m/(m-1)}} \quad (2.106)$$

and for $m < 1$

$$p(x) = \frac{1}{\sigma}\left[\frac{1-m}{\pi(3-m)}\right]^{1/2} \frac{\Gamma((3-m)/(2(1-m)))}{\Gamma(1/(1-m))}\left[1 - \frac{1-m}{3-m}\frac{x^2}{\sigma^2}\right]^{m/(1-m)} \quad (2.107)$$

if $x < \sigma[(3 - m)/(1 - m)]^{1/2}$ and 0 otherwise (cutoff is maintained even if $m \leq 0$). For $m < 3$ (and not only for $m < 5/3$), $P_m(x)$ has a finite second moment. For $m < 5/3$, the escort distribution is Gaussian $\propto \exp\{-[m(5 - 3m)/2(3 - m)]x^2/(\sigma^2 N)\}$, where $N$ = number of particles.

## REFERENCES

Abe, S. and Bagci, G.B. (2005). Necessity of q-expectation value in nonextensive statistical mechanics. *Physical Review E*, 71, 016139-1–016139-5.

Bashkirov, A.G. (2004). On maximum entropy principle, superstatistics, power-law distribution and Renyi parameter. *Physica A*, 340, 153–162.

Beck, C. (2000). Application of generalized thermostatistics to fully developed turbulence. *Physica A*, 277, 115–123.

Evans, N., Hastings, N., and Peacock, B. (2000). *Statistical Distributions*. John Wiley & Sons, New York.

Jaynes, E.T. (1957a). Information theory and statistical mechanics, I. *Physical Reviews*, 106, 620–630.

Jaynes, E.T. (1957b). Information theory and statistical mechanics, II. *Physical Reviews*, 108, 171–190.

Koutsoyiannis, D. (2004a). Statistics of extremes and estimation of rainfall: 1. Theoretical investigation. *Hydrological Sciences Journal*, 49(4), 575–590.

Koutsoyiannis, D. (2004b). Statistics of extremes and estimation of rainfall: 1. Empirical investigation of long rainfall records. *Hydrological Sciences Journal*, 49(4), 591–610.

Koutsoyiannis, D. (2005a). Uncertainty, entropy, scaling and hydrological stochastics. 1. Marginal distributional properties of hydrological processes and state scaling. *Hydrological Sciences Journal*, 50(3), 381–404.

Koutsoyiannis, D. (2005b). Uncertainty, entropy, scaling and hydrological stochastics. 2. Time dependence of hydrological processes and state scaling. *Hydrological Sciences Journal*, 50(3), 405–426.

Koutsoyiannis, D. (2006). An entropic-stochastic representation of rainfall intermittency: The origin of clustering and persistence. *Water Resources Research*, 42, W01401, doi: 10.1029/2005WR004175.

Papalexiou, S.M. and Koutsoyiannis, D. (2012). Entropy-based derivation of probability distributions: A case study to daily rainfall. *Advances in Water Resources*, 45, 51–57.

Papalexiou, S.M. and Koutsoyiannis, D. (2013). Battle of extreme value distributions: A global survey extreme daily rainfall. *Water Resources Research*, 49, 87–201, doi: 10.029/2012WR012557.

Prato, D. and Tsallis, C. (1999). Nonextensive foundation of Levy distributions. *Physical Review E*, 60(2), 2398–2401.

Tsallis, C., Mendes, R.S., and Plastino, A.R. (1998). The role of constraints within generalized nonextensive statistics. *Physica A*, 261, 534–554.

# Section II

---

## Hydraulic Engineering

# 3 One-Dimensional Velocity Distributions

Fundamental to determining flow discharge, scour around bridge piers, erosion and sediment transport, pollutant transport, energy and momentum distribution coefficients, hydraulic geometry, watershed runoff, and river behavior is velocity distribution. Consider, for example, discharge measurements that involve velocity sampling. In order for these measurements to be simple and efficient, the number of velocity samples to be taken must be sufficiently small so that sampling may be accomplished quickly and within the time frame of the particular flow and velocity regime being investigated. Translating a small number of velocity samples into the cross-sectional mean velocity requires a velocity distribution equation. In wide open channels, velocity increases monotonically from the channel bed toward the water surface and can be approximately considered as one-dimensional. The objective of this chapter is to present velocity distributions in one dimension in open channels using the Tsallis entropy.

## 3.1 PRELIMINARIES

Flow in an open channel at a given time and location can be laminar, turbulent, or mixed (transitional). The flow in open channels on alluvial sand beds as well as gravel beds is generally hydraulically rough, and therefore, turbulent flow prevails for most natural conditions. If the flow is laminar, then the velocity can be defined accurately. However, in turbulent flow, the velocity vector fluctuates both spatially and temporally and velocity is not stationary.

Velocity distributions have been derived using either experimental or deterministic hydrodynamic methods. The velocity distributions popular in hydraulics are the Prandtl–von Karman universal velocity distribution and power law velocity distribution (Karim and Kennedy, 1987). Limitations of these velocity distributions have been discussed by Chiu (1987), Singh (1996), among others. The Prandtl–von Karman universal velocity distribution was initially developed for pipe flow (von Karman, 1935) and was then applied to wide open channels (Vanoni, 1941). The universal velocity distribution does not predict the velocity near the bottom well, especially in sediment-laden flows (Einstein and Chien, 1955), and is also found to be inaccurate near the water surface. The power law velocity distribution was first developed for smooth pipes (Blasius, 1913) and then expanded to open channel flow (Sarma et al., 1983). This distribution is simple to apply but its accuracy is limited.

The velocity of flow in an open channel cross section varies along a vertical from zero at the bed to a maximum value that may or may not occur at the water surface.

At any point or in any cross section, the flow velocity varies with time, but this time variation does not follow a particular pattern and depends on the water and sediment influx. Chiu and his associates (Chiu and Chiou, 1986; Chiu, 1987, 1988, 1989, 1995; Chiu and Murray, 1992; Chiu and Said, 1995; Chiu and Tung, 2002; Chiu and Chen, 2003; Chiu et al., 2005) assumed that the time-averaged velocity at any point is a random variable and, therefore, has a probability distribution. This assumption has since been employed by a number of investigators for determining distributions of flow velocity and pollutant transport (Barbe et al., 1991; Xia, 1997; Araujo and Chaudhry, 1998; Choo, 2000; Chen and Chiu, 2004). Chiu (1987) was the first to derive a velocity distribution using the Shannon entropy (Shannon, 1948) and the principle of maximum entropy (POME) (Jaynes, 1957a,b) and discuss the uncertainty associated with the distribution. The Chiu distribution has been found to be more accurate than the power law and universal velocity distributions, has the advantage that it satisfactorily predicts the velocity near the bed, and can be employed for optimum velocity sampling.

On the other hand, Singh and Luo (2009, 2011), Luo and Singh (2011), and Cui and Singh (2013) employed the Tsallis entropy (Tsallis, 1988), a generalization of the Shannon entropy, to derive the velocity distribution. The Tsallis entropy–based velocity distribution has been shown to have an advantage over the Shannon entropy–based distribution. However, in these entropy-based velocity distributions, the cumulative probability distribution has been assumed to be linear, meaning the velocity is equally likely along the vertical from the channel bed to the water surface. This assumption is fundamental to the derivation of velocity distributions but has not been adequately scrutinized (Cui and Singh, 2012). Further, this assumption is weak and may partly explain the reason that these velocity distributions do not accurately describe the velocity near the channel bed.

## 3.2 ONE-DIMENSIONAL VELOCITY DISTRIBUTIONS

The procedure for deriving the velocity distribution entails the following steps: (1) formulation of a hypothesis for cumulative distribution of velocity as a function of flow depth, (2) defining the Tsallis entropy, (3) specification of constraints, (4) maximization of entropy, (5) entropy of velocity distribution, and (6) determination of Lagrange multipliers. Each step is discussed in the following.

### 3.2.1 HYPOTHESIS

The time-averaged velocity in a channel cross section is considered as a random variable under the premise that between 0 and channel depth $D$ all values of flow depth $y$ ($0 \leq y \leq D$) are equally likely. This also means that all values of velocity $u$ between zero and maximum are equally likely. In reality, this may not be necessarily true for all channel cross-sections (Cui and Singh, 2012). The probability of velocity being equal to or less than a given value $u$ is assumed to be $y/D$; thus, the cumulative probability distribution of velocity $F(u) = P(\text{velocity} \leq u)$, $P$ = probability, can be expressed in terms of flow depth as

$$F(u) = \frac{y}{D} \qquad (3.1)$$

where
  $y$ is the distance from the bed
  $D$ is the water depth

Differentiation of $F(u)$ in Equation 3.1 yields the probability density function (PDF) of $u$, $f(u)$, as

$$f(u) = \frac{dF(u)}{du} = \frac{1}{D}\frac{dy}{du} \quad \text{or} \quad f(u) = \left( D\frac{du}{dy} \right)^{-1} \qquad (3.2)$$

The term $dF(u) = f(u)du = F(u + du) - F(u)$ denotes the probability of velocity between $u$ and $u + du$. Equation 3.1 constitutes the fundamental hypothesis that is employed for deriving velocity distributions in this chapter.

### 3.2.2 TSALLIS ENTROPY

The objective is to determine the PDF of $u$, $f(u)$. This is accomplished by maximizing the Tsallis entropy of velocity, $H(u)$. If velocity is treated as a discrete random variable taking on $n$ values, then $H(u)$ can be defined as

$$H = \frac{1}{m-1}\left[ 1 - \sum_{i=0}^{n} p(u_i)^m \right] = \frac{1}{m-1}\sum_{i=0}^{n} p(u_i)[1 - p(u_i)^{m-1}] \qquad (3.3)$$

where
  $p(u_i)$ is the probability of $u = u_i$, $i = 1, 2, ..., n$
  $m$ is a real number

Equation 3.3 expresses a measure of uncertainty about $p(u)$ or the average information content of sampled $u$. For continuous nonnegative velocity ($u = u_D$, $y = D$),

$$H(U) = \frac{1}{m-1}\left\{ 1 - \int_0^{u_{max}} [f(u)]^m\, du \right\} = \frac{1}{m-1}\int_0^{u_{max}} f(u)\{1 - [f(u)]^{m-1}\}du \qquad (3.4)$$

where $u_{max}$ is the maximum velocity.

Maximizing $H(u)$ is equivalent to maximizing $f(u)\{1 - [f(u)]^{m-1}\}$. In order to maximize $H(u)$, certain constraints need to be satisfied. If $\{1 - [f(u)]^{m-1}\}/(m - 1)$ is considered as a measure of uncertainty, then Equation 3.4 represents the average uncertainty of $u$ or $f(u)$. More the uncertainty more the ignorance and more information will be needed to characterize $u$. In this sense, information and uncertainty are related. Thus, the key in Equation 3.4 is to derive the least-biased PDF $f(u)$, subject to given information on $u$. This means that the resulting distribution is least biased toward information not given and takes full advantage of the given information.

### 3.2.3 Specification of Constraints

Constraints can be specified in different ways, but whatever the way, they should be simple. The first constraint, $C_1$, is the total probability:

$$C_1 = \int_0^{u_{max}} f(u)du = 1 = b_1 \tag{3.5}$$

In actuality, $C_1$ is not a constraint, for $f(u)$ must always satisfy the total probability. Nevertheless, it is treated as a constraint for the sake of discussion.

Flow in open channels satisfies the laws of conservation of mass, momentum, and energy. These laws constitute additional constraints that entropy maximization is subjected to. The mass conservation constraint can be written as

$$C_2 = \int_0^{u_{max}} uf(u)du = u_m = \bar{u} \tag{3.6}$$

The momentum conservation constraint can be written as

$$C_3 = \int_0^{u_{max}} u^2 f(u)du = \beta u_m^2 = \beta \overline{u^2} \tag{3.7}$$

The energy conservation constraint can be written as

$$C_3 = \int_0^{u_{max}} u^3 f(u)du = \alpha u_m^3 = \overline{u^3} \tag{3.8}$$

where
$u_m$ is the mean velocity
$\alpha$ and $\beta$ are energy and momentum distribution coefficients

It may be interesting to note that in this case the momentum conservation constraint corresponds to the second moment of velocity about the mean velocity or variance and the energy conservation constraint corresponds to the third moment of velocity about the mean velocity and in turn skewness.

### 3.2.4 Maximization of Entropy

Jaynes (1957a,b) formulated the principle of maximum entropy (POME), according to which the PDF can be obtained by maximizing the uncertainty expressed by entropy, subject to given constraints. The probability so derived will be least biased toward the information that is not given about velocity. To maximize the entropy given by Equation 3.3, subject to the specified constraint equations (Equations 3.5 through 3.8), the Lagrange multiplier method is employed. For the method of Lagrange multipliers, for $m > 0$, the Lagrangian function $L$ is constructed as

$$L = \int_0^{u_{max}} \frac{1}{m-1} f(u)\{1-[f(u)]^{m-1}\}du + \lambda_0 \left[\int_0^{u_{max}} f(u)du - 1\right] + \lambda_1 \left[\int_0^{u_{max}} u f(u)du - u_m\right]$$

$$+ \lambda_2 \left[\int_0^{u_{max}} u^2 f(u)du - \beta u_m^2\right] + \lambda_3 \left[\int_0^{u_{max}} u^3 f(u)du - \alpha u_m^3\right]$$

$$(3.9)$$

where $\lambda_i$, $i = 0, 1, 2, 3$, are the Lagrange multipliers. Equation 3.9 can be written as

$$L = \int_0^{u_{max}} f(u)\left\{\frac{1-[f(u)]^{m-1}}{m-1} + \lambda_0 + \lambda_1 u + \lambda_2 u^2 + \lambda_3 u^3\right\}du - \left(\lambda_0 + \lambda_1 u_m + \lambda_2 u_m^2 + \lambda_3 u_m^3\right)$$

$$(3.10)$$

Differentiating Equation 3.9 or 3.10 with respect to $f(u)$ and equating the derivative to zero, one obtains the velocity PDF:

$$f(u) = \left[\frac{m-1}{m}\left(\frac{1}{m-1} + \lambda_0 + \lambda_1 u + \lambda_2 u^2 + \lambda_3 u^3\right)\right]^{1/(m-1)} \tag{3.11}$$

For simplicity, let $\lambda_* = 1/(m-1) + \lambda_0$. Then, Equation 3.11 can be cast as

$$f(u) = \left[\frac{m-1}{m}\left(\lambda_* + \lambda_1 u + \lambda_2 u^2 + \lambda_3 u^3\right)\right]^{1/(m-1)} \tag{3.12}$$

Equation 3.12 defines the general PDF of velocity.

One can also determine the PDF of dimensionless velocity obtained by dividing velocity by shear velocity as $(u/u_*)$, where $u_*$ is the shear velocity. Chiu (1987) noted that the probability distribution function can be described as

$$f\left(\frac{u}{u_*}\right) = u_* f(u) \tag{3.13}$$

where $f(u)$ is given by Equation 3.12. Equation 3.13 has the same parameters as Equation 3.12.

### 3.2.5 Determination of Lagrange Multipliers

Equation 3.12 contains unknown Lagrange parameters $\lambda_*$ and $\lambda_i$, $i = 0, 1, 2, 3$. Inserting Equation 3.12 in Equations 3.5 through 3.8 leads to a system of four equations that can be solved numerically for $\lambda_i$, $i = 0, 1, 2, 3$.

### 3.2.6 Entropy of Velocity Distribution

Substitution of Equation 3.12 in Equation 3.3 yields the maximum entropy as

$$H(U) = \frac{1}{m-1} - \frac{1}{m-1} \int_0^{u_{max}} \left[ \frac{m-1}{m}(\lambda_* + \lambda_1 u + \lambda_2 u^2 + \lambda_3 u^3) \right]^{m/(m-1)} du \quad (3.14)$$

### 3.2.7 General Velocity Distribution

Substitution of Equation 3.12 in Equation 3.2 yields the velocity distribution as

$$\int_0^u \left[ \frac{m-1}{m}[\lambda_* + \lambda_1 u + \lambda_2 u^2 + \lambda_3 u^3] \right]^{1/(m-1)} du = \frac{y}{D} + C \quad (3.15)$$

where $C$ is a constant of integration evaluated using $u = 0$ at $y = 0$. An analytical solution of Equation 3.15 is not tractable. Equation 3.15 has four parameters: Lagrange multipliers $\lambda_i$, $i = 0, 1, 2$, and $3$, that can be estimated using constraint equations (Equations 3.5 through 3.8). However, the method of parameter estimation for the general velocity distribution becomes cumbersome and hence will not be presented here. Therefore, velocity distributions are derived for simple cases.

## 3.3 ONE-DIMENSIONAL NO-CONSTRAINT VELOCITY DISTRIBUTION

In this case, there are no physical constraints, that is, $\lambda_1 = \lambda_2 = \lambda_3 = 0$, and the only constraint is the total probability equation (Equation 3.5). Then Equation 3.12 becomes

$$f(u) = \left[ \frac{m-1}{m}\lambda_* \right]^{1/(m-1)} = C \quad (3.16)$$

where $C$ is a constant. The velocity distribution from Equation 3.15 becomes

$$u = \frac{y}{CD} \quad (3.17)$$

The value of $C$ can be obtained by substituting Equation 3.16 in Equation 3.5 and solving

$$\int_0^{u_{max}} C du = 1 \quad \text{or} \quad C = \frac{1}{u_{max}} \tag{3.18}$$

Equating Equation 3.18 to Equation 3.16, the result is

$$\lambda_* = \frac{m}{m-1} u_{max}^{1-m} = \lambda_0 + \frac{1}{m-1} \quad \text{or} \quad \lambda_0 = -\frac{1}{m-1} + \frac{m}{m-1} u_{max}^{1-m} \tag{3.19}$$

Thus, the velocity distribution is given as

$$u = u_{max} \frac{y}{D} \tag{3.20}$$

Entropy of this velocity distribution can be expressed as

$$H = \frac{u_{max}}{m-1} \left[ 1 - \left( \frac{1}{u_{max}} \right)^m \right] \tag{3.21}$$

## Example 3.1

Consider a 5 m wide open channel with a flow depth of 0.75 m. The velocity at the surface is about 0.5 m/s. Plot the velocity as a function of flow depth, assuming zero velocity at the bottom. What is the average flow velocity? Comment on this velocity distribution. How realistic is it?

### Solution

According to the mass conservation expressed by Equation 3.6 in which $u_m$ is the average velocity and $u_{max}$ is the maximum velocity at the water surface. The velocity distribution is expressed as

$$u = u_{max} \frac{y}{D} = \frac{y}{D} = 0.5 \times \frac{y}{0.75} \text{ m/s}$$

Substituting $u$ in the mass conservation equation and solving for $u_m$

$$u_m = \frac{u_{max}}{2} = \frac{0.5}{2} = 0.25 \text{ m/s}$$

The velocity distribution is shown in Figure 3.1.

The velocity distribution, shown in Figure 3.1, is linear, that is, the flow velocity increases linearly from a value of zero at the channel bed to a maximum of 0.5 m/s at the water surface. In real world, the velocity distribution is significantly different from being linear.

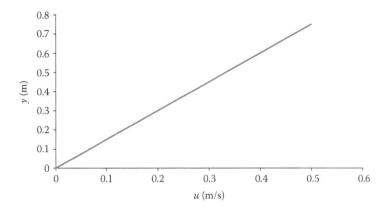

**FIGURE 3.1**    Velocity as a function of depth.

## Example 3.2

A set of experimental velocity measurements is given in Table 3.1. Compute the distribution of velocity for this data set using Equation 3.20 and comment on the goodness of this distribution. Also, compute the value of Lagrange multiplier.

### Solution

For the given data, the maximum velocity $u_{max} = 7.113$ ft/s observed at surface. Thus, according to Equation 3.20, the distribution of velocity can be obtained from

$$u = u_{max}\frac{y}{D} = \frac{y}{D} = 7.113 \times \frac{y}{0.12} \text{ ft/s}$$

## TABLE 3.1
## Velocity Distribution (From Einstein and Chien's [1955] Experiment)

| $y$ (ft) | Observed $u$ (ft/s) | Computed $u$ (ft/s) | $y$ (ft) | Observed $u$ (ft/s) | Computed $u$ (ft/s) |
|---|---|---|---|---|---|
| 0 | 0 | 0 | 0.04 | 4.831 | 2.371 |
| 0.01 | 2.221 | 0.593 | 0.04 | 5.075 | 2.371 |
| 0.01 | 2.497 | 0.593 | 0.05 | 5.298 | 2.964 |
| 0.01 | 2.72 | 0.593 | 0.05 | 5.522 | 2.964 |
| 0.01 | 2.858 | 0.593 | 0.06 | 5.806 | 3.557 |
| 0.01 | 2.964 | 0.593 | 0.07 | 6.09 | 4.149 |
| 0.02 | 3.329 | 1.186 | 0.08 | 6.293 | 4.742 |
| 0.02 | 3.573 | 1.186 | 0.09 | 6.516 | 5.335 |
| 0.02 | 3.898 | 1.186 | 0.1 | 6.699 | 5.928 |
| 0.03 | 4.519 | 1.778 | 0.12 | 7.113 | 7.113 |

*Source:* Einstein, H.A. and Chien, N., Effects of heavy sediment concentration near the bed on velocity and sediment distribution, Report No. 8, M.R.D. Sediment Series, U.S. Army Corps of Engineers, Omaha, Nebraska, August 1955.

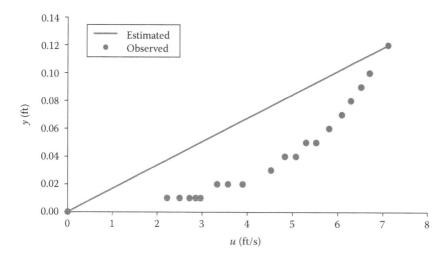

**FIGURE 3.2** Velocity distribution for the data set given in Table 3.1.

Thus, the velocity is computed from the previous equation, as given in Table 3.1. The mean velocity estimated is $u_m = u_D/2 = 7.113/2 = 3.51$ ft/s, which is smaller than observed $u_m$ of 4.39 ft/s. The computed as well as observed velocity distributions are plotted in Figure 3.2. Since Equation 3.20 is derived without any physical constraint, it is a linear distribution. However, the observed data do not exhibit a linear increase along the flow depth. This suggests that additional constraints are needed for obtaining a more accurate velocity distribution.

## 3.4 ONE-DIMENSIONAL ONE-CONSTRAINT VELOCITY DISTRIBUTION

For this case, $\lambda_2 = \lambda_3 = 0$ in Equation 3.12, and the physical constraint is the mass conservation constraint given in Equation 3.6. Equation 3.12 then reduces to

$$f(u) = \left[ \frac{m-1}{m} (\lambda_* + \lambda_1 u) \right]^{1/(m-1)} \tag{3.22}$$

Chiu (1987) expressed the probability distribution function of dimensionless velocity ($u/u_{max}$) as

$$f\left( \frac{u}{u_{max}} \right) = u_{max} f(u) \tag{3.23}$$

where $f(u)$ is given by Equation 3.22 that has the same parameters as does $f(u/u_{max})$ given by Equation 3.23.

### 3.4.1 DETERMINATION OF THE LAGRANGE MULTIPLIERS

The probability distribution of velocity, given by Equation 3.22, has two parameters $\lambda_1$ and $\lambda_*$, which can be determined using Equations 3.5 and 3.6. Substituting Equation 3.22 into Equation 3.5, one obtains

$$\int_0^{u_{max}} \left[ \frac{m-1}{m}(\lambda_* + \lambda_1 u) \right]^{1/(m-1)} du = 1 \tag{3.24}$$

On integration, Equation 3.24 yields

$$(\lambda_* + \lambda_1 u_{max})^{m/(m-1)} = \lambda_1 \left( \frac{m}{m-1} \right)^{m/(m-1)} + \lambda_*^{m/(m-1)} \tag{3.25}$$

Inserting Equation 3.22 in Equation 3.6 one gets

$$\int_0^{u_{max}} u \left[ \frac{m-1}{m}(\lambda_* + \lambda_1 u) \right]^{1/(m-1)} du = u_m \tag{3.26}$$

Integrating by parts, the solution of Equation 3.26 follows

$$u_{max}(\lambda_* + \lambda_1 u_{max})^{m/(m-1)} + \frac{m-1}{2m-1}\frac{1}{\lambda_1}(\lambda_* + \lambda_1 u_{max})^{(2m-1)/(m-1)} - \frac{m-1}{2m-1}\frac{1}{\lambda_1}(\lambda_*)^{(2m-1)/(m-1)}$$

$$= \lambda_1 \bar{u} \left( \frac{m}{m-1} \right)^{m/(m-1)} \tag{3.27}$$

Equations 3.25 and 3.27 can be used to numerically solve for $\lambda_*$ and $\lambda_1$.

### 3.4.2 VELOCITY DISTRIBUTION

In order to relate the entropy-based probability distribution to space domain, Equation 3.22 is inserted in Equation 3.1 and then is integrated with respect to $u$ using the condition that $F(u) = 0$ when $u = 0$. To that end, Equation 3.22 is integrated first:

$$F(u) = \int_0^u f(u) du = \int_0^u \left[ \frac{m-1}{m}(\lambda_* + \lambda_1 u) \right]^{1/(m-1)} du = \frac{1}{\lambda_1}\frac{m}{m-1}\left[ \frac{m-1}{m}(\lambda_* + \lambda_1 u) \right]^{m/(m-1)}$$

$$\tag{3.28}$$

Rearranging Equation 3.28, the relation between velocity and its cumulative probability is obtained as

$$u = -\frac{\lambda_*}{\lambda_1} + \frac{1}{\lambda_1}\frac{m}{m-1}\left[\lambda_1 F(u) + \left(\frac{m-1}{m}\lambda_*\right)^{m/(m-1)}\right]^{(m-1)/m} \tag{3.29}$$

Equation 3.29 specifies velocity at a specified probability and can be called as velocity quantile–probability relation.

Substituting Equation 3.22 in Equation 3.2, one can write

$$\left[\frac{m-1}{m}(\lambda_* + \lambda_1 u)\right]^{1/(m-1)} = \frac{1}{D}\frac{dy}{du} \tag{3.30}$$

Integrating both sides of Equation 3.30, one obtains

$$\int_0^u \left[\frac{m-1}{m}(\lambda_* + \lambda_1 u)\right]^{1/(m-1)} du = \frac{1}{D}\int_0^y dy = \frac{y}{D} \tag{3.31}$$

or

$$\left[\frac{m-1}{m}(\lambda_* + \lambda_1 u)\right]^{1/(m-1)} = \lambda_1\frac{y}{D}\left[\left(\frac{m-1}{m}\right)\lambda_*\right]^{m/(m-1)} \tag{3.32}$$

Now the velocity distribution can be obtained as a function of $y$ as

$$u = \left(\frac{m}{m-1}\right)\frac{1}{\lambda_1}\left\{\left(\lambda_1\frac{y}{D}\right) + \left[\left(\frac{m-1}{m}\right)\lambda_*\right]^{m/(m-1)}\right\}^{(m-1)/m} - \frac{\lambda_*}{\lambda_1} \tag{3.33}$$

which is the same as Equation 3.29 where $F(u) = y/D$.

For simplicity, let $m/(m-1)$ be denoted as $k$, Equation 3.33 can be written as

$$u = \frac{k}{\lambda_1}\left[\lambda_1\frac{y}{D} + \left(\frac{\lambda_*}{k}\right)^k\right]^{1/k} - \frac{\lambda_*}{\lambda_1} \tag{3.34}$$

Equation 3.34 is a Tsallis entropy–based velocity distribution for the flow in a wide open channel in which the velocity varies nonlinearly with the vertical distance from the channel bed.

**TABLE 3.2**

**Computation of Lagrange Multipliers**

| $m$ | $\lambda_1$ | $\lambda_*$ | $H$ |
|-----|-------------|-------------|-------|
| 2/3 | 0.389 | 0.473 | 6.272 |
| 3/4 | 0.184 | 0.301 | 5.479 |
| 1.5 | 0.054 | 0.122 | 1.761 |
| 2 | 0.014 | 0.231 | 0.854 |
| 3 | −0.012 | 0.010 | 0.514 |
| 4 | 0.001 | 0.001 | 0.332 |

## Example 3.3

For the velocity measurements in Table 3.1, compute the values of the Lagrange multipliers and the entropy value.

### Solution

It is known from the given data that $u_{max} = 7.113$ ft/s and $u_m = 4.391$ ft/s. The Lagrange multipliers are computed by solving Equations 3.25 and 3.27 numerically. For different $m$ values, $\lambda_1$ and $\lambda_*$ are computed and are listed in Table 3.2. Entropy can be computed from

$$H(U) = \frac{1}{m-1} - \frac{1}{m-1} \int_0^{u_{max}} \left[ \frac{m-1}{m}(\lambda_* + \lambda_1 u) \right]^{m/(m-1)} du = \frac{1}{m-1} - \frac{1}{m}(\lambda_* + \lambda_1 \bar{u})$$

and is given in Table 3.2.

## Example 3.4

Compute the Lagrange multipliers for different $m$ values (1/4, 1/3, 2/3, 3/4, 1.25, 3/2, 2.0, 3.0) for four sets of mean and maximum values of velocity given (ft/s) as $u_{max} = 9.194$, $u_m = 7.303$; $u_{max} = 11.42$, $u_m = 8.7$; $u_{max} = 0.535$, $u_m = 0.412$; and $u_{max} = 1.046$, $u_m = 0.890$.

### Solution

The Lagrange multipliers are computed for different $m$ values, as shown in Table 3.3.

## Example 3.5

For various values of $m$, compute the dimensionless velocity density function and plot it.

### Solution

With Lagrange multipliers computed in Example 3.4, the dimensionless velocity density function can be computed from

**TABLE 3.3**

**Parameters of Tsallis Entropy–Based Velocity Distribution (Maximum and Mean Velocities Are in ft/s)**

| $m$ | $u_{max} = 9.194,$ $u_m = 7.303$ | | $u_{max} = 11.42,$ $u_m = 8.7$ | | $u_{max} = 0.535,$ $u_m = 0.412$ | | $u_{max} = 1.046,$ $u_m = 0.890$ | |
|---|---|---|---|---|---|---|---|---|
| | $\lambda_1$ | $\lambda_*$ | $\lambda_1$ | $\lambda_*$ | $\lambda_1$ | $\lambda_*$ | $\lambda_1$ | $\lambda_*$ |
| 1/4 | 0.672 | −6.473 | 0.528 | −6.489 | 1.190 | −0.680 | 1.636 | −1.742 |
| 1/3 | 0.746 | −7.354 | 0.576 | −7.289 | 1.675 | −0.984 | 2.173 | −2.345 |
| 2/3 | 0.712 | −8.839 | 0.527 | −8.644 | 4.109 | −3.140 | 4.281 | −5.429 |
| 3/4 | −0.665 | 9.499 | −0.483 | 9.255 | 4.861 | −4.333 | 4.799 | −6.787 |
| 1.25 | −0.864 | 3.218 | 0.672 | −3.214 | −10.804 | −2.271 | 12.125 | −4.532 |
| 3/2 | −0.273 | 0.505 | −0.224 | 0.783 | −15.026 | 0.638 | −6.868 | 1.515 |
| 2.00 | 0.084 | −0.167 | 0.049 | −0.106 | 22.585 | −2.303 | 7.696 | −2.113 |
| 3.00 | 0.032 | 0.012 | 0.031 | 0.076 | 1.367 | 0.382 | 4.223 | 1.452 |

$$f(u) = u_{max}\left[\frac{m-1}{m}(\lambda_* + \lambda_1 u)\right]^{1/(m-1)}$$

This is shown in Figure 3.3. When $m$ is smaller than 2, the PDF decreases as dimensionless velocity increases. For $m$ equal to or larger than 2, it increases with dimensionless velocity. It is ensured that the area underlying each curve equals 1.

## Example 3.6

For the velocity measurements given in Table 3.1, compute the velocity distribution and compare it with the observed distribution.

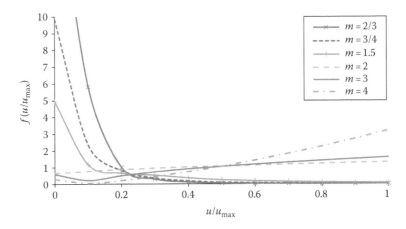

**FIGURE 3.3** Probability density function of dimensionless velocity for different values of $m$ for data given in Table 3.1.

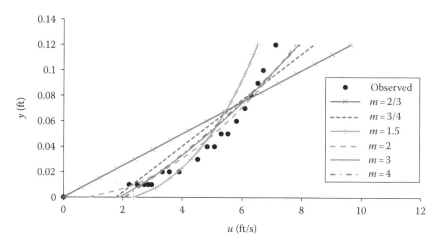

**FIGURE 3.4**   Velocity distribution for different values of $m$ for data set given in Table 3.2.

### Solution

The velocity distribution is computed using Equation 3.33 for different $m$ values and is plotted in Figure 3.4. The Lagrange multipliers have already been computed for these data as given in Table 3.2. It can be seen from the figure that the velocity distribution obtained for $m = 2/3$ or $m = 3/4$ fails to capture the observed velocity distribution pattern and the velocity distribution is best fitted with $m = 3$ and $m = 4$. However, when $m = 3$, the computation of parameters is much simpler than when $m = 4$. Thus, $m = 3$ is preferred.

### 3.3.3   REPARAMETERIZATION

Reparameterization reduces the computational difficulty and facilitates comparison of different velocity distributions (Chiu, 1988). In a manner similar to that in Chiu (1988), a dimensionless parameter $G$ can be defined as

$$G = \frac{\lambda_1 u_{max}}{\lambda_1 u_{max} + \lambda_*} \tag{3.35}$$

With the use of parameter $G$, the relationship between $f(0)$ and $f(u_{max})$ can be shown as

$$\frac{f(0)}{f(u_{max})} = \frac{[(m-1)/m(\lambda_*)]^{1/(m-1)}}{[(m-1)/m(\lambda_* + \lambda_1 u_{max})]^{1/(m-1)}} = \left(\frac{\lambda_*}{\lambda_* + \lambda_1 u_{max}}\right)^{1/(m-1)} = (1-G)^{1/(m-1)} \tag{3.36}$$

Thus, $G$ can be used as an index of the velocity distribution. Equation 3.36 shows that if $G = 0, f(0) = f(u_{max})$ and the PDF of velocity would tend to be uniform. If $G = 1$,

$f(0) > 0$ and $f(u_{max})$ would tend to infinity. This means that the PDF would be highly nonuniform.

When $u = u_{max}$ and $F(u) = 1$, Equation 3.29 becomes

$$u_{max} = -\frac{\lambda_*}{\lambda_1} + \frac{1}{\lambda_1}\frac{m}{m-1}\left[\lambda_1 + \left(\frac{m-1}{m}\lambda_*\right)^{m/(m-1)}\right]^{(m-1)/m} \qquad (3.37)$$

Dividing Equation 3.29 by $u_{max}$ and then using Equation 3.35, the dimensionless velocity equation is obtained as

$$\frac{u}{u_{max}} = \frac{\lambda_*}{\lambda_1 u_{max}} - \frac{1}{\lambda_1 u_{max}}\frac{m}{m-1}\left\{-\lambda_1\frac{m-1}{m}F(u) + \left[\frac{m-1}{m}\lambda_*\right]^{m/(m-1)}\right\}^{(m-1)/m}$$

$$= 1 - \frac{1}{G}\left(1 - \left(\left(\frac{m}{m-1}\right)^{m/(m-1)}\frac{G}{u_{max}}\left(\left(\frac{y}{D}\right)-1\right)+1\right)^{(m-1)/m}\right) \qquad (3.38)$$

Substituting $u = u_0 = 0$, at $y = 0$, Equation 3.38 reduces to

$$0 = 1 - \frac{1}{G}\left(1 - \left(1 - \left(\frac{m}{m-1}\right)^{m/(m-1)}\frac{G}{u_{max}}\right)^{(m-1)/m}\right) \qquad (3.39)$$

Rearranging Equation 3.39, one obtains

$$\left(\frac{m}{m-1}\right)^{m/(m-1)}\frac{G}{u_{max}} = 1 - (1-G)^{m/(m-1)} \qquad (3.40)$$

Now, Equation 3.38 can be simplified with the use of Equation 3.40 as

$$\frac{u}{u_{max}} = 1 - \frac{1}{G}\left(1 - \left[(1-G)^{m/(m-1)} + (1 - (1-G)^{m/(m-1)})\left(\frac{y}{D}\right)\right]^{(m-1)/m}\right) \qquad (3.41)$$

Equation 3.41 shows that for a given $m$ value, the velocity distribution can be obtained with only one parameter $G$. Figure 3.5 shows a family of dimensionless velocity distributions for different $G$ values for a fixed $m = 3$. It is seen from the figure that a bigger $G$ value tends to slow the increase in velocity from the channel bed to the water surface. For a lower value of $G$, the velocity distribution tends to linearize and for a higher value it tends to nonlinearize. The velocity distribution is, therefore, highly sensitive to $G$.

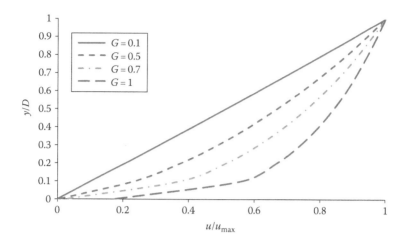

**FIGURE 3.5**  Dimensionless velocity distributions for different $G$ values for a fixed $m = 3$.

### Example 3.7

For the velocity measurements in Table 3.1, compute the value of parameter $G$ and then compute the dimensionless velocity distribution and compare it with the observed velocity distribution.

#### Solution

From Table 3.2, it is seen that for $m = 3$, $\lambda_1 = -0.012$ and $\lambda_* = 0.010$. Then,

$$G = \frac{\lambda_1 u_{max}}{\lambda_1 u_{max} + \lambda_*} = \frac{0.012 \times 7.113}{0.012 \times 7.113 + 0.010} = 0.895$$

Thus, using Equation 3.41, the velocity distribution can be computed and plotted as shown in Figure 3.6. The computed velocity is first overestimated and then underestimated.

### 3.4.4 ENTROPY

The maximum entropy can be obtained by substituting Equation 3.22 in Equation 3.4 as

$$H = \frac{1}{m-1} - \frac{1}{m}(\lambda_* + \lambda_1 \bar{u}) \tag{3.42}$$

The entropy of velocity distribution, given by Equation 3.42, is expressed in terms of given information. It can also be derived in terms of dimensionless parameter $G$. First, the Lagrange multipliers $\lambda_*$ and $\lambda_1$ are expressed in terms of $G$. By rearranging Equation 3.35, $\lambda_*$ can be expressed as

$$\lambda_* = \lambda_1 u_{max}\left(\frac{1}{G} - 1\right) \quad \text{or} \quad \frac{u_{max}}{\lambda_* + \lambda_1 u_{max}} = \frac{\lambda_1}{G} \tag{3.43}$$

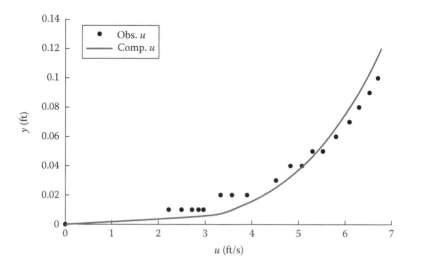

**FIGURE 3.6**   Velocity distribution with parameter $G$ (obs. = observed and comp. = computed).

Now, multiplying Equation 3.25 by $u_{\max}/(\lambda_* + \lambda_1 u_{\max})^{m/(m-1)}$ on both sides, we obtain

$$u_{\max} = \frac{\lambda_1 u_{\max}(m/(m-1))^{m/(m-1)}}{(\lambda_* + \lambda_1 u_{\max})^{m/(m-1)}} + \frac{\lambda_*^{m/(m-1)} u_{\max}}{(\lambda_* + \lambda_1 u_{\max})^{m/(m-1)}}$$

$$= \frac{\lambda_1 u_{\max}(m/(m-1))^{m/(m-1)}}{(\lambda_* + \lambda_1 u_{\max})^{m/(m-1)}} + (1-G)^{m/(m-1)} \tag{3.44}$$

Then, eliminating $\lambda_*$ in Equation 3.44, with the use of Equation 3.43 one can write $\lambda_1$ as

$$\lambda_1 = \left(\frac{m-1}{m}\right)^{-m}[1-(1-G)^{m/(m-1)}]^{-(m-1)}\left(\frac{u_{\max}}{G}\right)^{-m} \tag{3.45}$$

Now substituting Equations 3.43 and 3.45 into Equation 3.42, the entropy can be computed in terms of parameter $G$ instead of the Lagrange multipliers as

$$H = \frac{1}{m-1} - \frac{1}{m}\left(\lambda_1 u_{\max}\left(\frac{1}{G}-1\right) + \left(\frac{m-1}{m}\right)^{-m}[1-(1-G)^{m/(m-1)}]^{-(m-1)}\left(\frac{u_{\max}}{G}\right)^{-m}\bar{u}\right)$$

$$= \frac{1}{m-1} - \frac{1}{m}\left\{\left(\frac{m-1}{m}\right)^{-m}[1-(1-G)^{m/(m-1)}]^{-(m-1)}\left(\frac{u_{\max}}{G}\right)^{-m}\left[u_{\max}\left(1-\frac{1}{G}\right)+\bar{u}\right]\right\}$$

$$\tag{3.46}$$

It can be seen from Equation 3.46 that for known $u_{max}$ and $\bar{u} = u_{mean}$, the maximum entropy can be determined from parameter $G$ and for the $G$ value between 0 and 1, the maximum entropy monotonically increases with increasing $G$. Since entropy varies for different combinations of maximum and mean velocities, it is more convenient to use $G$ as an index for the entropy of a given data set. When a high value of $G$ is obtained, it will imply a larger value of entropy and vice versa. This means that the probability distribution will be more uncertain for greater $G$ values than for smaller $G$ values.

### Example 3.8

For $u_m = 2.25$ m/s, $u_{max} = 5.0$ m/s, and $m = 3$, compute entropy and plot maximum entropy as a function of $G$.

**Solution**

For $\bar{u} = 2.25$ m/s and $u_{max} = 5.0$ m/s, the entropy value is computed using Equation 3.46 for various $G$ values that are plotted in Figure 3.7. The maximum entropy decreases with increasing $G$ value.

### Example 3.9

For the velocity measurements in Table 3.1, compute the entropy value in two different ways and compare them.

**Solution**

For $m = 3$, the Lagrange multipliers $\lambda_1 = -0.012$ and $\lambda_* = 0.010$ are given in Table 3.2. Then, the Tsallis entropy is computed as

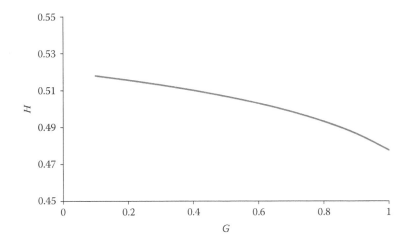

**FIGURE 3.7**  Maximum entropy as a function of $G$.

$$H = \frac{1}{m-1} - \frac{1}{m}(\lambda_* + \lambda_1 \bar{u}) = \frac{1}{3-1} - \frac{1}{3}(0.010 + 0.012 \times 4.391) = 0.514$$

Using $G = 0.895$ from Example 3.7,

$$H = \frac{1}{m-1} - \frac{1}{m}\left\{\left(\frac{m-1}{m}\right)^{-m}[1-(1-G)^{m/(m-1)}]^{-(m-1)}\left(\frac{u_{max}}{G}\right)^{-m}\left[u_{max}\left(1-\frac{1}{G}\right)+\bar{u}\right]\right\}$$

$$= \frac{1}{3-1} - \frac{1}{3}\left\{\left(\frac{3-1}{3}\right)^{-3}[1-(1-0.895)^{3/2}]^{-(3-1)}\left(\frac{7.113}{0.895}\right)^{-3}\left[7.113\left(1-\frac{1}{0.895}\right)+4.391\right]\right\}$$

$$= 0.574$$

Both entropy values are quite close each other.

## Example 3.10

For various combinations of $u_m$ and $u_{max}$, given in Table 3.4, compute the Lagrange multiplier $\lambda_1$ and $H$. Then, plot $H$ as a function of $u_{max}$ and establish a relation between $H$ and $\lambda_1$, using a simple regression. Similarly, establish a relation between $H$ and $u_{max}$, using regression analysis. Plot both figures on the same graph. What can be concluded from the plot?

### Solution

For given $u_D$ and $u_m$ in Table 3.4, $\lambda_1$ is solved from Equation 3.27 and $H$ is computed from Equation 3.42 as listed in Table 3.5.

Figure 3.8 gives the relation between entropy and $u_{max}$ as well as the relation between entropy and $\lambda_1$, and these relations are obtained by regression analysis. Entropy has a positive relation with $u_{max}$ and a negative relation with $\lambda_1$. Parameter $\lambda_1$ can be considered as a hydraulic parameter and can be used to characterize and classify open channel flows under the effect of coarseness of bed material and sediment concentration. The data points in Figure 3.8 at lower values of $\lambda_1$ and hence higher

### TABLE 3.4
### Mean and Maximum Velocity Values

| $u_D$ (ft/s) | $u_m$ (ft/s) | Data Source | $u_D$ (ft/s) | $u_m$ (ft/s) | Data Source |
|---|---|---|---|---|---|
| 1.755 | 1.352 | Iran | 8.246 | 6.430 | Einstein |
| 2.073 | 1.450 | Iran | 8.474 | 6.640 | Einstein |
| 3.432 | 2.920 | Iran | 8.810 | 6.550 | Einstein |
| 5.485 | 4.480 | Iran | 9.194 | 7.303 | Einstein |
| 7.591 | 6.050 | Einstein | 10.415 | 7.160 | Einstein |
| 7.940 | 6.130 | Einstein | 11.420 | 8.700 | Einstein |
| 8.021 | 6.190 | Einstein | | | |

**TABLE 3.5**
**Computed $\lambda_1$ and $H$ for Data in Table 3.4**

| $u_D$ (ft/s) | $u_m$ (ft/s) | $\lambda_1$ | $H$ |
|---|---|---|---|
| 1.755 | 1.352 | 2.01 | 0.18 |
| 2.073 | 1.45 | 1.83 | 0.57 |
| 3.432 | 2.92 | 1.98 | 0.52 |
| 5.485 | 4.48 | 0.81 | 1.31 |
| 7.591 | 6.05 | 0.79 | 1.89 |
| 7.94 | 6.13 | 0.68 | 2.08 |
| 8.021 | 6.19 | 0.63 | 2.10 |
| 8.246 | 6.43 | 0.66 | 2.08 |
| 8.474 | 6.64 | 0.68 | 2.14 |
| 8.81 | 6.55 | 0.52 | 2.39 |
| 9.194 | 7.303 | 0.68 | 2.18 |
| 10.415 | 7.16 | 0.33 | 2.90 |
| 11.42 | 8.7 | 0.49 | 2.70 |

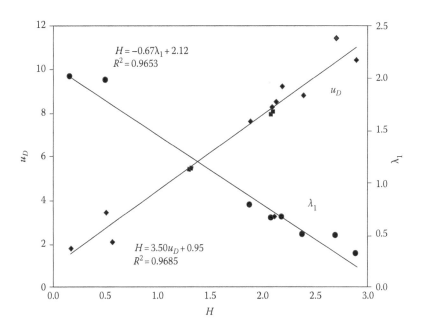

**FIGURE 3.8**    Relation of $H(u)$ to $u_D$ and $\lambda_1$ (based on data given in Table 3.4).

values of entropy represent flows over coarser channel bed and/or with higher levels of sediment concentration. A considerably wide range of $\lambda_1$ in the clear water flows manifests the marked trend of $\lambda_1$ to decrease with the coarseness of bed material. For sediment-laden flows, data points reflect the effect of both the coarseness of bed material and the sediment concentration.

## 3.5 RELATION BETWEEN MEAN VELOCITY AND MAXIMUM VELOCITY

The relationship between the mean velocity and the maximum velocity has been examined by Chiu (1995) and Xia (1997). If the numerical relationship between these two velocities can be established, then the mean velocity can be obtained using this relationship since the maximum velocity is measurable. The ratio between these two velocities is found to be constant for a given river reach. For 13 sets of velocity data, given in Table 3.4, representing different flow profiles (Einstein and Chien, 1955 and Iran data), the mean and maximum velocities obey a linear relationship with a very high $R^2$, as shown in Figure 3.9.

Following Chiu (1988), a relationship between the mean velocity and the maximum velocity can be established. To obtain an analytical solution of the mean velocity, Equation 3.41 can be integrated over the whole cross-sectional area (or depth):

$$\bar{u} = \frac{1}{A}\int u\,dA = \frac{1}{A}\int u_{\max}\left[1 - \frac{1}{G}\left\{1-(1-G)^{m/(m-1)} + \{1-(1-G)^{m/(m-1)}\}\left(\frac{y}{D}\right)^a\right\}^{(m-1)/m}\right]dA$$

(3.47)

Equation 3.47 can be recast as

$$\frac{\bar{u}}{u_{\max}} = \Psi(G) = \frac{1}{A}\int\left[1 - \frac{1}{G}\left[1-(1-G)^{m/(m-1)} + \{1-(1-G)^{m/(m-1)}\}\left(\frac{y}{D}\right)^a\right]^{(m-1)/m}\right]dA$$

(3.48)

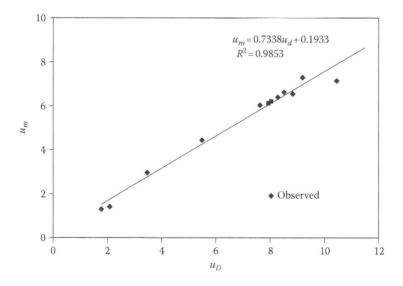

**FIGURE 3.9** Mean and maximum velocity values.

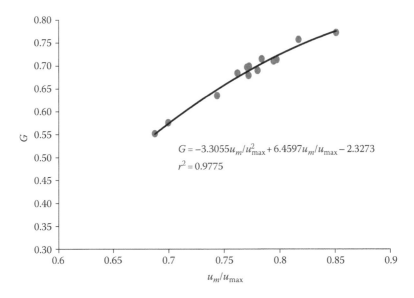

$$G = -3.3055 u_m/u_{max}^2 + 6.4597 u_m/u_{max} - 2.3273$$
$$r^2 = 0.9775$$

**FIGURE 3.10**   Relationship between $G$ and $u_m/u_{max}$.

In Equation 3.48, $G$ should be computed first from its definition (Equation 3.35) in terms of the Lagrange multipliers that are themselves obtained from Equations 3.25 and 3.27 with known mean and maximum velocity values. In this manner, the relation between the ratio of mean velocity to maximum velocity and the Lagrange multipliers can be transformed into a relationship with parameter $G$, which is plotted in Figure 3.10 with field data collected from Iranian rivers and Italian rivers. Applying polynomial regression, Cui and Singh (2013) obtained a numerical solution of $\Psi(G)$ as

$$\Psi(G) = 0.554G^2 - 0.077G + 0.568 \qquad (3.49)$$

Table 3.6 compares the values of $G$ computed from Equations 3.35 and 3.49. It is seen that the mean difference is about 0.003. Thus, it is reasonable to use Equation 3.49 instead of Equation 3.35 involving the Lagrange multipliers.

### Example 3.11

Show Equation 3.49 on a graph. How good is this equation?

**Solution**

With given $u_D(u_{max})$ and $u_m$, the Lagrange multipliers can be solved from Equations 3.25 and 3.27. Then, parameter $G$ is obtained from Equation 3.35 with computed $\lambda_1$ and $\lambda_*$. For example,

$$G = \frac{\lambda_1 u_{max}}{\lambda_1 u_{max} + \lambda_*} = \frac{0.217 \times 1.755}{0.217 \times 1.755 + 0.166} = 0.697$$

**TABLE 3.6**

**Computation of Entropy Parameter G**

| $u_D$(ft/s) | $u_m$ | Data Source | $\lambda_1$ | $\lambda_*$ | G | G' |
|---|---|---|---|---|---|---|
| 1.755 | 1.352 | Iran | 0.217 | 0.166 | 0.697 | 0.669 |
| 2.073 | 1.450 | Iran | 0.261 | 0.398 | 0.576 | 0.549 |
| 3.432 | 2.920 | Iran | 0.177 | 0.180 | 0.772 | 0.764 |
| 5.485 | 4.480 | Iran | 0.314 | 0.553 | 0.757 | 0.729 |
| 7.591 | 6.050 | Einstein | 0.086 | 0.263 | 0.713 | 0.706 |
| 7.940 | 6.130 | Einstein | 0.273 | 0.933 | 0.699 | 0.672 |
| 8.021 | 6.190 | Einstein | 0.188 | 0.714 | 0.679 | 0.671 |
| 8.246 | 6.430 | Einstein | 0.317 | 1.174 | 0.690 | 0.683 |
| 8.474 | 6.640 | Einstein | 0.407 | 1.373 | 0.715 | 0.688 |
| 8.810 | 6.550 | Einstein | 0.113 | 0.572 | 0.635 | 0.628 |
| 9.194 | 7.303 | Einstein | 0.146 | 0.549 | 0.710 | 0.702 |
| 10.415 | 7.160 | Einstein | 0.075 | 0.633 | 0.552 | 0.526 |
| 11.420 | 8.700 | Einstein | 0.216 | 1.140 | 0.684 | 0.657 |

$G'$ is the one computed from the relationship expressed by Equation 3.49. By solving $(1.352/1.755) = 0.554G^2 - 0.077G + 0.568$, $G' = 0.669$ is obtained.

The relationship between $G$ and $u_m/u_{max}$ from Equation 3.49 is examined in Figure 3.10. The solid line is the $G'$ from Equation 3.49 and dotted line is the one obtained directly from the Lagrange multipliers. It is seen that Equation 3.49 fits the $G$ values obtained from the Lagrange multipliers well with $r^2$ of 0.977.

## 3.6 SIMPLIFICATION OF VELOCITY DISTRIBUTION

With a fixed value of $m = 3$, the velocity distribution equation (Equation 3.41) can be simplified by using parameter $G$ as

$$\frac{u}{u_{max}} = 1 - \frac{1}{G}\left(1 - \left[(1-G)^{3/2} + (1-(1-G)^{3/2})\left(\frac{y}{D}\right)\right]^{2/3}\right) \tag{3.50}$$

Mathematically, equation $(1-G)^{3/2}$ is always smaller than 1, since the $G$ value is smaller than 1. Using an expansion method, $(1-G)^{3/2}$ can be approximated by $-0.5 \ln G$. Then Equation 3.50 can be simplified as

$$\frac{u}{u_{max}} = 1 - \frac{1}{G}\left\{1 - \left[(1+0.5\ln G)\left(\frac{y}{D}\right)^a - 0.5\ln G\right]^{2/3}\right\} \tag{3.51}$$

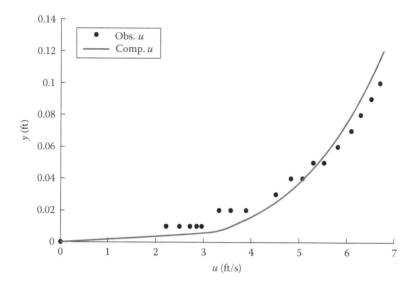

**FIGURE 3.11**    Velocity distribution from Equation 3.51.

### Example 3.12

Compute the velocity distribution using Equation 3.51 for the data listed in Table 3.1 and compare it with the observed distribution.

#### Solution

The velocity distribution of this data set is obtained with a value of $G$ of 0.895 and an entropy value as 0.514. Such a high value of $G$ represents that the maximum entropy obtains a high value within the possible range, which implies that the velocity is distributed most likely as uniform under the given maximum and mean velocity values. Figure 3.12 compares the velocity distribution from Equation 3.51 with observations. It is seen from the figure that the bed-affected region for this data set is very small, only about 0.01 ft, where the velocity increases slowly due to the high shear stress. Figure 3.11 shows a good agreement between computed and observed velocity values. The computed curve increases from the channel bed to the water surface across the observation set, even satisfactory in the region near channel bed.

## 3.7   ESTIMATION OF MEAN VELOCITY

Equation 3.49 can be used to compute the $G$ value for cross sections with known maximum and mean velocities. On the other hand, Equation 3.49 can be used to compute the mean velocity with known $G$ values since the $G$ value is supposed to be approximately constant for a given cross section.

### Example 3.13

Compute the values of mean velocity for the following values of maximum velocity obtained from Santa Lucia gauge, where $G = 0.761$ is given (Table 3.7).

**TABLE 3.7**

**Maximum Velocity Observed from Santa Lucia Gauge**

| $u_{max}$ (m/s) | $u_m$ (m/s) | $u_{max}$ (m/s) | $u_m$ (m/s) | $u_{max}$ (m/s) | $u_m$ (m/s) |
|---|---|---|---|---|---|
| 0.088 | 0.047 | 2.243 | 1.648 | 2.437 | 1.873 |
| 0.269 | 0.182 | 0.129 | 0.067 | 0.107 | 0.052 |
| 1.208 | 0.948 | 0.495 | 0.324 | 0.482 | 0.315 |
| 1.467 | 1.072 | 0.644 | 0.401 | 0.735 | 0.497 |
| 1.773 | 1.135 | 1.155 | 0.736 | 1.022 | 0.672 |
| 1.631 | 1.179 | 2.194 | 1.497 | 1.678 | 1.151 |
| 2.760 | 1.478 | | | | |

**Solution**

For given maximum velocities and $G$ value, the mean velocity is computed using Equation 3.49. The computed mean velocity is plotted against observations in Figure 3.12. It is seen from the figure that the computed mean velocity from Equation 3.49 fits observations well. When computed values are regressed against the observed values, the regression line has a slope of 0.926 and $r^2$ is as high as 0.958. However, the deviation grows with increasing mean velocity value, which can be seen from the figure that the data points are more spread for larger values of mean velocity than for smaller ones.

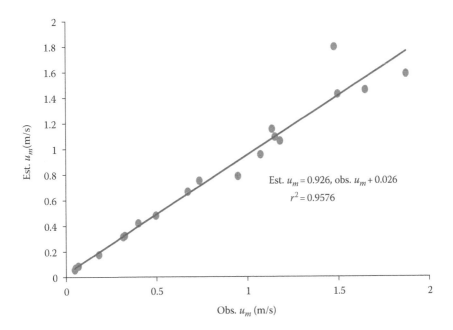

**FIGURE 3.12**   Estimated mean velocity from Equation 3.49.

## REFERENCES

Araujo, J. and Chaudhry, F. (1998). Experimental evaluation of 2-D entropy model for open-channel flow. *Journal of Hydraulic Engineering*, 124(10), 1064–1067.

Barbe, D.E., Cruise, J.F., and Singh, V.P. (1991). Solution of 3-constraint entropy-based velocity distribution. *Journal of Hydraulic Engineering*, 117(10), 1389–1396.

Blasius, H. (1913). Das¨ Ahnlichkeitsgesetz bei Reibungsvorgängen in Flüssigkeiten. *Forsch. Arb. Ing.*, 134.

Chen, Y.C. and Chiu, C.L. (2004). A fast method of flood discharge estimation. *Hydrological Processes*, 18, 1671–1683.

Chiu, C.-L. (1987). Entropy and probability concepts in hydraulics. *Journal of Hydraulic Engineering, ASCE*, 113(5), 583–600.

Chiu, C.-L. (1988). Entropy and 2-D velocity distribution in open channels. *Journal of Hydraulic Engineering, ASCE*, 114(7), 738–755.

Chiu, C.-L. (1989). Velocity distribution in open channel flows. *Journal of Hydraulic Engineering, ASCE*, 115(5), 576–594.

Chiu, C.L. (1995). Maximum and mean velocity and entropy in open channel flow. *Journal of Hydraulic Engineering, ASCE*, 121(1), 26–35.

Chiu, C.L. and Chen, Y.C. (2003). An efficient method of discharge estimation based on probability concept. *Journal of Hydraulic Research, IAHR*, 41(6), 589–596.

Chiu, C.-L. and Chiou, J.-D. (1986). Structure of 3-D flow and shear in open channels. *Journal of Hydraulic Engineering, ASCE*, 109(11), 1424–1440.

Chiu, C.L., Hsu, S.M., and Tung, N.C. (2005). Efficient methods of discharge measurements in rivers and streams based on the probability concept. *Hydrological Processes*, 19, 3935–3946.

Chiu, C.L. and Murray, D.W. (1992). Variation of velocity distribution along nonuniform open-channel flow. *Journal of Hydraulic Engineering*, 118(7), 989–1001.

Chiu, C.-L. and Said, A.A. (1995). Maximum and mean velocities in open-channel flow. *Journal of Hydraulic Engineering, ASCE*, 121(1), 26–35.

Chiu, C.L. and Tung, N.C. (2002). Maximum velocity and regularities in open-channel flow. *Journal of Hydraulic Engineering*, 128(8), 803.

Choo, T.H. (2000). An efficient method of the suspended sediment-discharge measurement using entropy concept. *Water Engineering Research*, 1(2), 95–105.

Cui, H. and Singh, V.P. (2012). On the cumulative distribution function for entropy-based hydrologic modeling. *Transactions of ASABE*, 55(2), 429–438.

Cui, H. and Singh, V.P. (2013). Two-dimensional velocity distribution in open channels using the Tsallis entropy. *Journal of Hydrologic Engineering*, 18(3), 331–339.

Einstein, H.A. and Chien, N. (1955). Effects of heavy sediment concentration near the bed on velocity and sediment distribution. Report No. 8, M.R.D. Sediment Series, U.S. Army Corps of Engineers, Omaha, Nebraska, August.

Jaynes, E.T. (1957a). Information theory and statistical mechanics, I. *Physical Reviews*, 106, 620–630.

Jaynes, E.T. (1957b). Information theory and statistical mechanics, II. *Physical Reviews*, 108, 171–190.

Karim, M.-F. and Kennedy, J.F. (1987). Velocity and sediment concentration profiles in river flows. *Journal of Hydraulic Engineering, ASCE*, 113(2), 159–178.

Luo, H. and Singh, V.P. (2011). Entropy theory for two-dimensional velocity distribution. *Journal of Hydrologic Engineering*, 16(4), 303–315.

Sarma, K.V.N., Lakshminarayana, P., and Rao, N.S.L. (1983). Velocity distribution in smooth rectangular open channels. *Journal of Hydraulic Engineering, ASCE*, 109(2), 270–289.

Shannon, C.E. (1948). The mathematical theory of communications, I and II. *Bell System Technical Journal*, 27, 379–423.

Singh, V.P. (1996). *Kinematic Wave Modeling in Water Resources: Surface Water Hydrology.* John Wiley, New York.

Singh, V.P. and Luo, H. (2009). Derivation of velocity distribution using entropy. *Proceedings of IAHR Congress*, Vancouver, British Columbia, Canada, pp. 31–38.

Singh, V.P. and Luo, H. (2011). Entropy theory for distribution of one-dimensional velocity in open channels. *Journal of Hydrologic Engineering*, 16(9), 725–735.

Tsallis, C. (1988). Possible generalization of Boltzmann-Gibbs statistics. *Journal of Statistical Physics*, 52, 479–487.

Vanoni, V.A. (1941). Velocity distribution in open channels. *Civil Engineering*, 11(6), 356–357.

von Karman, T. (1935). Some aspects of the turbulent problem. *Mechanical Engineering*, 57(7), 407–412.

Xia, R. (1997). Relation between mean and maximum velocities in a natural river. *Journal of Hydraulic Engineering*, 123(8), 720–723.

# 4 Two-Dimensional Velocity Distributions

In natural open channels, velocity varies in two directions, and therefore, the velocity distribution should be considered in two dimensions. The existing velocity distribution laws and equations that are described in the 1-D case are, however, applicable only to wide channels in which velocity is assumed to increase monotonically in the vertical direction from the channel bed to the water surface. They cannot be regarded as general or universal laws governing velocity distributions in open channels. The objective of this chapter is to present the derivation of 2-D velocity distributions using the Tsallis entropy.

## 4.1 PRELIMINARIES

In natural open channels, velocity varies in both vertical ($y$) and transverse ($x$) directions. In the transverse direction, the velocity is near zero at the boundaries and is maximum somewhere in the middle of the channel but not necessarily in the center. In the vertical direction, the velocity increases from 0 at the channel boundary to the maximum at or below the water surface near the channel center. The phenomenon in which the velocity reaches the maximum value below the water surface is called dip phenomenon.

The occurrence of maximum velocity below the water surface is an important feature of open-channel flow. Therefore, a 2-D analysis and modeling of velocity distribution is needed to be able to deal with the geometry of isovels (lines of equal velocity) in a cross section. Classical laws like the power law and the Prandtl–von Karman universal velocity distribution are 1-D distributions that are satisfactory for wide rectangular channels in which the variation in velocity is dominant in the vertical direction.

Another characteristic that distinguishes the 2-D velocity distribution is the dip phenomenon, which was reported more than a century ago by Stearns (1883) and Murphy (1904). The dip phenomenon is caused by secondary currents (Nezu and Nakagawa 1993) that result in the circulation in the transverse channel cross section. In the velocity field, the longitudinal flow component is called primary flow and the secondary motion transports the low-momentum fluids from the near-bank region to the center and the high-momentum fluids from the free surface toward the bed. Yang et al. (2004) investigated the mechanism of dip phenomenon in relation to secondary currents in open-channel flow. In their study, a dip-modified log-law for the velocity distribution in smooth uniform open channel was developed. This modified velocity distribution was capable of describing the dip phenomenon and is applicable to the velocity profile in the region from near the bed to just below the free surface and transversely from the centerline to the near-wall region of the channel. From experiments, they showed that the location of the maximum velocity was related to the lateral portion of the measured velocity profile. It was concluded that the dip may even occur in a very wide channel, not on the centerline but in the sidewall region. Absi (2011) derived an ordinary differential equation for velocity distribution using a simple dip-modified log-wake law and found his approach to yield improved velocity predictions for arbitrary open-channel flow.

## 4.2   2-D VELOCITY DISTRIBUTIONS

The entropy theory permits the development of an efficient method to describe velocity in both one and two dimensions. The 2-D velocity distribution should be valid regardless of the location of maximum velocity. It should allow deriving equations for the location (on a vertical) of mean velocity in a channel cross section and derive other equations that can be used to provide additional descriptions of velocity distribution.

Plastino and Plastino (1999), among others, have discussed the advantages of the Tsallis entropy. Luo and Singh (2011) and Cui and Singh (2013) employed the Tsallis entropy to derive the 2-D velocity distribution. The Tsallis entropy (Tsallis, 1988), $H(u)$, of velocity, $u$, can be expressed as

$$H(u) = \frac{1}{m-1} \int_0^{u_{max}} f(u)\{1 - [f(u)]^{m-1}\} du \tag{4.1}$$

where
  $m$ is the entropy index
  $f(u)$ is the probability density function (PDF)
  $u_{max}$ is the maximum velocity

The 2-D velocity distributions have been derived by two methods: (1) using curvilinear coordinates (Chiu, 1988) and (2) regular coordinates (Marini et al., 2011). Both methods are discussed using the Tsallis entropy. The procedure for deriving a 2-D velocity distribution using the Tsallis entropy comprises the following steps: (1) setting up a coordinate system, (2) hypothesizing cumulative probability distribution function (CDF) of velocity, (3) specification of constraints, (4) determination of entropy, (5) derivation of velocity distribution, and (6) determination of parameters.

## 4.3   ONE-CONSTRAINT VELOCITY DISTRIBUTION USING THE CHIU COORDINATE SYSTEM

Using the coordinate system of Chiu (1988), Luo and Singh (2011) employed the Tsallis entropy. The Tsallis entropy–based approach of Luo and Singh (2011) was either superior or comparable to Chiu's distribution for the data sets used for testing.

However, due to the complexity of the coordinate system and a large number of parameters involved, application of the velocity distribution with the Chiu coordinate system may be limited.

### 4.3.1   2-D COORDINATES

In an open channel that is not "wide," the (time-averaged) velocity varies in both the vertical ($y$) and transverse ($z$) directions (Chiu, 1988). The isovels curve up toward the water surface under the influence of the two sides of the channel, among other factors. For modeling the velocity distribution, it is, therefore, logical to first transform the Cartesian $y$- and $z$-coordinates into another coordinate system, say, the $r$–$s$ coordinate system, in which $r$ has a unique, one-to-one relation with a value of velocity, and $s$ (coordinate) curves are their orthogonal trajectories, as shown in Figure 4.1. The idea of using the $r$–$s$ coordinates is similar to that of using the cylindrical coordinates in studying flows in a pipe. The time-averaged velocity $u$ that varies from 0 to $u_{max}$ is assigned to the $r$ value varying from $r_0$ to $r_{max}$. The time-averaged velocity ($u$) is almost zero along an isovel that has an $r$ value equal to $r_0$, which has a small value representing the channel bed (including the bottom and sides). In addition, $u$ is $u_{max}$, the maximum value of $u$, at $r$ equal to $r_{max}$, which may occur on or below the water surface. The velocity $u$ increases monotonically with the spatial coordinate $r$ from $r_0$ to $r_{max}$, although it may not increase monotonically with $y$, the elevation from the channel bed.

Referring to Figure 4.1, Chiu and Lin (1983) and Chiu and Chiou (1986) represented different features of isovels in a channel cross section as

$$r = Y(1-Z)^{\beta_i} \exp(\beta_i Z - Y + 1) \tag{4.2}$$

with its orthogonal trajectories as

$$s = \pm \frac{1}{Z}(|1-Y|)^{\beta_i[(D+\delta_y+h)/(B_i+\delta_i)]^2} \exp\left[Z + \beta_i\left(\frac{D+\delta_y+h}{B_i+\delta_i}\right)^2 Y\right] \tag{4.3}$$

in which

$$Y = \frac{y+\delta_y}{D+\delta_y+h} \tag{4.4}$$

$$Z = \frac{|z|}{B_i+\delta_i} \tag{4.5}$$

It is noted that $s$ takes on the negative sign only when $y > D + h$ and $h < 0$. In other cases, $s$ takes on the positive sign. Term $B_i$ for $i$ equal to either 1 or 2 is the transverse distance on the water surface between the $y$-axis and either the left or the right

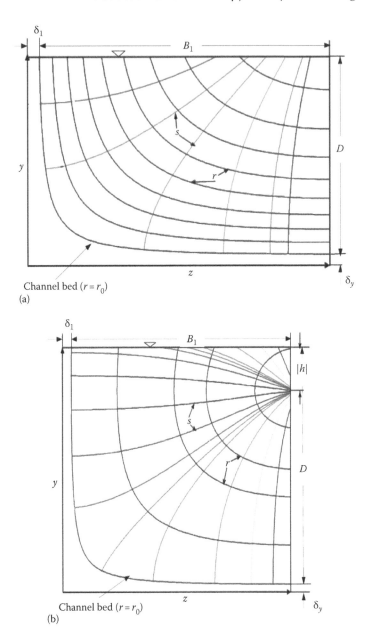

**FIGURE 4.1**   Velocity distribution and curvilinear coordinate system: (a) $h > 0$ and (b) $h < 0$.

bank of a channel cross section; $z$ is the coordinate in the transverse direction; $y$ is the coordinate in the vertical direction ($y$ is selected such that it passes through the point of maximum velocity); $h$, $\delta_y$, $\delta_i$, and $\beta_i$ are the coefficients characterizing the isovel geometry. Coefficient $h$ may vary from a negative value—$D$ to a positive value up to $+\infty$. When greater than zero, $h$ does not have any special physical meaning; it is simply a coefficient instrumental in shaping the pattern of isovels shown in Figure 4.1a, in which $u_{max}$ occurs on the water surface. However, when $h$ is less than or equal to zero, its magnitude $|h|$ represents the actual depth to $u_{max}$ below the water surface, as shown in Figure 4.1b. These figures also show the coordinates chosen, along with other variables that appear in the preceding equation. If the magnitude of $h$ is very large, isovels are parallel horizontal lines such that velocity varies only with $y$ and $r$ approaches $y/D$. Such a situation tends to occur in wide channels.

For a particular vertical along the $y$-axis where $z = 0$ and $\delta y = 0$, Equation 4.2 gives

$$r = Y \exp(1 - Y) = \frac{y}{D+h} \exp\left(1 - \frac{y}{D+h}\right) \tag{4.6}$$

Quantities $\delta_y$ and $\delta_i$ are usually small (especially for a rectangular channel) and $r_0 = 0$. For estimating flow velocity during high floods, it can be assumed that Equation 4.6, written for the vertical where the maximum velocity occurs ($z = 0$), holds.

If $h < 0$, $r_{max}$ and $u_{max}$ occur at $y = D + h$, so that $r_{max} = 1$ from Equation 4.2. Then, Equation 4.6 gives

$$\frac{r - r_0}{r_{max} - r_0} = \frac{r - 0}{1 - 0} = r = \frac{y}{D+h} \exp\left(1 - \frac{y}{D+h}\right) \tag{4.7}$$

If $h \geq 0$, $r_{max}$ and $u_{max}$ occur at the water surface where $y = D$ and

$$r_{max} = \frac{D}{D+h} \exp\left(1 - \frac{D}{D+h}\right) \tag{4.8}$$

Then, using Equations 4.6 and 4.8, one obtains

$$\frac{r - r_0}{r_{max} - r_0} = \frac{(y/(D+h))\exp(1 - (y/(D+h)))}{(D/(D+h))\exp(1 - (D/(D+h)))} = \frac{y}{D} \exp\left(\frac{D-y}{D+h}\right) \tag{4.9}$$

### 4.3.2 Formulation of a Hypothesis on Cumulative Probability Distribution

Chiu (1987, 1989) expressed the cumulative probability distribution (CDF) of velocity in a channel cross section in terms of the cylindrical coordinate system as

$$F(u) = \int_0^u f(u)du = \frac{r - r_0}{r_{max} - r_0} \tag{4.10}$$

From Equation 4.10 and the definition of PDF follows:

$$f(u) = \frac{dF(u)}{du} = \frac{dF(u)}{dr}\frac{dr}{du} = \left[(r_{max} - r_0)\frac{du}{dr}\right]^{-1}$$

(4.11)

### 4.3.3 SPECIFICATION OF CONSTRAINTS

The constraints can be defined variously but for illustrative purposes they are simply defined as

$$\int_0^{u_{max}} f(u)du = 1$$

(4.12)

$$\int_0^{u_{max}} uf(u)du = \bar{u}$$

(4.13)

where $\bar{u}$ is the cross-sectional mean velocity.

### 4.3.4 ENTROPY MAXIMIZING

Following Jaynes (1957a,b), the least-biased estimation of $f(u)$ can be obtained by maximizing the entropy subject to the specified constraints. Using the Lagrange multipliers, the Lagrangian function, subjected to Equations 4.12 and 4.13, can be written as

$$L = \int_0^{u_{max}} \frac{1}{m-1}f(u)\{1-[f(u)]^{m-1}\}du + \lambda_0\left[\int_0^{u_{max}} f(u)du - 1\right] + \lambda_1\left[\int_0^{u_{max}} uf(u)du - u_m\right]$$

(4.14)

where $\lambda_0$ and $\lambda_1$ are the Lagrange multipliers. Now, differentiating Equation 4.14 with respect to $f(u)$, the velocity PDF is obtained as

$$f(u) = \left[\frac{m-1}{m}\left(\frac{1}{m-1} + \lambda_0 + \lambda_1 u\right)\right]^{1/(m-1)}$$

(4.15)

Let $\lambda_* = (1/(m-1)) + \lambda_0$ be replaced with $\lambda_*$. Then, Equation 4.15 can be recast as

$$f(u) = \left[\frac{m-1}{m}(\lambda_* + \lambda_1 u)\right]^{1/(m-1)}$$

(4.16)

Equation 4.16 expresses the entropy-based probability distribution of velocity.

### 4.3.5 VELOCITY DISTRIBUTION

Combining the PDF of $u$ (Equation 4.16) with Equation 4.10, one obtains the velocity distribution as

$$u = \left(\frac{m}{m-1}\right)\frac{1}{\lambda_1}\left\{\lambda_1\left(\frac{r-r_0}{r_{max}-r_0}\right)+\left[\left(\frac{m}{m-1}\right)\lambda_*\right]^{(m/(m-1))}\right\}^{(m-1)/m} - \frac{\lambda_*}{\lambda_1} \qquad (4.17)$$

Equation 4.17 is in terms of the $r$-coordinate and has two unknown parameters $\lambda_1$ and $\lambda_*$ and exponent $m$ which need to be determined. Figure 4.2 shows the velocity distribution for Tiber River, Italy, at two different verticals for different $m$ values. The velocity observations are given in Table 4.A.1.

### 4.3.6 DETERMINATION OF PARAMETERS

Based on the analysis of both field and experimental velocity data, the feasible range of $m$ has been found to be between 0 and 2. Figure 4.2 shows the velocity distribution at two different verticals in a cross section of Tiber River, Italy, for different $m$ values. It can be seen that the velocity distribution changes a little with the variation of $m$ under the coordinate transformation; this means that the velocity distribution is not highly sensitive to exponent $m$ within the feasible range. In other words, the velocity curves derived from different $m$ values do not have significant differences between each other. Furthermore, it was found that for fixing $m = 2$, parameters $\lambda_1$ and $\lambda_*$ have simple analytical expressions as shown later.

Following the same procedure as in Chapter 3, one can write for $m = 2$

$$\int_0^{u_{max}} f(u) = \int_0^{u_{max}} \frac{1}{2}(\lambda_* + \lambda_1 u)du = 1 \qquad (4.18)$$

$$\int_0^{u_{max}} uf(u)du = u_m = \int_0^{u_{max}} u\left[\frac{1}{2}(\lambda_* + \lambda_1 u)\right]du = u_m \qquad (4.19)$$

Integration of Equations 4.18 and 4.19 yields, respectively,

$$\left(\frac{1}{2}\lambda_* u + \frac{1}{4}\lambda_1 u^2\right)_0^{u_{max}} = \frac{1}{2}\lambda_* u_{max} + \frac{1}{4}\lambda_1 u_{max}^2 = 1 \qquad (4.20)$$

$$u\left[\frac{1}{2}(\lambda_* + \lambda_1 u)\right]_0^{u_{max}} = \frac{1}{4}\lambda_* u_{max}^2 + \frac{1}{6}\lambda_1 u_{max}^3 = u_m \qquad (4.21)$$

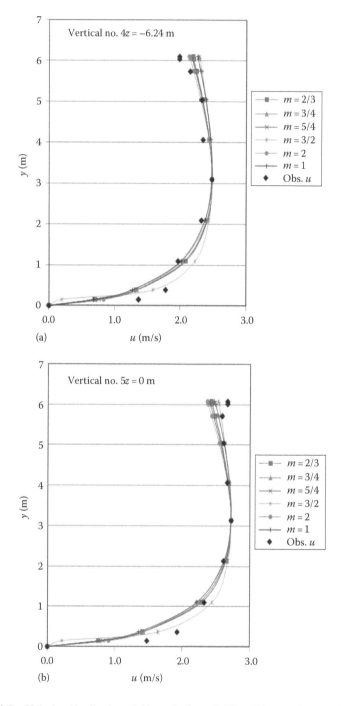

**FIGURE 4.2** Velocity distribution of (a) vertical no. 4, Tiber River and (b) vertical no. 5, Tiber River.

Therefore, parameters $\lambda_1$ and $\lambda_*$ have a simple analytical solution:

$$\lambda_1 = -\frac{12}{u_{max}^3}(u_{max} - 2u_m) \tag{4.22}$$

$$\lambda_* = \frac{4 - \lambda_1 u_{max}^2}{2u_{max}} = \frac{8u_{max} - 12u_m}{u_{max}^2} \tag{4.23}$$

With $u_{max}$ and $u_m$ known from observations, parameters $\lambda_1$ and $\lambda_*$ can be obtained from Equations 4.22 and 4.23.

## Example 4.1

Twelve sets of observed mean and maximum velocity data collected from straight rectangular reaches of Ghamasiab River in western Iran are listed in Table 4.1. Compute the Lagrange multipliers for the velocity data sets.

### Solution

The Lagrange multipliers can be computed from Equations 4.22 and 4.23. For example, taking $u_{max} = 0.524$ m/s and $u_m = 0.348$ m/s, one gets

$$\lambda_1 = -\frac{12}{u_{max}^3}(u_{max} - 2u_m) = -\frac{12}{0.524^3}(0.524 - 2 \times 0.348) = 14.346$$

$$\lambda_* = \frac{8u_{max} - 12u_m}{u_{max}^2} = \frac{8 \times 0.524 - 12 \times 0.348}{0.524^2} = -2.914$$

Thus, the Lagrange multipliers for other sets are computed in the same way and listed in Table 4.2.

### TABLE 4.1
### Observed Maximum and Mean
### Velocity from Ghamasiab River

| Run No. | $u_{max}$ (m/s) | $u_m$ (m/s) |
|---------|-----------------|-------------|
| A0-1 | 0.524 | 0.348 |
| A0-2 | 0.491 | 0.335 |
| A0-3 | 0.358 | 0.308 |
| A0-4 | 0.421 | 0.323 |
| A1-1 | 0.582 | 0.421 |
| A1-2 | 0.578 | 0.424 |
| A1-3 | 0.575 | 0.345 |
| A1-4 | 0.607 | 0.378 |
| A2-1 | 1.071 | 0.708 |
| A2-2 | 0.885 | 0.584 |
| A2-3 | 0.774 | 0.516 |
| A2-4 | 0.682 | 0.493 |

**TABLE 4.2**

**Computation of Lagrange Multipliers**

| Run No. | $\lambda_1$ | $\lambda_*$ |
|---------|-------------|-------------|
| A0-1 | 14.346 | −2.914 |
| A0-2 | 18.146 | −8.922 |
| A0-3 | 67.476 | −139.811 |
| A0-4 | 36.184 | −46.939 |
| A1-1 | 15.827 | −5.925 |
| A1-2 | 16.779 | −7.454 |
| A1-3 | 7.259 | 6.338 |
| A1-4 | 7.995 | 4.978 |
| A2-1 | 3.370 | 5.390 |
| A2-2 | 4.899 | 5.386 |
| A2-3 | 6.677 | 4.585 |
| A2-4 | 11.500 | −0.459 |

### 4.3.7 DEFINING AN ENTROPY PARAMETER

It is possible to simplify the velocity distribution equation by introducing a new dimensionless parameter $M$ that can be defined as

$$M = \lambda_1 u_{max}^2 \tag{4.24}$$

The mathematical range of $M$ is in the interval (0, 12). For most rivers in the United States, the maximum velocity is 25%–50% larger than the mean, so the $M$ values for these rivers are between 4 and 7.2. The $M$ value can play an important part in understanding and controlling open-channel flows and hence as a new key hydraulic parameter. Parameter $M$ combines mean and maximum velocities and in turn the influence of the Lagrange multiplier $\lambda_1$. It also serves to succinctly express a range of flow characteristics.

### Example 4.2

Compute the values of $M$ for data sets listed in Table 4.1.

**Solution**

Using Equation 4.24, the entropy parameter $M$ can be computed as

$$M = \lambda_1 u_{max}^2 = 14.346 \times 0.524^2 = 3.939$$

The PDF of $u$, $f(u)$, can also be expressed in terms of $M$ as

$$f(u) = \frac{4 - M}{4u_{max}} + \frac{M}{2u_{max}^2} u \tag{4.25}$$

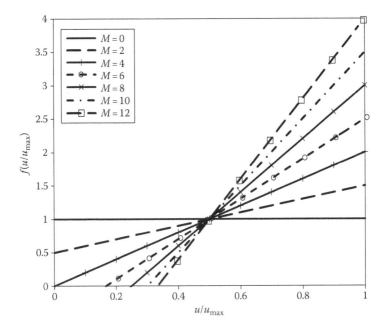

**FIGURE 4.3**   Parameter $M$ and probability density of dimensionless velocity $f(u/u_{max})$.

The PDF of dimensionless velocity $f(u/u_{max})$ can also be expressed with $M$ as a parameter:

$$f\left(\frac{u}{u_{max}}\right) = u_{max} f(u) = \frac{4-M}{4} + \frac{M}{2}\left(\frac{u}{u_{max}}\right) \tag{4.26}$$

Equation 4.26 shows that the probability density of dimensionless velocity $u/u_{max}$ follows a simple linear relationship, and the slope of the line is determined by parameter $M$. Figure 4.3 plots the function $f(u/u_{max})$ for various $M$ values.

## Example 4.3

Compute the PDF of $u/u_{max}$.

### Solution

It can be seen from Equation 4.26 that the PDF of dimensionless velocity is defined by entropy parameter $M$. For different $M$ values, the PDF of $u/u_{max}$ is obtained from Equation 4.26 and plotted in Figure 4.3. It is seen from the figure that when $M$ equals 0, the PDF is constant at 1 for any velocity; thus, it is a uniform distribution. As $M$ value increases, the PDF gets skewed and the slope increases with the $M$ value.

## 4.3.8   Maximum Entropy

Substitution of Equations 4.24 and 4.25 in Equation 4.1 gives, respectively, the maximum entropy of velocity and dimensionless velocity $u/u_{max}$ as

$$H(u) = \int_0^{u_{max}} f(u)\{1 - [f(u)]\} du = 1 - \int_0^{u_{max}} f(u)^2 du$$

$$= 1 - \int_0^{u_{max}} \left( \frac{4 - M}{4u_{max}} + \frac{M}{2u_{max}^2} u \right)^2 du = 1 - \frac{48 + M^2}{48u_{max}} \tag{4.27}$$

or

$$H\left( \frac{u}{u_{max}} \right) = \int_0^1 f(w)\{1 - [f(w)]\} dw = 1 - \int_0^{u_{max}} \left( \frac{4 - M}{4} + \frac{M}{2} w \right)^2 dw = -\frac{M^2}{48} \tag{4.28}$$

where $w = u/u_{max}$. Equation 4.27 shows that the maximum entropy increases with increasing maximum velocity.

## Example 4.4

Compute the maximum entropy for the data in Table 4.3.

### Solution

Using Equation 4.27, the maximum entropy can be obtained as shown in Table 4.4. As an illustration,

$$H(u) = 1 - \frac{48 + M^2}{48u_{max}} = 1 - \frac{48 + 3.939^2}{48 \times 0.524} = -1.525$$

Equation 4.27 shows that the maximum entropy is determined by parameter $M$. The maximum entropy of dimensionless velocity computed using Equation 4.27 is plotted in Figure 4.4 for various $M$ values, and a unique relation is observed. It is seen that $H(u)$ decreases with the increase in $M$ in its feasible range. Entropy

**TABLE 4.3**

**Computation of M Values**

| Run No. | M | Run No. | M |
|---------|-------|---------|-------|
| A0-1 | 3.939 | A1-3 | 2.400 |
| A0-2 | 4.375 | A1-4 | 2.946 |
| A0-3 | 8.648 | A2-1 | 3.866 |
| A0-4 | 6.413 | A2-2 | 3.837 |
| A1-1 | 5.361 | A2-3 | 4.000 |
| A1-2 | 5.606 | A2-4 | 5.349 |

**TABLE 4.4**
**Computed Values of Maximum Entropy**

| Run No. | H | Run No. | H |
|---------|--------|---------|--------|
| A0-1 | −1.525 | A1-3 | −0.948 |
| A0-2 | −1.849 | A1-4 | −0.945 |
| A0-3 | −6.146 | A2-1 | −0.224 |
| A0-4 | −3.411 | A2-2 | −0.477 |
| A1-1 | −1.747 | A2-3 | −0.723 |
| A1-2 | −1.863 | A2-4 | −1.340 |

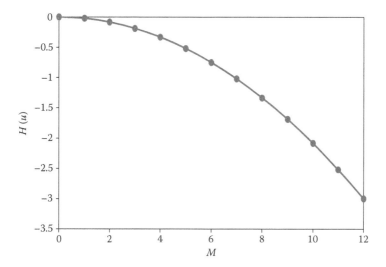

**FIGURE 4.4** Maximum entropy with various $M$.

has the largest value at $M = 0$, which corresponds to the result from Figure 4.3, where the PDF is uniform for $M = 0$. It may, however, be noted that the value of entropy must be positive, and therefore, the value of entropy reported in the table can be taken with respect to a selected benchmark so all values of entropy are positive.

### 4.3.9 VELOCITY DISTRIBUTION IN TERMS OF PARAMETER $M$

Replacing the two parameters in the velocity distribution of Equation 4.17 with parameter $M$, the velocity distribution equation becomes

$$u = \frac{2u_{max}}{M}\left[M\left(\frac{r - r_0}{r_{max} - r_0}\right) + \frac{(4 - M)^2}{16}\right]^{1/2} - \frac{(4 - M)u_{max}}{2M} \qquad (4.29)$$

Therefore, the dimensionless velocity can be expressed as

$$\frac{u}{u_{max}} = \frac{2}{M}\left[M\left(\frac{r-r_0}{r_{max}-r_0}\right)+\frac{(4-M)^2}{16}\right]^{1/2} - \frac{4-M}{2M} \tag{4.30}$$

The cross-sectional mean $M$ can be used to compute the velocity distribution that can be compared with observed data in Figure 4.3.

## Example 4.5

Compute and plot the dimensionless velocity distribution for various values of $M$ with $h/D = -0.4$ and 0.05.

### Solution

When $h/D = -0.4$, $u_{max}$ occurs at $y/D = 0.6$ and $(r - r_0)/(r_{max} - r_0)$ is given by Equation 4.7. Thus, the dimensionless velocity is computed from

$$\frac{u}{u_{max}} = \frac{2}{M}\left[M\frac{y}{D+h}\exp\left(1-\frac{y}{D+h}\right)+\frac{(4-M)^2}{16}\right]^{1/2} - \frac{4-M}{2M}$$

When $h/D = 0.05$, $u_{max}$ occurs at water surface and $(r - r_0)/(r_{max} - r_0)$ is given by Equation 4.9. Now the dimensionless velocity is estimated from

$$\frac{u}{u_{max}} = \frac{2}{M}\left[M\frac{y}{D}\exp\left(\frac{D-y}{D+h}\right)+\frac{(4-M)^2}{16}\right]^{1/2} - \frac{4-M}{2M}$$

Figure 4.5 shows the velocity distribution curve changes gradually according to the change in parameter $M$, but not so dramatically; during the reasonable mathematical range of $M$, profiles with different $M$ values tend to the same shape. For $h < 0$, the intersection of velocity profiles with different $M$ values is determined by $h/D$. The differences focus on the region close to the channel bottom.

## Example 4.6

Estimate the velocity distribution of Tiber River for the data given in Appendix A using the $M$ value and plot it.

### Solution

For given $u_{max} = 2.72$ m/s and $u_m = 2.20$ m/s, the Lagrange multipliers are determined first

$$\lambda_1 = -\frac{12}{u_{max}^3}(u_{max} - 2u_m) = 1.002$$

$$\lambda_* = \frac{8u_{max} - 12u_m}{u_{max}^2} = 2.643$$

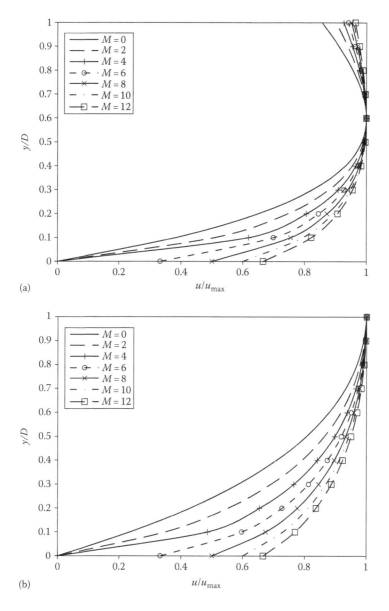

**FIGURE 4.5** (b) Dimensionless velocity distributions with various $M$ values at (a) $h/d = -0.4$ and (b) $h/D = 0.05$.

Thus, $M = \lambda_1 u_{max}^2 = 7.412$

As given $h = -2.7$, $u_{max}$ occurs at $D + h = 3.61$ m, and Equation 4.7 is used. By inputting Equation 4.7 into Equation 4.30, the velocity for each location can be determined as in the previous example. The velocity estimated for no. 5 vertical line (the centerline) and no. 4 vertical line is plotted in Figure 4.6.

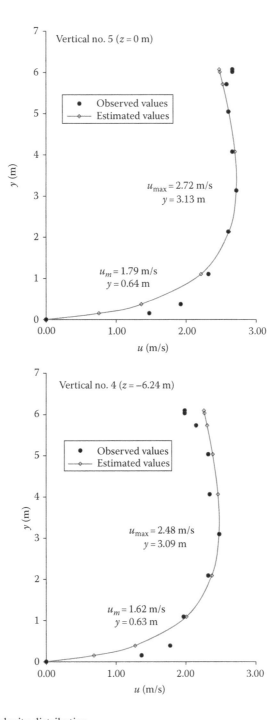

**FIGURE 4.6**   Velocity distribution.

### 4.3.10 CONSTRUCTION OF ISOVELS

For a given cross section, isovels can be constructed following the steps shown in Example 4.7.

### Example 4.7

Construct the isovels of the velocity estimated from Example 4.6 with $n = 0.015$, $B = 20.8$ m, $D = 6.31$ m, $h = -2.7$ m and $u_{max} = 2.72$ m/s, $u_m = 2.20$ m/s.

**Solution**

For $z = (-20.8, 20.8)$ and $y = (0, 6.31)$, we first transfer $z$ and $y$ into $Z$ and $Y$ using Equations 4.4 and 4.5 with $B_i = 20.8$ m, $D = 6.31$ m, and $h = -7.2$ m using the given information. Then, $(z, y)$ coordinates are meshed with the 2-D plane and changed into the isovel coordinate $r$ using Equation 4.2, where $\beta_i$ is assumed to be 1. With the $M$ value computed from Example 4.6, the velocity distribution is estimated for each $r$ value using Equation 4.30. Thus, for each combined value of $(z, y)$ a corresponding velocity value is estimated. Therefore, the contour plot of $(z, y, u)$ is obtained, as shown in Figure 4.7.

### 4.3.11 MEAN VELOCITY

In open-channel hydraulics, the mean velocity is needed in the governing equations for the transport of mass, momentum, and energy through a channel cross section. In comparison with mean velocity, the maximum velocity in a channel cross section has

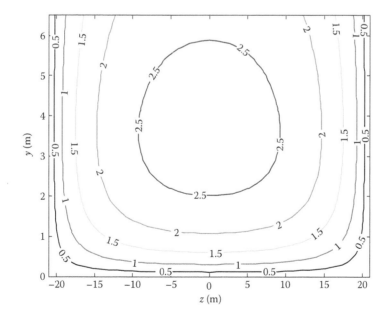

**FIGURE 4.7** Cross-sectional velocity distribution for a section in Tiber River, Italy.

not been considered important enough to receive special attention until Chiu (1991) explored the relationship between the mean velocity and the maximum velocity using the entropy parameter $M_c$, which can be expressed as

$$\frac{u_m}{u_{max}} = \frac{e^{M_c}}{e^{M_c} - 1} - \frac{1}{M_c} \tag{4.31}$$

It is considered worthwhile to make special efforts for measurement, analysis and modeling of maximum velocity, and determining the relation of maximum velocity to the mean velocity (Chiu and Said, 1994). On the other hand, based on the definition of $M$ and its relationship with the two Lagrange multipliers, it is found that

$$\frac{u_m}{u_{max}} = \frac{12 + M}{24} \tag{4.32}$$

The relation found based on the Tsallis-based 2-D velocity distribution follows a linear distribution with $M$. The ratio of mean and maximum velocity is plotted against various $M$ values for the Tsallis entropy–based velocity distribution in Figure 4.8.

### Example 4.8

Compute the mean velocity for data given in Example 4.1.

**Solution**
Following Equation 4.32, the mean velocity can be computed from

$$u_m = u_{max}\left(\frac{12+M}{24}\right) = 0.524 \times \left(\frac{12+3.939}{24}\right) = 0.348$$

which is the same as the observed value (Table 4.5).

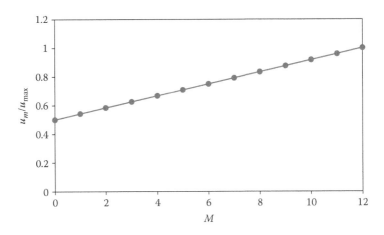

**FIGURE 4.8**   $u_m/u_{max}$ versus various $M$.

**TABLE 4.5**
**Estimation of Mean Velocity from Parameter $M$**

| Run No. | $u_{max}$(m/s) | Observed $u_m$(m/s) | Estimated $u_m$(m/s) |
|---------|---------------|---------------------|----------------------|
| A0-1 | 0.524 | 0.348 | 0.348 |
| A0-2 | 0.491 | 0.335 | 0.335 |
| A0-3 | 0.358 | 0.308 | 0.308 |
| A0-4 | 0.421 | 0.323 | 0.323 |
| A1-1 | 0.582 | 0.421 | 0.421 |
| A1-2 | 0.578 | 0.424 | 0.424 |
| A1-3 | 0.575 | 0.345 | 0.345 |
| A1-4 | 0.607 | 0.378 | 0.378 |
| A2-1 | 1.071 | 0.708 | 0.708 |
| A2-2 | 0.885 | 0.584 | 0.584 |
| A2-3 | 0.774 | 0.516 | 0.516 |
| A2-4 | 0.682 | 0.493 | 0.493 |

## 4.3.12 LOCATION OF MEAN VELOCITY

On an isovel where $r = r_m$ and $u = u_m$, Equation 4.30 becomes

$$\frac{u_m}{u_{max}} = \frac{2}{M}\left[M\left(\frac{r_m - r_0}{r_{max} - r_0}\right) + \frac{(4-M)^2}{16}\right]^{\frac{1}{2}} - \frac{4-M}{2M} \tag{4.33}$$

Combining Equations 4.33 and 4.32, the location of mean velocity can be obtained as

$$\frac{r_m - r_0}{r_{max} - r_0} = \frac{1}{M}\left[\left(\frac{M^2}{48} - 1\right)^2 - \frac{(4-M)^2}{16}\right] \tag{4.34}$$

### Example 4.9

Compute the location of mean velocity.

**Solution**

For $M$ varying from 0 to 12, the location of mean velocity can be determined from Equation 4.34 and is plotted in Figure 4.9. For example, when $M = 1$

$$\frac{r_m - r_0}{r_{max} - r_0} = \frac{1}{M}\left[\left(\frac{M^2}{48} - 1\right)^2 - \frac{(4-M)^2}{16}\right] = \left[\left(\frac{1}{48} - 1\right)^2 - \frac{(4-1)^2}{16}\right] = 0.396$$

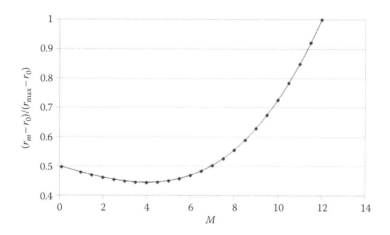

**FIGURE 4.9** $(r_m - r_0)/(r_{max} - r_0)$ versus $M$.

## 4.4 VELOCITY DISTRIBUTION USING GENERAL FRAMEWORK

To reduce the many parameters in the curvilinear coordinate system, Marini et al. (2011) developed a new method for deriving 2-D velocity distribution using the Shannon entropy, where the CDF was hypothesized using the $x$–$y$ coordinates and was continuous and differentiable. The 2-D velocity distribution, so developed, has been shown to have an advantage over Chiu's velocity distribution. However, the velocity with lower values is not captured accurately.

### 4.4.1 CUMULATIVE DISTRIBUTION FUNCTION

An idealized rectangular channel is shown in Figure 4.10, which is the half size of the cross section. The coordinate is set in this way such that $(0, 0)$ represents the center of the channel bed, $y$ represents the depth from the channel bed, and $x$ represents the distance from the center. Thus, $y$ is always positive and is measured from the channel bed ($y = 0$) up, while $x$ increases positively toward the right bank and takes on negative values in the left half of the cross section. Since the velocity is assumed to be 0 at the boundary, the velocity isovel $I(0) = \{x = B \text{ or } y = 0\}$. For the maximum velocity that is assumed to occur at the axis passing through the center of the water surface, the isovel $I(u_{max}) = \{x = 0 \text{ and } y = D - h\}$, $h$ is the distance to the maximum velocity below the water surface. For the rest of the points within the flow cross section, velocity isovels monotonically increase from 0 to $u_{max}$.

The velocity is shown to be monotonically increasing from isovel $I(0)$ to isovel $I(u_{max})$. Thus, the cumulative distribution function will have a value of 0 at $I(0)$ and a value of 1 at $I(u_{max})$. With the relation between the $x$ (transverse) and $y$ (vertical) coordinates and isovel $I(u)$, the cumulative distribution function can be expressed in the space domain. Thus, the CDF needs to be 0 at $x = B$ or $y = 0$, which corresponds to $I(0)$, and needs to be 1 at $x = 0$ and $y = D$, which corresponds to $I(u_{max})$. Here, $D$ is the flow depth. Cui and Singh (2012) have shown that a nonlinear cumulative distribution function of the type

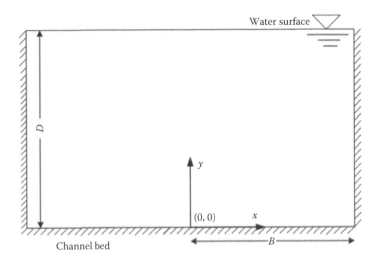

**FIGURE 4.10** Idealized rectangular cross section.

$$F(u) = \left[1 - \left(\frac{x}{B}\right)^2\right]^b \left(\frac{y}{D}\right)^a \quad \text{for all } (x, y) \text{ on } I(u) \qquad (4.35)$$

is satisfactory for the probabilistic description of 2-D velocity distribution in open channels. In Equation 4.35, $a$ and $b$ are both shape parameters and are related to the width-depth ratio.

### 4.4.2 2-D VELOCITY DISTRIBUTION

With the hypothesis on the cumulative distribution function expressed by Equation 4.35, the next step is to compute velocity profiles using the entropy-based PDF given by Equation 4.16. To that end, consider the PDF in 2-D domain as $(x, y)$, where $y$ represents the depth from the channel bed and $x$ represents the transverse distance from the centerline. Following Marini et al. (2011), since $u$ is a function of $x$ and $y$, $f(u)$ can be written as $f(u(x, y))$. Since $f(u)$ is the derivative of the cumulative distribution function $F(u)$, taking the partial derivatives of $F(u)$ with respect to $x$ and $y$, the following two equations are obtained:

$$\frac{\partial F(u)}{\partial x} = \frac{\partial F(u)}{\partial u}\frac{\partial u}{\partial x} = f(u)\frac{\partial u}{\partial x} = \left[\frac{m-1}{m}(\lambda_* + \lambda_1 u)\right]^{\frac{1}{m-1}}\frac{\partial u}{\partial x} \qquad (4.36)$$

$$\frac{\partial F(u)}{\partial y} = \frac{\partial F(u)}{\partial u}\frac{\partial u}{\partial y} = f(u)\frac{\partial u}{\partial y} = \left[\frac{m-1}{m}(\lambda_* + \lambda_1 u)\right]^{\frac{1}{m-1}}\frac{\partial u}{\partial y} \qquad (4.37)$$

Now defining a new variable

$$w = \left[ \frac{m-1}{m}(\lambda_* + \lambda_1 u) \right]^{\frac{m}{m-1}} \tag{4.38}$$

and taking partial derivatives of $w$ with respect to $x$ and $y$, one obtains

$$\frac{\partial w}{\partial x} = \frac{\partial w}{\partial u}\frac{\partial u}{\partial x} = \lambda_1 \left[ \frac{m-1}{m}(\lambda_* + \lambda_1 u) \right]^{\frac{1}{m-1}} \frac{\partial u}{\partial x} \tag{4.39}$$

$$\frac{\partial w}{\partial y} = \frac{\partial w}{\partial u}\frac{\partial u}{\partial y} = \lambda_1 \left[ \frac{m-1}{m}(\lambda_* + \lambda_1 u) \right]^{\frac{1}{m-1}} \frac{\partial u}{\partial y} \tag{4.40}$$

Comparing Equations 4.39 and 4.40 with Equations 4.36 and 4.37, the relationship between $F(u)$ and $w$ can be written as

$$\frac{\partial w}{\partial x} = \lambda_1 \frac{\partial F(u)}{\partial x} \tag{4.41}$$

$$\frac{\partial w}{\partial y} = \lambda_1 \frac{\partial F(u)}{\partial y} \tag{4.42}$$

Equations 4.41 and 4.42 can be seen as a system of linear differential equations that can be integrated using the Leibniz rule:

$$\int_{(0,0)}^{(x,y)} \frac{\partial w}{\partial x}dx + \frac{\partial w}{\partial y}dy = w(x,y) - w(0,0) \tag{4.43}$$

Since the point with coordinates (0, 0) lies in the middle of channel floor, which has the velocity $u = 0$, $w(0, 0)$ on the right-hand side of Equation 4.43 equals $[((m-1)/m)\lambda_*]^{m/(m-1)}$ from Equation 4.38. Hence, the right-hand side of Equation 4.43 now becomes

$$w(x,y) - w(0,0) = w(x,y) - \left[ \frac{m-1}{m}\lambda_* \right]^{\frac{m}{m-1}} \tag{4.44}$$

The definite integral on the left-hand side of Equation 4.43 can be calculated at a generic point of coordinates, say $(\bar{x},\bar{y})$, which is identified by means of a polygonal curve that starts from the origin of axes (0, 0), passing through the point $(\bar{x},0)$ and ends at $(\bar{x},\bar{y})$. The cumulative distribution function $F(u)$ is constantly 0 at point (0, 0) to $(\bar{x}, 0)$. Thus, using Equations 4.39 and 4.40, the integral of Equation 4.43 yields

$$\int_{(0,0)}^{(x,y)} \frac{\partial w}{\partial w} dx + \frac{\partial w}{\partial y} dy = \int_{(0,0)}^{(\bar{x},\bar{y})} \lambda_1 \frac{\partial F(u)}{\partial u} dx + \lambda_1 \frac{\partial F(u)}{\partial y} dy = \int_0^{\bar{y}} \lambda_1 \frac{\partial F(u)}{\partial y} dy = \lambda_1 F(u) \quad (4.45)$$

Replacing the left side of Equation 4.43 with Equation 4.45 and the right side with Equation 4.44, $w(x, y)$ can be obtained as

$$w(x,y) = \lambda_1 F(u) + \left[ \frac{m-1}{m} \lambda_* \right]^{\frac{m}{m-1}} \quad (4.46)$$

Substituting the definition of $w$ from Equation 4.38 into Equation 4.46, the velocity distribution function is obtained as

$$u = -\frac{\lambda_*}{\lambda_1} + \frac{1}{\lambda_1} \frac{m}{m-1} \left[ \lambda_1 F(u) + \left( \frac{m-1}{m} \lambda_* \right)^{m/(m-1)} \right]^{(m-1)/m} \quad (4.47)$$

Equation 4.47 is the 2-D velocity distribution equation based on the Tsallis entropy. It is interesting to note that Equation 4.47 reduces to 1-D velocity distribution:

$$u = -\frac{\lambda_*}{\lambda_1} + \frac{1}{\lambda_1} \frac{m}{m-1} \left[ \lambda_1 \left( \frac{y}{D} \right) + \left( \frac{m-1}{m} \lambda_* \right)^{m/(m-1)} \right]^{(m-1)/m} \quad (4.48)$$

with the CDF hypothesized as $F(u) = y/D$ (Cui, 2011). Equation 4.48 can also be derived directly (Cui, 2011).

### Example 4.10

Determine the 2-D velocity distribution using the general framework for data listed in Table 4.6.

### Solution

It can be seen from the table that the cross section is 25.58 m wide from center to the left bank and 16.47 m wide from center to the right bank with a depth of 6.15 m at the center. $u_{max} = 3.36$ m/s occurred at $y = 6.09$ m and $u_m = 2.206$ m/s from observations. Thus, following the steps from Chapter 3, for different $m$ values, the Lagrange multipliers are computed, as given in Table 4.7.

Thus, the velocity distribution can be determined from Equation 4.47 with fitted $a = 0.2$ and $b = 1$. Figure 4.11 plots the dimensionless velocity distribution of centerline for various $m$ values.

## 4.4.3 DIMENSIONLESS PARAMETER G

As defined in the case of 1-D velocity distribution (Cui, 2011) in Chapter 3, the dimensionless entropy parameter $G$ can also be used as follows:

**TABLE 4.6**

**2-D Velocity Profile Obtained from Tiber River, Italy**

| z (m) | Depth (m) | u obs. (m/s) | z (m) | Depth (m) | u obs. (m/s) |
|-------|-----------|--------------|-------|-----------|--------------|
| −25.58 | 0 | 0 | 0 | 0.06 | 3.36 |
| −18.86 | 0 | 0.83 | 0 | 0.2 | 3.16 |
| −18.86 | 0.06 | 0.83 | 0 | 1 | 3.20 |
| −18.86 | 0.2 | 0.74 | 0 | 2 | 3.28 |
| −18.86 | 1 | 0.64 | 0 | 3 | 2.91 |
| −18.86 | 2 | 1.15 | 0 | 4 | 2.78 |
| −18.86 | 3 | 0.96 | 0 | 5 | 2.32 |
| −18.86 | 3.9 | 0.74 | 0 | 5.8 | 2.03 |
| −18.86 | 4.1 | 0.71 | 0 | 6 | 1.86 |
| −18.86 | 4.25 | 0.00 | 0 | 6.15 | 0.00 |
| −14.66 | 0 | 1.21 | 3.78 | 0 | 3.16 |
| −14.66 | 0.06 | 1.21 | 3.78 | 0.06 | 3.16 |
| −14.66 | 0.2 | 1.21 | 3.78 | 0.2 | 3.11 |
| −14.66 | 1 | 1.18 | 3.78 | 1 | 3.28 |
| −14.66 | 2 | 1.56 | 3.78 | 2 | 3.20 |
| −14.66 | 3 | 1.09 | 3.78 | 2.88 | 2.61 |
| −14.66 | 3.8 | 0.83 | 3.78 | 3.88 | 2.57 |
| −14.66 | 4 | 0.71 | 3.78 | 4.88 | 2.53 |
| −14.66 | 4.15 | 0.00 | 3.78 | 5.7 | 2.03 |
| −10.46 | 0 | 2.06 | 3.78 | 5.94 | 1.86 |
| −10.46 | 0.06 | 2.06 | 3.78 | 6.09 | 0.00 |
| −10.46 | 0.2 | 2.34 | 7.34 | 0 | 2.36 |
| −10.46 | 1 | 2.31 | 7.34 | 0.06 | 2.36 |
| −10.46 | 1.93 | 2.44 | 7.34 | 0.2 | 2.61 |
| −10.46 | 2.88 | 2.19 | 7.34 | 1 | 2.70 |
| −10.46 | 3.88 | 2.06 | 7.34 | 2 | 2.74 |
| −10.46 | 4.28 | 1.65 | 7.34 | 2.93 | 2.61 |
| −10.46 | 4.48 | 1.59 | 7.34 | 3.88 | 2.53 |
| −10.46 | 4.63 | 0.00 | 7.34 | 4.88 | 2.32 |
| −6.29 | 0 | 2.99 | 7.34 | 5.38 | 0.98 |
| −6.29 | 0.06 | 2.99 | 7.34 | 5.7 | 1.19 |
| −6.29 | 0.2 | 2.66 | 7.34 | 5.85 | 0.00 |
| −6.29 | 0.8 | 2.82 | 10.49 | 0 | 1.78 |
| −6.29 | 1.71 | 2.66 | 10.49 | 0.06 | 1.78 |
| −6.29 | 3 | 2.61 | 10.49 | 0.2 | 1.44 |
| −6.29 | 4 | 2.36 | 10.49 | 1 | 2.15 |
| −6.29 | 5 | 1.95 | 10.49 | 2 | 2.32 |
| −6.29 | 5.2 | 1.60 | 10.49 | 3 | 2.57 |
| −6.29 | 5.45 | 1.46 | 10.49 | 3.5 | 2.32 |
| −6.29 | 5.6 | 0.00 | 10.49 | 4.6 | 1.95 |
| | | | 10.49 | 5 | 1.61 |
| | | | 10.49 | 5.7 | 0.00 |
| | | | 16.47 | 0 | 0 |

**TABLE 4.7**

**Lagrange Multipliers Computed for Various $m$ Values**

| $M$ | 1.5 | 2 | 3 | 4 | 6 |
|---|---|---|---|---|---|
| $\lambda_1$ | 0.101 | 0.055 | 0.016 | 0.005 | 0.001 |
| $\lambda_*$ | 0.464 | 0.5028 | 0.1064 | 0.027 | 0.0013 |

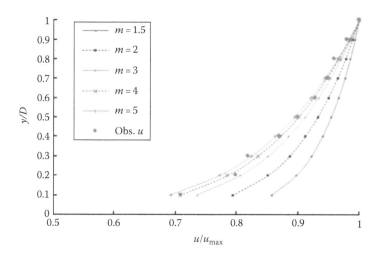

**FIGURE 4.11**   Dimensionless velocity distribution for various $m$ values.

$$G = \frac{\lambda_1 u_{max}}{\lambda_1 u_{max} + \lambda_*} \tag{4.49}$$

Note that in Equation 4.49 the Lagrange multiplier $\lambda_1$ has dimensions $[T/L]$ and $\lambda_*$ is dimensionless; accordingly, $G$ is dimensionless. Parameter $G$ is found to be related to the ratio of mean and maximum velocity and a quadratic relation is obtained from observed mean and maximum velocity values as (Cui, 2011):

$$\frac{\bar{u}}{u_{max}} = \Psi(G) = 0.554G^2 - 0.077G + 0.568 \tag{4.50}$$

Now with the use of entropy parameter $G$, the general velocity distribution equation (Equation 4.47) can be developed using the same steps as presented in Chapter 3. From Equation 4.47, the maximum velocity, where $y = D$, $F(u_{max}) = 1$, can be obtained as

$$u_{max} = -\frac{\lambda_*}{\lambda_1} + \frac{1}{\lambda_1}\frac{m}{m-1}\left[\lambda_1 + \left(\frac{m-1}{m}\lambda_*\right)^{m/(m-1)}\right]^{(m-1)/m} \tag{4.51}$$

Dividing Equation 4.47 by Equation 4.51 and using Equation 4.49, one obtains

$$\frac{u}{u_{max}} = \frac{\lambda_*}{\lambda_1 u_{max}} - \frac{1}{\lambda_1 u_{max}} \frac{m}{m-1} \left\{ -\lambda_1 \frac{m-1}{m} F(u) + \left[ \frac{m-1}{m} \lambda_* \right]^{m/(m-1)} \right\}^{(m-1)/m}$$

$$= 1 - \frac{1}{G} \left( 1 - \left( \left( \frac{m}{m-1} \right)^{m/(m-1)} \frac{G}{u_{max}} \left( \left( \frac{y}{D} \right) - 1 \right) + 1 \right)^{(m-1)/m} \right) \tag{4.52}$$

Substituting $u = u_0 = 0$, at $y = 0$, Equation 4.52 reduces to

$$0 = 1 - \frac{1}{G} \left( 1 - \left( 1 - \left( \frac{m}{m-1} \right)^{m/(m-1)} \frac{G}{u_{max}} \right)^{(m-1)/m} \right) \tag{4.53}$$

Rearranging Equation 4.53, one obtains

$$\left( \frac{m}{m-1} \right)^{m/(m-1)} \frac{G}{u_{max}} = 1 - (1-G)^{m/(m-1)} \tag{4.54}$$

Using the right side of Equation 4.54, Equation 4.52 can be written as

$$\frac{u}{u_{max}} = 1 - \frac{1}{G} (1 - [(1-G)^{m/(m-1)} + (1-(1-G)^{m/(m-1)})F(u)]^{(m-1)/m}) \tag{4.55}$$

Equation 4.55 is the general 2-D velocity distribution equation in terms of parameter $G$ and maximum velocity.

## Example 4.11

Determine the velocity distribution for the data in Table 4.6.

### Solution
The Lagrange multiplier values are given in Table 4.7. For given $u_{max} = 3.36$ m/s and $u_m = 2.206$ m/s, parameter $G$ is estimated by solving Equation 4.50, which gives $G = 0.336$. Thus, the velocity distribution can be computed from Equation 4.55 for each vertical line. The velocity profiles at $x = 0$ m, 7.34 m, and −18.86 m are plotted in Figure 4.12.

It can be seen from Figure 4.12 that the velocity can be divided into three portions. In the first region from the channel bed to some depth of about 0.2 m, the velocity distribution on each vertical line increases slowly. Then it starts to increase faster to the maximum velocity, which is the second region. Thereafter, the velocity decreases from the maximum to some value at the water surface, where the rate of decrease is

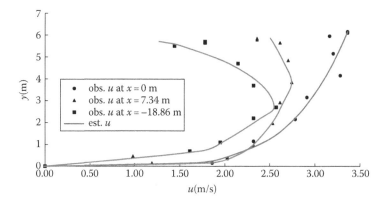

**FIGURE 4.12** Velocity distribution observed in Tiber River.

similar to the rate of increase in the second region. It can be concluded that the slow increase in the first region is caused by the resistance due to the bed shear stress, so that beyond the bed effective region the velocity grows faster. In the third portion, it is the secondary currents that retard the velocity from growing to the water surface.

The highlighted point is the maximum velocity observed from each vertical. The computed values capture the point well. Comparing the curves in Figure 4.12, it is found that the farther away from the centerline the location of the maximum velocity is occurring lower, which is in accord with the analysis of Yang et al. (2004). Overall, the estimated values fit the observed values reasonably well with the coefficient of determination between computed and observed values as high as 0.977.

### 4.4.4 VELOCITY ENTROPY

With entropy-based probability distribution obtained from Equation 4.15, the maximum entropy can be expressed by substituting Equation 4.15 in Equation 4.1 as

$$H = \frac{1}{m-1} - \frac{1}{m}(\lambda_* + \lambda_1 \bar{u}) \tag{4.56}$$

The Lagrange multipliers from Equation 4.22 to 4.23 can be replaced by parameter $G$ using the same steps as presented in Chapter 3. Then, the entropy can be computed in terms of parameter $G$ instead of the Lagrange multipliers as

$$H = \frac{1}{m-1} - \frac{1}{m} \left\{ \left( \frac{m-1}{m} \right)^{-m} [1 - (1-G)^{m/(m-1)}]^{-(m-1)} \left( \frac{u_{\max}}{G} \right)^{-m} \left[ u_{\max} \left( 1 - \frac{1}{G} \right) + \bar{u} \right] \right\} \tag{4.57}$$

which is the same as derived by Cui (2011). However, because the velocity distribution is 2-D, the value of $G$ will be different from that corresponding to 1-D case. Equation 4.57 is a measure of uncertainty associated with the 2-D velocity distribution.

**Example 4.12**

Compute the entropy value for Example 4.11.

**Solution**

For $m = 3$, $\lambda_1 = 0.016$, $\lambda_* = 0.106$, and $G = 0.336$ are obtained in Example 4.11. Therefore,

$$H = \frac{1}{m-1} - \frac{1}{m}(\lambda_* + \lambda_1 \bar{u}) = \frac{1}{3-1} - \frac{1}{3}(0.106 + 0.016 \times 2.027) = 0.454$$

$$H = \frac{1}{m-1} - \frac{1}{m}\left\{\left(\frac{m-1}{m}\right)^{-m}[1-(1-G)^{m/(m-1)}]^{-(m-1)}\left(\frac{u_{max}}{G}\right)^{-m}\left[u_{max}\left(1-\frac{1}{G}\right)+\bar{u}\right]\right\}$$

$$= \frac{1}{3-1} - \frac{1}{3}\left\{\left(\frac{3-1}{3}\right)^{-3}[1-(1-0.336)^{3/2}]^{-(3-1)}\left(\frac{3.36}{0.336}\right)^{-3}\left[3.36\left(1-\frac{1}{0.336}\right)+2.027\right]\right\}$$

$$= 0.474$$

Equation 4.57 implies that the value of maximum entropy increases with entropy parameter $G$, which implies that the velocity distribution with larger $G$ values tends to be more uniformly distributed. Though not quite obvious, it is found from Cui and Singh (2012) that the flow with larger $G$ values results in smaller Reynolds number (Re). The velocity is distributed more possibly uniformly and entropy tends to be bigger when Re is small than when Re is large.

### 4.4.5 LOCATION OF MAXIMUM VELOCITY

The maximum velocity over the cross section occurs some distance below the water surface near the center in natural open channels. Due to the transverse change in the velocity distribution, the distribution along each vertical is different and the location of the maximum velocity along each vertical may also differ. It has been stated that the depth of maximum velocity is mainly related to the lateral position of velocity profiles in natural channels (Yang et al., 2004). The flow near the wall is more affected by the boundary shear and vegetation if any than near the center area and is dominated by secondary currents. To determine the location of the maximum velocity below the water surface, one may begin with the consideration of bed shear stress as

$$\tau = \rho g S_f (D - h) = \rho \varepsilon_0 \left.\frac{\partial u}{\partial y}\right|_{y=0} \tag{4.58}$$

where
  $S_f$ is the friction slope
  $\varepsilon_0$ is the momentum transfer coefficient at the channel bed, which is equal to the kinematic viscosity of the fluid
  $\partial u/\partial y$ can be computed from Equation 4.52

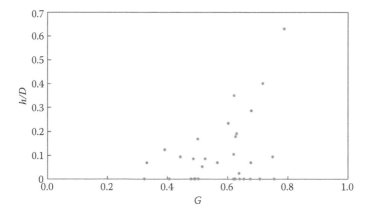

**FIGURE 4.13**   Location of maximum velocity.

Thus,

$$\frac{\partial u}{\partial y} = \frac{u_{\max}}{G}\frac{m-1}{m}(1-(1-G)^{m/(m-1)})((1-G)^{m/(m-1)}+(1-(1-G)^{m/(m-1)})F(u))^{-1/m}\frac{\partial F(u)}{\partial y}$$

(4.59)

It is seen from Equation 4.59 that the possible factors impacting the shear stress are the dimensionless parameter $G$ and the cumulative density function $F(u)$, which may further impact the location of maximum velocity. Figure 4.13 plots the measured depth of maximum velocity below the water surface versus the entropy parameter $G$, from which it can be seen that the dip phenomenon is more likely to occur when the $G$ value is bigger than 0.5 and the larger the $G$ is the lower the maximum velocity may locate.

In order to be able to describe the change in depth, the cumulative distribution function needs to be modified as

$$F(u) = \left[1-\left(\frac{x}{B}\right)^2\right]^b\left(\frac{y+c}{D(x)-h}\right)^a \quad \text{for all } (x,y) \text{ on } I(u)$$

(4.60)

where $h$ is a variable, representing the distance of the maximum velocity from the water surface along each line and $c$ is a parameter that ensures the CDF to be within the range between 0 and 1. The flow depth $D$ in Equation 4.35 is changed to be $D(x)$ so that it can represent the real geometry of the channel boundaries, since natural channels are not ideally rectangular shaped. Then, with the use of Equations 4.58 and 4.60, Equation 4.59 changes to

$$\frac{\partial u}{\partial y}\bigg|_{y=0} = \frac{1}{(D-h)^2}\frac{\varepsilon_0}{v}\frac{2a}{3G}[(1-G)^{-1/2}-(1-G)]\left[1-\left(\frac{x}{B}\right)^2\right]^{2/3}$$

(4.61)

Recall the Darcy–Weisbach equation

$$\frac{\bar{u}}{u_*} = \sqrt{\frac{8}{f_d}}$$

(4.62)

where $f_d$ is the Darcy friction factor and the shear velocity is defined as $u_* = \sqrt{\tau/\rho}$. Thus, the shear stress can be written as

$$\tau = \rho u_*^2 = \rho f_d \frac{\bar{u}^2}{8}$$

(4.63)

Substituting Equations 4.61 and 4.63 into Equation 4.58 with $m = 3$, one obtains

$$\tau = \rho f_d \frac{\bar{u}^2}{8} = \frac{1}{(D-h)^2} \frac{\varepsilon_0}{\nu} \frac{2a}{3G} [(1-G)^{-1/2} - (1-G)] \left[ 1 - \left( \frac{x}{B} \right)^2 \right]^{2/3}$$

(4.64)

Dividing both sides of Equation 4.64 by $f_d$, we obtain

$$\frac{\rho \bar{u}^2}{D} = \frac{1}{(D-h)^2} \frac{\varepsilon_0}{\nu} \frac{16a}{3Gf_d} [(1-G)^{-1/2} - (1-G)] \left[ 1 - \left( \frac{x}{B} \right)^2 \right]^{2/3}$$

(4.65)

It is noted that the left side is the inverse of the Reynold number. Thus,

$$\frac{1}{R_e} = \frac{1}{(D-h)^2} \frac{\varepsilon_0}{\nu} \frac{16a}{3Gf_d} [(1-G)^{-1/2} - (1-G)] \left[ 1 - \left( \frac{x}{B} \right)^2 \right]^{2/3}$$

(4.66)

Moving $(D - h)$ to the left side of equation, one obtains

$$(D-h)^2 = \frac{\varepsilon_0}{\nu} \frac{16 \, \mathrm{Re} \, a}{3Gf_d} [(1-G)^{-1/2} - (1-G)] \left[ 1 - \left( \frac{x}{B} \right)^2 \right]^{2/3}$$

(4.67)

Thus, the location of maximum velocity turns out to be

$$h = D - \sqrt{\frac{\varepsilon_0}{\nu} \frac{16}{3f_d} \mathrm{Re} \frac{a}{G} [(1-G)^{-1/2} - (1-G)] \left[ 1 - \left( \frac{x}{B} \right)^2 \right]^{2/3}}$$

(4.68)

For $x = 0$, Equation 4.68 reduces to

$$h = D - \sqrt{\frac{\varepsilon_0}{\nu} \frac{16}{3f_d} \mathrm{Re} \frac{a}{G} [(1-G)^{-1/2} - (1-G)]}$$

(4.69)

which represents the location of the maximum velocity at the centerline.

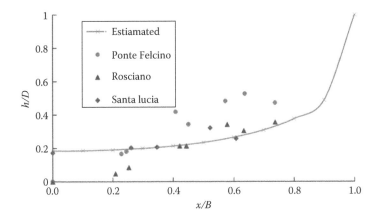

**FIGURE 4.14** Location of maximum velocity across the bank.

Equation 4.68 yields the depth of maximum velocity of the whole cross section. Figure 4.14 shows the computed location of maximum velocity $h/D$ on several verticals for field data from Italian rivers. The estimated $h/D$ represents the mean value of observed values. It can be seen from Figure 4.14 that the maximum velocity occurs further below the water surface as $x$ increases, where it is more affected by the boundary shear, and at the boundary it may possibly be as low as the channel bed. Results for the Ponte Felcino are higher than the estimated values with a standard deviation as 0.157, while results from other two sections are similar to the estimated values.

Figure 4.15 plots the relationship between the location of maximum velocity and parameter $G$ at the centerline of the Iranian rivers using Equation 4.69. The $h$–$G$ relationship curve is only valid when the dip phenomenon occurs, but does not provide the probability of the dip phenomenon and fails to describe when maximum velocity occurs at the water surface.

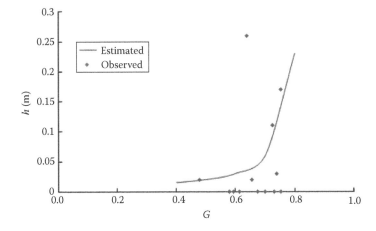

**FIGURE 4.15** Computed location of maximum velocity at centerline.

### 4.4.6 CONSTRUCTION OF ISOVELS

The procedure for construction of isovels is similar to that explained in Example 4.7. The difference is that the $(x, y)$ coordinates no longer need to be transferred to the $r$ coordinate first. In Example 4.7, the velocity value is one-to-one correlated to the $r$ value and linked to the $(x, y)$ coordinates. Here, the velocity can be computed directly for each location described with $(x, y)$.

#### Example 4.13

Construct the isovels for the velocity obtained from Tiber River in Table 4.6.

#### Solution

For given 2-D velocity distribution listed in Table 4.6, with $u_{max} = 3.36$ m/s and $u_m = 2.206$ m/s, we first compute $G = 0.336$ as in Example 4.11. Then, using $F(u)$ of Equation 4.60 with the modification of flow depth $D(x)$ and the location of maximum velocity $h$, the velocity distribution is obtained from Equation 4.55, as shown in Example 4.11. Thus, there will be one velocity value associated with each $(x, y)$. At last, meshing the grid of $(x, y)$ coordinates and using the contour plot on $(x, y, u)$, the isovels are plotted, as shown in Figure 4.16.

Although the velocity is not computed from point to point, the overall trend of isovels provides a qualitative idea. The location and value of the maximum velocity and the boundary condition were correctly determined. Compared to observations, the computed isovels are more uniformly distributed, because the entropy-based velocity distribution is the distribution with the maximum entropy, which tends toward the uniform distribution under given constraints.

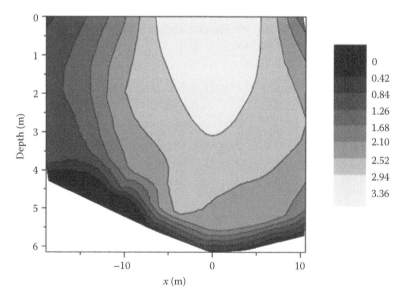

**FIGURE 4.16**   Velocity isovels for Tiber River.

## 4.A   APPENDIX A

**TABLE 4.A.1**

**Velocity Observations at Pointe Nuovo at Tiber River, Italy, June 3, 1997**

| Vertical 1 | | Vertical 2 | | Vertical 3 | | Vertical 4 | |
|---|---|---|---|---|---|---|---|
| y (m) | u (m/s) | y (m) | u (m/s) | y (m) | u (m/s) | y (m) | u (m/s) |
| 4.71 | 0.52 | 6.31 | 0.99 | 6.21 | 1.54 | 6.09 | 1.98 |
| 4.65 | 0.52 | 6.25 | 0.99 | 6.15 | 1.54 | 6.03 | 1.98 |
| 4.35 | 0.63 | 5.95 | 1.25 | 5.85 | 1.74 | 5.73 | 2.15 |
| 3.65 | 0.92 | 5.25 | 1.51 | 5.15 | 1.87 | 5.03 | 2.32 |
| 2.65 | 1.02 | 4.25 | 1.81 | 4.15 | 2.13 | 4.06 | 2.34 |
| 1.85 | 0.97 | 3.25 | 1.83 | 3.15 | 2.08 | 3.09 | 2.48 |
| 1.15 | 0.74 | 1.25 | 1.65 | 2.18 | 2.06 | 2.09 | 2.32 |
| 0.45 | 0.67 | 0.45 | 0.97 | 1.18 | 1.92 | 1.09 | 1.97 |
| 0.15 | 0.34 | 0.15 | 0.81 | 0.48 | 1.47 | 0.39 | 1.78 |
| 0.00 | 0.00 | 0.00 | 0.00 | 0.15 | 1.28 | 0.15 | 1.37 |
| | | | | 0.00 | 0.00 | 0.00 | 0.00 |

| Vertical 5 | | Vertical 6 | | Vertical 7 | | Vertical 8 | | Vertical 9 | |
|---|---|---|---|---|---|---|---|---|---|
| y (m) | u (m/s) | y (m) | u (m/s) | y (m) | u (m/s) | y (m) | u (m/s) | y (m) | u (m/s) |
| 6.07 | 2.66 | 5.89 | 2.37 | 5.76 | 1.97 | 5.66 | 1.42 | 5.36 | 0.88 |
| 6.01 | 2.66 | 5.83 | 2.37 | 5.70 | 1.97 | 5.60 | 1.42 | 5.30 | 0.88 |
| 5.71 | 2.58 | 1.89 | 2.41 | 5.4 | 2.03 | 5.30 | 1.40 | 5.00 | 0.87 |
| 5.04 | 2.61 | 0.89 | 1.91 | 4.7 | 1.98 | 4.6 | 1.63 | 4.30 | 1.16 |
| 4.07 | 2.66 | 0.39 | 1.53 | 3.7 | 2.39 | 3.6 | 1.97 | 3.30 | 1.49 |
| 3.13 | 2.72 | 0.15 | 1.49 | 2.7 | 2.22 | 2.6 | 1.92 | 2.30 | 1.71 |
| 2.13 | 2.61 | 0.00 | 0.00 | 1.9 | 2.37 | 1.8 | 1.81 | 1.30 | 1.19 |
| 1.10 | 2.32 | | | 1.2 | 2.06 | 1.1 | 1.73 | 0.40 | 0.91 |
| 0.37 | 1.92 | | | 0.5 | 1.51 | 0.5 | 1.36 | 0.15 | 0.80 |
| 0.15 | 1.47 | | | 0.15 | 1.42 | 0.15 | 0.71 | 0.00 | 0.00 |
| 0.00 | 0.00 | | | 0.00 | 0.00 | 0.00 | 0.00 | | |

| Vertical | D (m) | Maximum u (m/s) | h (m) | z (m) |
|---|---|---|---|---|
| 1 | 4.71 | 1.02 | 2.7 | −20.2 |
| 2 | 6.31 | 1.83 | 2.7 | −2.64 |
| 3 | 6.21 | 2.13 | 2.7 | −11.44 |
| 4 | 6.09 | 2.48 | 2.7 | −6.24 |
| 5 | 6.07 | 2.72 | 2.7 | 0.0 |
| 6 | 5.89 | 2.41 | 2.7 | 6.24 |
| 7 | 5.76 | 2.39 | 2.7 | 11.44 |
| 8 | 5.66 | 1.97 | 2.7 | 15.60 |
| 9 | 5.36 | 1.72 | 2.7 | 19.76 |

*Note:* $y$ is the vertical distance (m) of each sampled point from the channel bed; $u$ is the observed velocity (m/s); $D$ is the water depth (m) along the vertical; maximum $u$ is the maximum sampled velocity (m/s) along the vertical; $h$ is the vertical distance (m) below the water surface where the maximum velocity occurs; $z$ is the horizontal distance from the vertical where the maximum velocity is sampled.

## 4.B   APPENDIX B

### TABLE 4.B.1
### Values of $M$ Based on the Data of Italy's Rivers

| $u_{max}$(m/s) | $u_m$(m/s) | $\lambda_1$ | $\lambda_v$ | $M$ | |
|---|---|---|---|---|---|
| 0.38 | 0.27 | 36.08 | −1.59 | 2.679 | |
| 1 | 0.77 | 6.55 | −1.27 | 3.912 | |
| 1.36 | 1.11 | 4.12 | −1.33 | 4.090 | |
| 1.36 | 1.06 | 3.60 | −0.98 | 4.028 | February 13, 1985 Pro |
| 1.33 | 1.00 | 3.38 | −0.75 | 3.361 | $M$ (Chiu) 1.945 |
| 1.57 | 1.11 | 2.02 | −0.31 | 2.483 | |
| 1.8 | 1.32 | 1.73 | −0.44 | 3.012 | |
| 1.71 | 1.19 | 1.60 | −0.20 | 2.259 | |
| 1.17 | 0.89 | 4.56 | −0.96 | 3.600 | |
| 1.18 | 0.93 | 4.03 | −1.27 | 4.393 | |
| 1.08 | 0.87 | 6.31 | −1.55 | 4.787 | |
| 0.56 | 0.40 | 14.94 | −0.89 | 2.509 | |
| | | | Mean | 3.509 | |
| 0.80 | 0.56 | 7.97 | −0.65 | 2.538 | |
| 1.26 | 0.86 | 2.85 | −0.20 | 2.121 | |
| 1.35 | 1.02 | 3.33 | −0.77 | 3.449 | |
| 1.42 | 1.10 | 3.30 | −0.93 | 4.015 | March 27, 1991 Pro |
| 1.44 | 1.17 | 3.64 | −1.24 | 4.018 | $M$ (Chiu) 1.945 |
| 1.49 | 1.17 | 3.13 | −0.98 | 4.281 | |
| 1.80 | 1.29 | 1.58 | −0.31 | 2.606 | |
| 1.80 | 1.41 | 2.08 | −0.76 | 4.119 | |
| 1.44 | 1.10 | 3.13 | −0.86 | 3.822 | |
| 1.37 | 1.04 | 3.31 | −0.80 | 3.561 | |
| 1.32 | 0.89 | 2.41 | −0.08 | 1.920 | |
| 1.07 | 0.79 | 4.09 | −0.83 | 3.162 | |
| 0.56 | 0.35 | 8.62 | 1.13 | 1.030 | |
| | | | Mean | 3.203 | |
| 1.02 | 0.63 | 2.71 | 0.58 | 1.077 | |
| 1.83 | 1.18 | 1.04 | 0.14 | 1.445 | |
| 2.13 | 1.60 | 1.33 | −0.48 | 3.401 | June 3, 1997 Tiber |
| 2.48 | 1.88 | 1.01 | −0.44 | 3.559 | $M$ (Chiu) 2.005 |
| 2.72 | 2.20 | 1.00 | −0.63 | 4.849 | |
| 2.41 | 1.73 | 0.90 | −0.25 | 2.694 | |
| 2.39 | 1.81 | 1.08 | −0.46 | 3.542 | |
| 1.97 | 1.40 | 1.30 | −0.27 | 2.553 | |
| 1.71 | 0.99 | 0.65 | 0.62 | 0.635 | |
| | | | Mean | 2.640 | |
| 0.86 | 0.63 | 7.55 | −0.92 | 2.995 | |
| 1.81 | 1.03 | 0.51 | 0.65 | 0.537 | |

(*Continued*)

**TABLE 4.B.1 (*Continued*)**
**The Values of *M* Based on the Data of Italy's Rivers**

| $u_{max}$(m/s) | $u_m$(m/s) | $\lambda_1$ | $\lambda_v$ | $M$ | |
|---|---|---|---|---|---|
| 1.98 | 1.46 | 1.45 | −0.43 | 3.098 | November 18, 1996 Tiber |
| 2.13 | 1.67 | 1.50 | −0.66 | 4.190 | $M$ (Chiu) 2.005 |
| 2.6 | 2.05 | 1.02 | −0.56 | 4.304 | |
| 2.45 | 1.93 | 1.15 | −0.59 | 4.286 | |
| 2.1 | 1.58 | 1.37 | −0.49 | 3.429 | |
| 1.71 | 1.12 | 1.27 | 0.08 | 1.596 | |
| 1.49 | 0.87 | 0.91 | 0.67 | 0.686 | |
| | | | Mean | 2.791 | |

# REFERENCES

Absi, R. (2011). An ordinary differential equation for velocity distribution and dip-phenomenon in open channel flows. *Journal of Hydraulic Research*, 49(1), 82–89.

Chiu, C.-L. (1987). Entropy and probability concepts in hydraulics. *Journal of Hydraulic Engineering*, ASCE, 113(5), 583–600.

Chiu, C.-L. (1988). Entropy and 2-D Velocity Distribution in open channels. *Journal of Hydraulic Engineering*, ASCE, 114(7), 738–754.

Chiu, C.-L. (1989). Velocity distribution in open channel flows. *Journal of Hydraulic Engineering*, ASCE, 115(5), 576–594.

Chiu, C.L. (1991). Application of entropy concept in open-channel flow. *Journal of Hydraulic Engineering*, 117(5), 615–628.

Chiu, C.-L. and Chiou, J.-D. (1986). Structure of 3-D flow and shear in open channels. *Journal of Hydraulic Engineering*, ASCE, 109(11), 424–1440.

Chiu, C.L. and Lin, G.F. (1983). Computation of 3-D flow and shear in open channels. *Journal of Hydraulic Engineering*, 112(11), 1424–1440.

Chiu, C.-L. and Said, A.A. (1994). Maximum and mean velocities in open-channel flow. *Journal of Hydraulic Engineering*, ASCE, 121(1), 26–34.

Cui, H. (2011). Estimation of velocity distribution and suspended sediment discharge in open channels using entropy. Unpublished M.S. thesis, Texas A&M University, College Station, TX.

Cui, H. and Singh, V.P. (2012). On the cumulative distribution function for entropy-based hydrologic modeling. *Transactions of the ASABE*, 55(2), 429–438.

Cui, H. and Singh, V.P. (2013). Two-dimensional velocity distribution in open channels using the Tsallis entropy. *Journal of Hydrologic Engineering*, 18(3), 331–339.

Jaynes, E.T. (1957a). Information theory and statistical mechanics 1. *Physical Review*, 106(4), 620–630.

Jaynes, E.T. (1957b). Information theory and statistical mechanics 2. *Physical Review*, 108(2), 171–190.

Luo, H. and Singh, V.P. (2011). Entropy theory for two-dimensional velocity distribution. *Journal of Hydrologic Engineering*, 16(4), 303–315.

Marini, G., Martino, G.D., Fontana, N., Fiorentino, M., and Singh, V.P. (2011). Entropy approach for 2D velocity distribution in-open channel flow. *Journal of Hydraulic Research*, 49(6), 784–790.

Murphy, C. (1904). Accuracy of stream measurements. Water supply and irrigation paper, 95, Department of the Interior, U.S. Geological Survey, Washington, D.C.

Nezu, I. and Nakagawa, H. (1993). *Turbulence in Open-Channel Flows*. Balkema, Rotterdam, the Netherlands.

Plastino, A. and Plastino, A.R. (1999). Tsallis entropy and Jaynes' information theory formalism. *Brazilian Journal of Physics*, 29(1), 50–60.

Stearns, F.P. (1883). On the current meter, together with a reason why the maximum velocity of water flowing in open channel is below the surface. *Transactions of American Society of Civil Engineers*, 12, 301–338.

Tsallis, C. (1988). Possible generalization of Boltzmann-Gibbs Statistics. *Journal of Statistical Physics*, 52, 479–487.

Yang, S.Q., Tan, S.K., and Lim, S.Y. (2004). Velocity distribution and dip-phenomenon in smooth uniform open channel flows. *Journal of Hydraulic Engineering*, 130(12), 1179–1186.

# 5 Suspended Sediment Concentration

Concentration of suspended sediment is of fundamental importance in environmental management, assessment of best management practices, water quality evaluation, reservoir ecosystem integrity, and fluvial hydraulics (Stewart et al., 1976a,b). Assuming time-averaged sediment concentration along a vertical as a random variable, this chapter discusses suspended sediment concentration profiles based on the Tsallis entropy and their analogy with a random walk model. The sediment concentration profiles so obtained are parsimonious as well as reasonably accurate.

## 5.1 METHODS FOR DETERMINING SEDIMENT CONCENTRATION

Sediment carries away chemicals adsorbed on soil particles and thus pollutes the environment. Since sediment occurs widely and in large volumes, it is a major pollutant. Determination of erosion and sediment transport is vital for the development of pollution abatement measures (Stewart et al., 1976a,b). Fundamental to determining sediment discharge and load are the velocity distribution and sediment concentration, as illustrated by Einstein (1950), who determined suspended sediment discharge by integrating the product of local sediment concentration and flow velocity over the zone of suspension.

Concentration of suspended sediment depends on particle size, settling velocity, fluid density, sediment density, and turbulent stress. Depending on the consideration of these factors, approaches to predicting suspended sediment concentration can be classified as (1) empirical, (2) hydraulic, and (3) entropy based. Empirical approaches to sediment concentration are of either exponential or power type (Singh, 1996) or linear (Simons and Senturk, 1992). Examples of popular hydraulic approaches include the O'Brien–Christiansen equation (O'Brien, 1933; Christiansen, 1935) that employs the exchange theory; Rouse equation (Rouse, 1937) that is based on the theory of turbulence; the Lane–Kalinske equation (Lane and Kalinske, 1941) that also uses the exchange theory; Einstein equation (Einstein, 1950) that is based on the bed load theory; the Ackers–White equation (Ackers and White, 1973) that employs the stream power concept of Bagnold (1966) and dimensional analysis; the Yang equation (Yang, 1973) that combines the unit stream power and dimensional analysis; the Chang–Simons–Richardson equation (Chang et al., 1967); and the Molinas and Wu equation (Molinas and Wu, 2001) that employs energy balance. Methods for determining flow velocity have been discussed in Chapters 3 and 4.

Entropy-based approaches to sediment concentration are based on the Shannon entropy theory or the Tsallis entropy theory. Chiu et al. (2000) and Choo (2000) employed the Shannon entropy and showed that the entropy-based equations predicted sediment concentration better than did empirical or hydraulics-based equations. Luo and Singh (2011) and Cui and Singh (2013, 2014) used the Tsallis entropy for deriving sediment concentration profiles and showed that the Tsallis entropy had an advantage over the Shannon entropy.

Classical methods of determining sediment concentration start from the mechanism of suspended sediment transport that occurs in turbulent flow, where turbulent velocity fluctuations in the vertical direction transport sediment upward. An equilibrium distribution of suspended sediment concentration develops due to the balance between turbulent diffusion of grains upward and gravitational settling of the grains downward. There are no vertical changes in the sediment concentration profile in the flow direction, if there are no changes in channel boundaries (Sturm, 2010). The equality between turbulent flux and gravitational settling flux leads to the following differential equation that governs the sediment concentration distribution

$$-\varepsilon_s \frac{dc}{dy} = \omega_s c \qquad (5.1)$$

where
    $c$ is the sediment concentration at a given point $y$
    $y$ is the vertical distance measured from the channel bed
    $\varepsilon_s$ is the diffusion coefficient for sediment transfer
    $\omega_s$ is the settling velocity of sediment particle

Since $c$ decreases with increasing $y$, $dc/dy$ is negative in Equation 5.1. The diffusion coefficient is not constant in alluvial channel flow, particularly near the bed where turbulence characteristics change with distance above the bed. Thus, $\varepsilon_s$ is often estimated as $\beta\varepsilon$, where $\beta$ is the coefficient of proportionality and $\varepsilon$ is the turbulent eddy viscosity.

The settling velocity, also called the fall velocity, $\omega_s$, plays an important role in distinguishing suspended sediment load and bed load. It is related to the particle size and shape, submerged specific weight, viscosity of water, and sediment concentration. In the laminar settling region, where the Reynolds number is smaller than 1, by solving the Navier–Stokes equations without inertia terms, Stokes (1851) derived the well-known Stokes law for the settling velocity of spherical particles and determined the drag force thereon. However, sediment particles in natural rivers are usually irregular shaped and have rough surfaces whose settling velocity is different from that of spherical particles. Rubey (1933) derived a formula for settling velocity of natural sediment particles. However, van Rijn (1984) suggested using the Stokes law for computing the velocity of sediment particles smaller than 0.1 mm and using Zanke's (1977) formula for particles of size from 0.1 to 1 mm; he also derived a formula for particles larger than 1 mm.

### 5.1.1 ROUSE EQUATION

The Rouse equation (Rouse, 1937) for determination of sediment concentration results from Equation 5.1. A classical model is derived from this equation with the use of the Prandtl–von Karman logarithmic velocity equation and linear shear stress distribution assumption. From the Prandtl–von Karman universal velocity distribution (see Chow, 1959), the gradient of the velocity distribution can be computed as

$$\frac{du}{dy} = \frac{u_*}{\kappa y} \tag{5.2}$$

where
  $u$ is the velocity at a point $y$ along the vertical from the bed
  $\kappa$ is the von Karman universal constant
  $u_*$ is the shear velocity defined as $u_* = \sqrt{\tau_0/\rho}$ in which $\tau_0$ is the bed shear stress
    and $\rho$ is the mass density of water

The vertical shear stress distribution in steady, uniform open-channel flow can be considered as linear:

$$\tau = \tau_0 \frac{(y_0 - y)}{y_0} \tag{5.3}$$

where
  $\tau$ is the shear stress at $y$
  $\tau_0$ is the shear stress at the bed at $y = 0$
  $y_0$ is the depth of uniform flow

The shear stress $\tau$ also equals

$$\tau = \rho \varepsilon \frac{du}{dy} \tag{5.4}$$

Recalling that $u_* = \sqrt{\tau_0/\rho}$ and substituting Equations 5.2 and 5.3 into Equation 5.4, $\varepsilon_s = \beta \varepsilon$ can be obtained as

$$\varepsilon_s = \beta \kappa u_* \frac{y}{y_0}(y_0 - y) \tag{5.5}$$

Substituting Equation 5.5 for $\varepsilon_s$ in Equation 5.1 and integrating, the result is the Rouse equation for concentration $c$ at a distance $y$ from the bed:

$$\frac{c}{c_a} = \left( \frac{(y_0 - y)}{y} \frac{a}{(y_0 - a)} \right)^{R_0} \tag{5.6}$$

in which $c_a$ is the reference concentration at a distance $y = a$, which is arbitrarily taken as $0.05y_0$ above the bed, and $R_0$ is referred to as the Rouse number, which is defined as $R_0 = \omega_s/(\beta \kappa u_*)$. The Rouse number is a measure of the relative contribution of settling velocity and turbulent stress. The reference concentration defined in the Rouse equation (5.6) compares the concentration at any distance with that at a reference level. Equation 5.6 does not account for bed load transport. The Rouse equation has been compared favorably with observed suspended sediment concentration distribution but it is not valid for sediment concentration near the channel bed or water surface (Simons and Senturk, 1992).

### 5.1.2 Chang–Simons–Richardson Equation

Applying the velocity distribution over the flow depth $D$, we can write

$$\frac{du}{dy} = \frac{u_*}{\kappa y}\sqrt{\frac{D-y}{D}} \tag{5.7}$$

Chang et al. (1967) derived a sediment concentration equation as

$$\frac{c}{c_a} = a_1\left[\frac{\sqrt{y/D}}{1-\sqrt{(D-y)/D}}\right]^{a_2} \tag{5.8}$$

where

$$a_1 = \left[\frac{1-\sqrt{(D-b)/D}}{\sqrt{b/D}}\right]^{a_2} \tag{5.9}$$

$$a_2 = \frac{2w_s}{\beta u_* \kappa} \tag{5.10}$$

The value of $b$ can be computed as

$$b = J\frac{\tau-\tau_0}{(1-n)(\gamma_s - \gamma)\tan\phi} \tag{5.11}$$

where
  $J = 10$
  $n$ is the porosity of the bed material
  $\phi$ is the angle of repose of the submerged bed material
  $\gamma$ is the weight density of water
  $\gamma_s$ is the weight density of sediment

The difference between Equations 5.8 and 5.6 arises from the use of different velocity distributions and consequent velocity gradients.

### 5.1.3 O'BRIEN–CHRISTIANSEN EQUATION

Using the exchange theory entailing the continuous exchange of sediment particles across any arbitrary layer in steady and uniform flow, O'Brien (1933) and Christiansen (1935) derived a sediment concentration equation:

$$c = c_a \exp\left(-w\int_a^y \frac{dy}{\varepsilon_s}\right) \tag{5.12}$$

where $c_a$ is the sediment concentration with fall velocity $w$ at level $y = a$ above the bed. Equation 5.12 states that the concentration is the greatest at the bottom and smallest at the water surface. For determining $c$, the variation of $\varepsilon_s$ must be prescribed.

### 5.1.4 LANE–KALINSKE EQUATION

Assuming that term $\varepsilon_s$ is equal to the kinematic eddy viscosity or the diffusion coefficient for momentum $(\varepsilon_m)$ which is the same as $\varepsilon$ defined earlier, that is, $\varepsilon_s = \varepsilon_m$, Lane and Kalinske (1941) derived

$$c = c_a \exp\left[-\frac{15w}{u_*}\left(\frac{y-a}{D}\right)\right] \tag{5.13}$$

where
  $w$ denotes the fall velocity of the representative grain size (e.g., sediment diameter $d_{35}$ or $d_{50}$) which in reality varies with $y$
  $a$ is the value of $y$ at which $c = c_a$
  $D$ is the flow depth

### 5.1.5 CHIU EQUATION

Instead of using the Prandtl–von Karman velocity distribution, Chiu (2000) employed the Shannon entropy–based velocity distribution (Chiu, 1987, 1988, 1989). Following the same method as earlier, integrating Equation 5.1 he obtained

$$\frac{c}{c_0} = \left(\frac{1-\dfrac{y}{D}}{1+(e^M-1)\dfrac{y}{D}}\right)^{z'} \tag{5.14}$$

where

$$z' = \frac{\omega_s u_{max}(1-e^{-M})}{\beta u_*^2 M}$$

$\omega_s$ is the settling velocity
$\beta$ is the coefficient relating to the viscosity
$u_*$ is the shear velocity
$u_{max}$ is the maximum velocity along the vertical or for the cross section
$D$ is the flow depth
$M$ is the entropy parameter defined in Chiu's (1988) velocity distribution, which equals $\lambda_1 u_{max}$, where $\lambda_1$ is the Lagrange multiplier used to maximize the Shannon entropy

In order to better account for shear stress, Chiu et al. (2000) refined Equation 5.14 for sediment concentration as

$$\frac{c}{c_0} = \exp\left[-z'I\left(\frac{y}{D},\frac{h}{D},M\right)\right] \tag{5.15}$$

where

$$I\left(\frac{y}{D},\frac{h}{D},M\right) = \frac{e^M}{\xi_{max}}\int_0^y \left\{h_\xi\left[1+(e^M-1)\frac{\xi}{\xi_{max}}\right]\frac{\tau}{\tau_0}\right\}^{-1} d\xi$$

$\xi$ is the variable that increases with velocity $u$
$e$ is the exponential base
$\xi_{max}$ is the maximum value of $\xi$ where $u_{max}$ occurs
$h$ is the distance from the flow surface at which the flow velocity is maximum
$h_\xi$ is the scale parameter of $\xi$

However, one of the disadvantages of Equation 5.15 is that it contains a large number of parameters and its practical use is therefore limited. It may be recalled that the basis of sediment concentration distributions, given by Equations 5.6, 5.8, 5.14, and 5.15, is Equation 5.1. These equations differ from one another in the use of different velocity distributions.

### 5.1.6 CHOO EQUATION

Skipping Equation 5.1, Choo (2000) developed a full Shannon entropy–based method for modeling sediment concentration. Considering time-averaged sediment concentration along a vertical as a random variable, Choo (2000) maximized the Shannon entropy and obtained the least-biased probability distribution of sediment concentration, and then using a linear hypothesis on the cumulative distribution of concentration he derived a sediment concentration profile as

$$c = \frac{c_0}{N} \ln \left\{ \exp(N') - \left[ \exp(N') - \exp\left(\frac{N'}{k}\right) \right] \frac{y}{D} \right\}$$  (5.16)

where

$N' = \lambda_1 c_0$ is a parameter

$\lambda_1$ is the Lagrange multiplier introduced when maximizing entropy

$c_0$ is the sediment concentration at the bed $y = 0$

$k = c_0/c_D$ in which $c_D$ is the sediment concentration at the water surface $y = D$, which can be considered as 0

When $k \approx \infty$, Equation 5.16 reduces to

$$c = \frac{c_0}{N} \ln \left[ \exp(N') - (\exp(N') - 1) \frac{y}{D} \right]$$  (5.17)

Equation 5.17 is fully entropy-based sediment concentration distribution (Choo, 2000).

### 5.1.7  SUMMATION

Although the Rouse equation provides a simple way to compute the sediment concentration distribution, it is based on the Prandtl–von Karman velocity distribution and has been found to poorly predict sediment concentration (Chiu et al., 2000; Choo, 2000). This may be because the Prandtl–von Karman velocity distribution does not predict the velocity near the bottom well, especially in sediment-laden flows (Einstein and Chien, 1955). The method (Equation 5.8) by Chang et al. (1967) has too many parameters and is not convenient to use in practice. Chiu's (2000) method has been shown to better predict the concentration than the Rouse equation; however, as can be seen from Equation 5.15, the Chiu equation contains too many parameters that are not convenient to determine. Equation 5.17 by Choo (2000) has been found to predict the sediment concentration along the vertical more accurately than does the Rouse equation (Equation 5.6) and the Chiu equation (Equation 5.15). However, the linear cumulative distribution function (CDF) hypothesis employed to derive Equation 5.16 has not been verified; hence, the validation of the sediment concentration profile is less than complete.

## 5.2  DERIVATION OF ENTROPY-BASED SUSPENDED SEDIMENT CONCENTRATION

Derivation of the vertical distribution of suspended sediment concentration using the Tsallis entropy comprises the following steps: (1) definition of the Tsallis entropy, (2) specification of constraints, (3) maximization of entropy, (4) hypothesizing the cumulative probability distribution function (CDF) of sediment concentration, (5) derivation of sediment concentration, (6) determination of distribution parameters, and (7) determination of the maximum entropy. Each of these steps is now discussed.

### 5.2.1 TSALLIS ENTROPY

Consider the time-averaged sediment concentration $C$ along a vertical from the bed as a random variable. Unlike the velocity in open channels, suspended sediment concentration has its maximum value ($c_0$) at the channel bed and decreases with increasing distance above the channel bed. If the sediment concentration is assumed 0 at the water surface, then $C$ varies from $c_0$ to 0. Therefore, the Tsallis entropy (Tsallis, 1988) of the time-averaged sediment concentration can be written as

$$H(C) = \frac{1}{m-1}\left\{1 - \int_{c_0}^{0}[f(c)]^m\,dc\right\} = \frac{1}{m-1}\int_{c_0}^{0}f(c)\left\{1-[f(c)]^{m-1}\right\}dc \qquad (5.18)$$

where
> $H(C)$ is the Tsallis entropy of the time-averaged sediment concentration $C$ at a specified point or elevation
> $c$ is a specific value of random variable $C$
> $c_0$ is the maximum sediment concentration of the profile
> $f(c)$ is the probability density function (PDF) of $C$
> $m$ is a real number

The objective is to obtain the least-biased probability distribution $f(c)$, which can be done by maximizing the entropy, subject to specified information expressed as constraints obtained from observations.

### 5.2.2 CONSTRAINTS

If some information on the sediment concentration is known from empirical observations or theory, then it can be codified in the form of what is called constraints. In this discussion, two constraints are applied: one arises from the satisfaction of the total probability theorem and the other is obtained from the conservation of mass of sediment (continuity equation). These are stated as

$$\int_{c_0}^{0} f(c)\,dc = 1 \qquad (5.19)$$

$$\int_{c_0}^{0} cf(c)\,dc = \bar{c} = c_m \qquad (5.20)$$

where $\bar{c}$ or $c_m$ is the mean suspended sediment concentration in the vertical flow profile, which equals $Q_s/Q$, where $Q_s$ is the suspended sediment discharge, and $Q$ is the flow discharge. Thus, Equation 5.20 is equivalent to satisfying the condition that $C$ must be distributed so that $\bar{c}Q = Q_s$. In actuality, Equation 5.19 is not a constraint, for all PDFs must satisfy the total probability theorem. Essentially what is being assumed here is that some information about the mean concentration is known. Specification of constraints determines the type of PDF but for practical usefulness they should be as few and simple as possible (Singh, 2011).

### 5.2.3 Maximization of Entropy

According to the principle of maximum entropy (POME) (Jaynes, 1957a,b), the least-biased PDF of sediment concentration can be obtained by maximizing the uncertainty expressed by entropy, subject to given constraints. The maximization of entropy given by Equation 5.18 can be achieved by using the method of Lagrange multipliers, subject to Equations 5.19 and 5.20. To that end, the Lagrangian function $L$ can be written as

$$L = \int_{c_0}^{0} \frac{f(c)}{m-1}\left\{1-\left[f(c)\right]^{m-1}\right\}dc + \lambda_0\left[\int_{c_0}^{0} f(c)dc - 1\right] + \lambda_1\left[\int_{c_0}^{0} cf(c)dc - \overline{c}\right] \quad (5.21)$$

in which $\lambda_0$ and $\lambda_1$ are the Lagrange multipliers. Differentiating Equation 5.21 with respect to $f(c)$ and equating the derivative to zero, the result is the PDF, which can be expressed as

$$f(c) = \left[\frac{m-1}{m}\left(\frac{1}{m-1} + \lambda_0 + \lambda_1 c\right)\right]^{1/(m-1)} \quad (5.22)$$

Let

$$\lambda_* = \frac{1}{m-1} + \lambda_0 \quad (5.23)$$

Equation 5.22, with the use of Equation 5.23, reduces to

$$f(c) = \left[\frac{m-1}{m}(\lambda_* + \lambda_1 c)\right]^{1/(m-1)} \quad (5.24)$$

Equation 5.24 is the least-biased entropy-based PDF of sediment concentration, which is fundamental to determining the sediment concentration distribution.

### 5.2.4 Cumulative Distribution Function

In order to determine the sediment concentration distribution in terms of vertical distance $y$ from the bed, a CDF, $F(c)$, of sediment concentration is hypothesized as

$$F(c) = 1 - \left(\frac{y}{D}\right)^{\eta} \quad (5.25)$$

where $\eta$ is a parameter related to sediment particle characteristics (Cui and Singh, 2012), which can be computed by fitting Equation 5.25 to observations using the least square method. Equation 5.25 shows that $F(c)$ is 1 at the channel bed, where $y = 0$, and is 0 at the water surface, where $y = D$. The validity of Equation 5.25 can be seen from Figures 5.1 and 5.2 for both experimental and field data. The $\eta$ value does not equal 1 for either of the two cases, suggesting that cumulative distribution hypothesis is not linear.

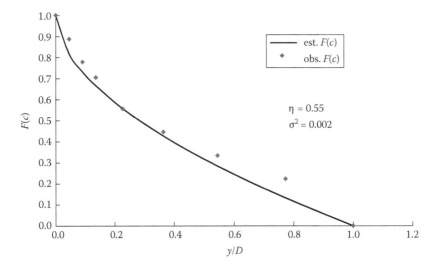

**FIGURE 5.1** Validation of cumulative distribution function for Run S10 of Einstein and Chien's (1955) data (obs. = observation, est. = estimation, and $\sigma^2$ = variance).

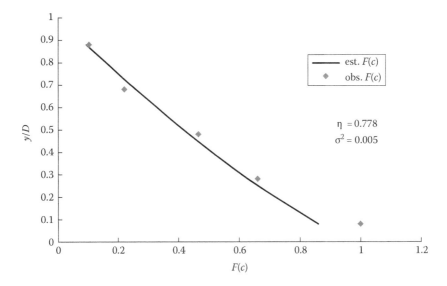

**FIGURE 5.2** Validation of cumulative distribution function or data collected from the Mississippi.

### 5.2.5  SEDIMENT CONCENTRATION

Combining the entropy-based PDF (Equation 5.24) with the CDF (Equation 5.25), the result is

$$F(c) = \int_0^c f(c)dc = \int_0^c \left[ \frac{m-1}{m}(\lambda_* + \lambda_1 c) \right]^{1/(m-1)} dc = 1 - \left( \frac{y}{D} \right)^\eta \quad (5.26)$$

From Equation 5.26, one obtains the sediment concentration distribution as

$$c = -\frac{\lambda_*}{\lambda_1} + \frac{1}{\lambda_1}\frac{m}{m-1}\left[ -\lambda_1\left[ 1 - \left( \frac{y}{D} \right)^\eta \right] + \left( \frac{m-1}{m}\lambda_* \right)^{m/(m-1)} \right]^{(m-1)/m} \quad (5.27)$$

Equation 5.27 is the Tsallis entropy–based sediment concentration distribution.

### 5.2.6  DETERMINATION OF THE LAGRANGE PARAMETERS

Equation 5.27 contains three parameters, $\lambda_*$, $\lambda_1$, and $m$. The Lagrange multipliers $\lambda_*$ and $\lambda_1$ can be determined using constraint equations (Equations 5.19 and 5.20). Substitution of Equation 5.24 in Equation 5.19 yields

$$\int_{c_0}^0 \left[ \frac{m-1}{m}(\lambda_* + \lambda_1 c) \right]^{1/m-1} dc = 1 \quad (5.28)$$

Integration of Equation 5.28 leads to

$$\frac{1}{\lambda_1}\left( \frac{m-1}{m} \right)^{\frac{m}{m-1}}\left[ (\lambda_*)^{\frac{m}{m-1}} - (\lambda_* + \lambda_1 c)^{\frac{m}{m-1}} \right] = 1 \quad (5.29)$$

Likewise, substitution of Equation 5.24 in Equation 5.20 yields

$$\int_{c_0}^0 c\left[ \frac{m-1}{m}(\lambda_* + \lambda_1 c) \right]^{\frac{1}{m-1}} dc = \bar{c} \quad (5.30)$$

Equation 5.30 can be integrated by parts as

$$\left( \frac{m-1}{m} \right)^{\frac{m}{m-1}}\left[ \frac{c}{\lambda_1}(\lambda_* + \lambda_1 c)^{\frac{m}{m-1}} - \frac{m-1}{2m-1}\frac{1}{\lambda_1^2}(\lambda_* + \lambda_1 c)^{\frac{2m-1}{m-1}} \right]\Big|_{c_0}^0 = \bar{c} \quad (5.31)$$

or

$$\left(\frac{m-1}{m}\right)^{\frac{1}{m-1}}\left\{-\frac{c_0}{\lambda_1}(\lambda_* + \lambda_1 c_0)^{\frac{m}{m-1}} + \frac{m-1}{2m-1}\frac{1}{\lambda_1^2}\left[(\lambda_* + \lambda_1 c_0)^{\frac{2m}{m-1}} - (\lambda_*)^{\frac{2m}{m-1}}\right]\right\} = \bar{c} \quad (5.32)$$

Equations 5.29 and 5.32 constitute a system of two nonlinear equations having two unknown parameters $\lambda_*$ and $\lambda_1$. However, as the value of $m$ increases, the difficulty of solving these nonlinear equations increases significantly. Similar to velocity distribution discussed in Chapters 3 and 4, Cui and Singh (2014) found a value of $m = 3$ as a good balance between convenience and accuracy.

### Example 5.1

A set of data on sediment concentration is given in Table 5.1. Compute the PDF of sediment concentration for these data and plot it. Also, show the computed Lagrange parameters.

### Solution

For given $c_0 = 58$ g/L and $\bar{c} = 31.05$ g/L, the Lagrange multipliers are obtained by solving Equations 5.29 and 5.32 that yield $\lambda_1 = 1.346$ and $\lambda_* = 80.14$. Then, the PDF of sediment concentration is obtained from

$$f(c) = \left[\frac{m-1}{m}(\lambda_* + \lambda_1 c)\right]^{1/(m-1)} = \left[\frac{2}{3}(80.14 + 1.346c)\right]^{1/2}$$

and is plotted in Figure 5.3.

---

**TABLE 5.1**

**Observed Sediment Concentration Data from Experimental Run S1 of Einstein and Chien (1955)**

| $y$ (mm) | $c$ (g/L) |
|---|---|
| 5.4 | 58 |
| 6.0 | 54 |
| 6.6 | 49.7 |
| 7.2 | 44.3 |
| 8.4 | 32.6 |
| 10.2 | 20.7 |
| 12.6 | 11.1 |
| 15.6 | 6.01 |
| 18.6 | 3.05 |

---

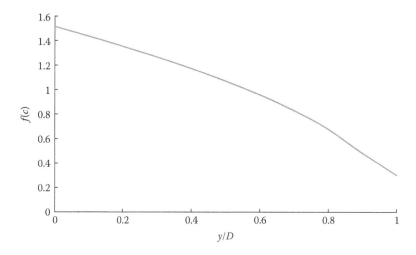

**FIGURE 5.3** Probability density function of sediment concentration.

However, a simpler way to estimate the Lagrange multipliers is by introducing a dimensionless parameter $N$ as a function of maximum concentration and the Lagrange multipliers as

$$N = \frac{\lambda_1 c_0}{\lambda_1 c_0 + \lambda_*} \tag{5.33}$$

Equation 5.33 simplifies computation and helps with analysis. It can be noted that the Lagrange multiplier $\lambda_1$ is of the dimensions of [L/g], which is the inverse of the dimensions of $c_0$, and $\lambda_*$ is dimensionless, thus $N$ is also dimensionless. It can then be stated that the Lagrange multipliers are related to the maximum concentration as

$$\lambda_* = \frac{\lambda_1 c_0 (1 - N)}{N} \tag{5.34}$$

and

$$\lambda_1 = \frac{N \lambda_*}{c_0 (1 - N)} \tag{5.35}$$

The sediment concentration distribution in Equation 5.27 can now be simplified. The maximum sediment concentration $c_0$ occurs at the bed; thus, substituting $y = 0$ into Equation 5.27 one obtains

$$c_0 = -\frac{\lambda_*}{\lambda_1} + \frac{1}{\lambda_1} \frac{m}{m-1} \left[ -\lambda_1 + \left( \frac{m-1}{m} \lambda_* \right)^{m/(m-1)} \right]^{(m-1)/m} \tag{5.36}$$

The non dimensional sediment concentration ($c/c_0$) can now be written as

$$\frac{c}{c_0} = \frac{-\lambda_* + m/(m-1)\left[-\lambda_1 F(c) + \left((m-1)/m\ \lambda_*\right)^{m/(m-1)}\right]^{(m-1)/m}}{-\lambda_* + m/(m-1)\left[-\lambda_1 + \left((m-1)/m\ \lambda_*\right)^{m/(m-1)}\right]^{(m-1)/m}} \qquad (5.37)$$

Replacing the Lagrange multipliers with parameter $N$ with the use of Equations 5.33 and 5.34, the dimensionless sediment concentration is obtained as

$$\frac{c}{c_0} = 1 - \frac{1}{N}\left(1 - \left\{(1-N)^{m/(m-1)} + \left[1-(1-N)^{m/(m-1)}\right]F(c)\right\}^{(m-1)/m}\right) \qquad (5.38)$$

If $m = 3$, Equation 5.38 reduces to

$$\frac{c}{c_0} = 1 - \frac{1}{N}(1-((1+0.5\ln N)F(c) - 0.5\ln N)^{2/3}) \qquad (5.39)$$

where $F(c)$ is given by Equation 5.25. The sediment concentration can be obtained with only parameter $N$.

## Example 5.2

Compute sediment concentration for data in Table 5.1. What is the best value of entropy index?

### Solution

With Lagrange multipliers computed from Example 5.1, the entropy index can be obtained from

$$N = \frac{\lambda_1 c_0}{\lambda_1 c_0 + \lambda_*} = \frac{1.346 \times 58}{1.346 \times 58 + 80.14} = 0.493$$

Thus, sediment concentration is now computed from

$$\frac{c}{c_0} = 1 - \frac{1}{N}(1-((1+0.5\ln N)F(c) - 0.5\ln N)^{2/3})$$

$$= 1 - \frac{1}{0.493}(1-((1+0.5\ln 0.493)F(c) - 0.5\ln 0.493)^{2/3})$$

With $F(c) = 1-(y/D)^{0.5}$, the concentration distribution is plotted in Figure 5.4.

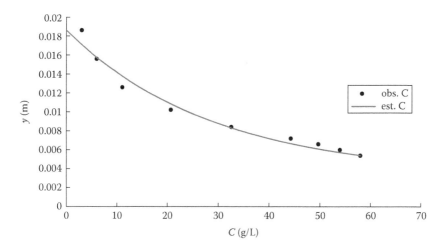

**FIGURE 5.4** Sediment concentration distributions for Run S1 of Einstein and Chien's (1955) data.

Following Chiu (1988) and Cui and Singh (2014), the entropy index is used as an index for characterizing the pattern of distribution. To that end, the mean sediment concentration is derived as

$$\overline{c} = \frac{1}{A}\int cdA = \frac{1}{A}\int c_0 \left[1 - \frac{1}{N}(1 - ((1 + 0.5\ln N)F(c) - 0.5\ln N)^{2/3})\right]dA \quad (5.40)$$

where $A$ is the cross-sectional area. The ratio of sediment concentration to the maximum sediment concentration can be written by dividing Equation 5.40 as

$$\frac{\overline{c}}{c_0} = \Psi(N) = \frac{1}{A}\int c_0 \left[1 - \frac{1}{N}(1 - ((1 + 0.5\ln N)F(c) - 0.5\ln N)^{2/3})\right]dA \quad (5.41)$$

Equation 5.41 shows the relationship between mean and maximum sediment concentration values as a function of entropy index $N$. Using empirical observations, the Lagrange multipliers are first computed by solving Equations 5.29 and 5.31, then the $N$ value is obtained by Equation 5.33. By plotting the empirical observations of the mean over the maximum concentration values and the corresponding $N$ values, Cui and Singh (2014) derived an explicit function by regression as

$$\frac{\overline{c}}{c_0} = \psi(N) = 0.176 + 0.5083N - 0.1561N^2 \quad (5.42)$$

**TABLE 5.2**
**Mean and Maximum Sediment**
**Concentration and N Values**

| $c_0$ | $\bar{c}$ | $\dfrac{\bar{c}}{c_0}$ | $N$ |
|---|---|---|---|
| 81.6 | 27.2 | 0.333 | 0.378 |
| 122.1 | 66.6 | 0.545 | 1.145 |
| 82.6 | 36.9 | 0.447 | 0.682 |
| 360.3 | 136.4 | 0.379 | 0.423 |
| 144 | 81.0 | 0.563 | 1.15 |
| 58 | 15.8 | 0.272 | 0.15 |
| 204.6 | 50.8 | 0.249 | 0.1 |
| 352 | 85.8 | 0.244 | 0.09 |
| 386 | 86.7 | 0.225 | 0.05 |
| 64.5 | 33.3 | 0.517 | 0.85 |
| 152 | 80.3 | 0.529 | 0.9 |

## Example 5.3

A number of mean and maximum sediment concentration values are shown in Table 5.2. Compute and tabulate the values of $N$ and plot Equation 5.42.

### Solution

For given mean and maximum sediment concentration values, parameter $N$ is computed from the Lagrange multipliers following Example 5.2, and its values are tabulated in Table 5.2.

Then, the relationship between the ratio of mean and maximum sediment concentration with $N$ given in Equation 5.42 is plotted in Figure 5.5.

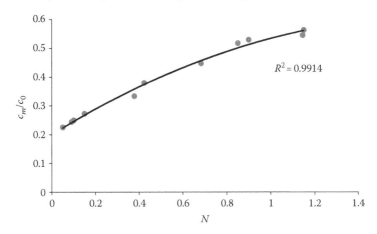

**FIGURE 5.5** Regression of mean/maximum sediment concentration by parameter $N$.

Equation 5.42, shown in Figure 5.5, has a coefficient of determination as 0.99. Thus, $N$ can be used for deriving sediment concentration distribution instead of solving nonlinear equations for $\lambda_1$ and $\lambda_*$. Once entropy index $N$ is determined, the mean concentration can be determined from Equation 5.42 for a given maximum concentration value. If Equation 5.42 can be validated, then the whole vertical profile of sediment concentration can be easily obtained from Equation 5.39 based on the Tsallis entropy approach.

## Example 5.4

Compute the Lagrange parameters for the data listed in Table 5.3 (Coleman, 1981) and sediment concentration.

### Solution

For given data $c_0 = 230$ g/L and $\bar{c} = 52.5$ g/L, the Lagrange multipliers are obtained by solving Equations 5.29 and 5.32 that yield $\lambda_1 = 0.515$ and $\lambda_* = 89.12$. Then,

$$N = \frac{\lambda_1 c_0}{\lambda_1 c_0 + \lambda_*} = \frac{0.515 \times 230}{0.515 \times 230 + 89.12} = 0.571$$

The sediment concentration is now computed from Equation 5.39 with $\eta = 0.1$ and is plotted in Figure 5.6.

## 5.2.7 ENTROPY OF SEDIMENT CONCENTRATION

Substituting Equation 5.24 into Equation 5.18 and then integrating, the result is the Tsallis entropy of sediment concentration as

$$H = \frac{1}{m-1} - \frac{1}{m}(\lambda_* + \lambda_1 \bar{c}) \tag{5.43}$$

---

**TABLE 5.3**
**Observed Sediment Concentration Data**

| Y (mm) | c (g/L) | y (mm) | C (g/L) |
|--------|---------|--------|---------|
| 6 | 230 | 69 | 26 |
| 12 | 120 | 91 | 16 |
| 18 | 82 | 122 | 7.6 |
| 24 | 61 | 137 | 4 |
| 30 | 48 | 152 | 2 |
| 46 | 33 | 162 | 1.1 |

*Source:* Coleman, N.L., *J. Hydr. Res.*, 19(3), 211, 1981.

---

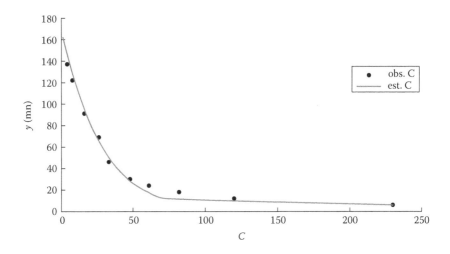

**FIGURE 5.6** Sediment concentration distribution for Coleman's data.

Replacing the Lagrange multipliers with $N$ using Equations 5.33 and 5.34, the entropy for the sediment concentration can be written as

$$H = \frac{1}{m-1} - \frac{1}{m}\left\{\left(\frac{m-1}{m}\right)^{-m}\left[1-(1-N)^{m/(m-1)}\right]^{-(m-1)}\left(\frac{c_0}{N}\right)^{-m}\left[c_0\left(1-\frac{1}{N}\right)+\bar{c}\right]\right\} \quad (5.44)$$

It can be seen from Equation 5.44 that for a given set of data, with known $c_0$ and $\bar{c}$, the entropy value is monotonically increasing with $N$. Thus, $N$, as an index of the concentration distribution pattern, yields the entropy value of the given distribution, which is always higher for bigger $N$ and lower for smaller $N$. This means that the probability distribution will have more uncertainty or uniformity for greater $N$ values than for smaller $N$ values.

## Example 5.5

Compute maximum entropy for data used in Examples 5.1 and 5.4.

**Solution**

For data in Example 5.1,

$$H = \frac{1}{m-1} - \frac{1}{m}\left\{\left(\frac{m-1}{m}\right)^{-m}\left[1-(1-N)^{m/(m-1)}\right]^{-(m-1)}\left(\frac{c_0}{N}\right)^{-m}\left[c_0\left(1-\frac{1}{N}\right)+\bar{c}\right]\right\}$$

$$= \frac{1}{2} - \frac{1}{3}\left\{\left(\frac{2}{3}\right)^{-3}\left[1-(1-0.493)^{3/2}\right]^{-2}\left(\frac{58}{0.493}\right)^{-3}\left[58\left(1-\frac{1}{0.493}\right)+31.05\right]\right\} = 0.78$$

For data in Example 5.4,

$$H = \frac{1}{m-1} - \frac{1}{m}\left\{\left(\frac{m-1}{m}\right)^{-m}\left[1-(1-N)^{m/(m-1)}\right]^{-(m-1)}\left(\frac{C_0}{N}\right)^{-m}\left[C_0\left(1-\frac{1}{N}\right)+\bar{c}\right]\right\}$$

$$= \frac{1}{2} - \frac{1}{3}\left\{\left(\frac{2}{3}\right)^{-3}[1-(1-0.571)^{3/2}]^{-2}\left(\frac{230}{0.571}\right)^{-3}\left[230\left(1-\frac{1}{0.571}\right)52.5\right]\right\} = 0.08$$

## 5.2.8 ESTIMATION OF MEAN SEDIMENT CONCENTRATION

The dimensionless entropy index $N$ is an index characterizing the distribution of sediment concentration. Once $N$ is known, the vertical distribution of sediment concentration can be obtained. One significant use of $N$ is to estimate the mean sediment concentration. As shown in Equation 5.42, the ratio of the mean and maximum sediment concentration values follows a quadratic function of $N$. With known maximum sediment concentration, the mean value is computed using Equation 5.42.

## Example 5.6

Estimate the mean sediment concentration for given maximum sediment concentration and entropy index $N$ listed in Table 5.4. Plot the estimated mean sediment concentration against the observed value.

**TABLE 5.4**
**Maximum Sediment Concentration and Entropy Parameter $N$**

| Maximum (g/L) | N | Observed Mean (g/L) | Estimated Mean (g/L) |
|---|---|---|---|
| 58 | 0.412 | 31.051 | 33.252 |
| 121 | 0.383 | 54.308 | 55.257 |
| 150.5 | 0.390 | 70.620 | 62.805 |
| 194 | 0.400 | 97.166 | 74.169 |
| 328 | 0.392 | 156.470 | 125.649 |
| 28 | 0.408 | 14.642 | 12.715 |
| 64.5 | 0.387 | 29.834 | 26.842 |
| 83.4 | 0.384 | 37.850 | 29.326 |
| 152 | 0.378 | 66.251 | 54.285 |
| 216 | 0.380 | 95.268 | 85.836 |
| 31.4 | 0.437 | 19.349 | 18.096 |
| 204.6 | 0.357 | 77.267 | 62.924 |
| 352 | 0.361 | 136.741 | 101.123 |
| 386 | 0.357 | 145.879 | 118.897 |
| 601 | 0.371 | 249.375 | 185.686 |
| 618 | 0.376 | 265.437 | 217.735 |

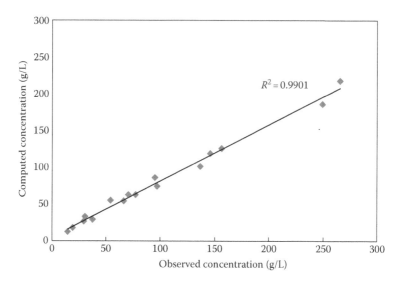

**FIGURE 5.7** Comparison of computed and observed mean sediment concentration.

### Solution

The mean sediment concentration can be estimated using Equation 5.42. For example,

$$\bar{c} = c_0(0.176 + 0.5083N - 0.1561N^2)$$

$$= 58(0.176 + 0.5083 \times 0.412 - 0.1561 \times 0.412^2) = 33.252\,\text{g/L}$$

that is comparable to the observed value of 31.051 g/L. The rest of mean sediment concentration values are computed in the same way and are tabulated in Table 5.4 and plotted in Figure 5.7.

Figure 5.7 shows that the estimated mean sediment concentration is a little bit smaller than the observed mean concentration as the slope of the straight line is smaller than 1:1. However, the coefficient of determination ($R^2$ = 1-residual sum of squares/total sum of squares) is higher than 0.99, thus showing the adequacy of this method.

### 5.2.9 COMPARISON WITH OTHER CONCENTRATION DISTRIBUTIONS

### Example 5.8

Compare entropy-based distribution with other concentration distributions.

### Solution

The sediment concentration using the Tsallis entropy is the same as in Examples 5.2 and 5.4, and thus, it will not be repeated here.

For given $a = 0.002$ mm and $c_a = 400$ g/L and $R_0 = 1.741$, using Equation 5.6 the sediment concentration with the Rouse equation is computed from

$$c = c_a \left( \frac{(y_0 - y)}{y} \frac{a}{(y_0 - a)} \right)^{R_0} = 400 \left( \frac{(0.01 - y)}{y} \frac{0.002}{(0.01 - 0.002)} \right)^{1.741}$$

To use Equation 5.15, $z'$ needs to be computed first. For given $\omega_s = 0.004$ ft/s, $u_{max} = 7.13$ ft/s, $u* = 0.342$ ft/s, and $M = 0.68$

$$z' = \frac{\omega_s u_{max}(1 - e^{-M})}{\beta u_*^2 M} = \frac{0.004 \times 7.13(1 - e^{-0.68})}{1.003 \times 0.342^2 \times 0.68} = 0.643$$

and the sediment concentration is computed from

$$c = c_0 \exp\left[ -z' \frac{e^M}{\xi_{max}} \int_0^y \left\{ h_\xi \left[ 1 + (e^M - 1)\frac{\xi}{\xi_{max}} \right] \frac{\tau}{\tau_0} \right\}^{-1} d\xi \right]$$

$$= 352 \exp\left[ -0.643 \times \frac{e^{0.643}}{\xi_{max}} \int_0^y \left\{ h_\xi \left[ 1 + (e^M - 1)\frac{\xi}{\xi_{max}} \right] \frac{\tau}{\tau_0} \right\}^{-1} d\xi \right]$$

For Choo's method with $N' = 5.647$, using Equation 5.17, the sediment concentration can be estimated from

$$c = \frac{c_0}{N} \ln\left[ \exp(N') - (\exp(N') - 1)\frac{y}{D} \right] = \frac{352}{5.647} \ln\left[ \exp(5.647) - (\exp(5.647) - 1)\frac{y}{D} \right].$$

The values of sediment concentration determined by all the previous methods are plotted in Figure 5.8.

It is seen from Figure 5.8 that the last three entropy-based methods (Equations 5.15, 5.17, and 5.36) lead to much smaller error than does the Rouse equation on average, which demonstrates the disadvantage of using the Rouse equation (Equation 5.6). The Rouse equation yields much lower values than observed values from the water surface down to 0.007 m in Figure 5.8 and the curve is farthest apart from observations in comparison with other equations. It can also be seen that Choo's method (Equation 5.17) and the Tsallis entropy–based equation (Equation 5.37) yield similar results that are closer to observations than other methods.

## 5.3 HYDRAULIC METHOD

Let there be a 3-D fluid medium in which flow is unsteady and nonuniform, that is, the velocity $u$ is varying in all three directions $x$, $y$, and $z$ as well as in time $t$, $u(x, y, z; t)$. Here, $x$ represents the horizontal direction, $y$ the vertical direction, and $z$ the transverse direction. It is assumed that sediment is released into a channel from a single source.

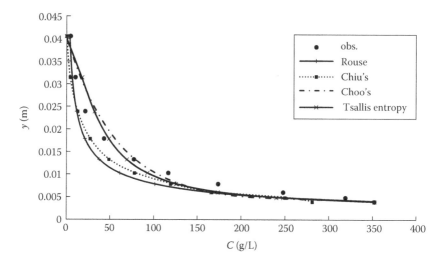

**FIGURE 5.8** Comparison for different methods for Run S13 of Einstein and Chien's (1955).

It has been shown (Chiu, 1967, 1987) that the movement of sediment particles fol-lows a "random walk." Then the position the sediment particle occupies during its movement can be considered as a random variable having a PDF, $f(x, y, z)$; the PDF describes the random walk. This suggests that there is potential for employing entropy in dealing with sediment movement. In order to simplify the probabilistic treatment of sediment movement using the entropy theory, it is assumed that the flow is steady (i.e., $u$ is independent of $t$.) and so is sediment movement and that the PDF does not vary in the longitudinal ($x$) and transverse ($z$) directions; thus, $f(x, y, z) = f(y)$ and $y$ can be taken as the distance the particle travels. The objective is to derive the sediment concentration distribution using the Tsallis entropy that requires the derivation of PDF $f(y)$. The methodology for deriving suspended sediment concentration distribu-tion using the Tsallis entropy is the same as earlier.

### 5.3.1 PROBABILITY DISTRIBUTION OF TRAVEL DISTANCE OF SEDIMENT

Let the flow depth in the channel be denoted by $D$. Then $0 \leq y \leq D$. The Tsallis entropy for $y$ or $f(y)$, $H[f(y)]$ or $H(y)$ can be written as

$$H[f(y)] = H(y) = \frac{1}{m-1} \int_0^D f(y)\{1 - [f(y)]^{m-1}\} dy \tag{5.45}$$

where $m$ is the Tsallis entropy index. The PDF of $y$ is derived by maximizing the Tsallis entropy subject to the following constraints:

$$\int_0^D f(y) dy = 1 \tag{5.46}$$

$$\int_0^D yf(y)dy = E[y] = \bar{y} \tag{5.47}$$

where $E$ is the expectation operator and $\bar{y}$ is the mean value of $y$. For maximizing the Tsallis entropy given by Equation 5.45, subject to Equations 5.45 and 5.46, the method of Lagrange multipliers is employed here, where the Lagrangian $L$ can be written as

$$L = \frac{1}{m-1}\int_0^D f(y)\{1-[f(y)]^{m-1}\}dy - \lambda_0\left[\int_0^D f(y)dy - 1\right] - \lambda_1\left[\int_0^D yf(y)dy - \bar{y}\right] \tag{5.48}$$

in which $\lambda_0$ and $\lambda_1$ are the Lagrange multipliers. Differentiating Equation 4.47 with respect to $f$ and equating the derivative to zero yield the maximum Tsallis entropy–based PDF of $y$:

$$f(y) = m^{1/(1-m)}[1-(m-1)(\lambda_0 + \lambda_1 y)]^{1/(m-1)} \tag{5.49}$$

The Tsallis entropy of Equation 5.49 can be written as

$$H(y) = \frac{1}{m-1} + \frac{m^{m/(1-m)}}{(2m-1)\lambda_1}$$

$$\times\left\{[(1-(m-1)(\lambda_0 + \lambda_1 D)]^{(2m-1)/(m-1)} - [1-(m-1)\lambda_0]^{(2m-1)/(m-1)}\right\} \tag{5.50}$$

The Lagrange multipliers $\lambda_0$ and $\lambda_1$ are now determined using Equations 5.45 and 5.46. Substituting Equation 5.49 in Equation 5.46 yields

$$\int_0^D \left(\frac{1}{m}\right)^{1/(m-1)}[1-(m-1)(\lambda_0 + \lambda_1 y)]^{1/(m-1)}dy = 1 \tag{5.51}$$

Equation 5.51 simplifies to

$$[1-(m-1)\lambda_0]^{m/(m-1)} - [1-(m-1)(\lambda_0 + \lambda_1 D)]^{m/(m-1)} = m^{m/(m-1)}\lambda_1 \tag{5.52}$$

Substituting Equation 5.49 into Equation 5.47 results in

$$\int_0^D y\left\{\frac{1}{m}[1-(m-1)(\lambda_0 + \lambda_1 y)]\right\}^{1/(m-1)}dy = \bar{y} \tag{5.53}$$

**TABLE 5.5**
**Lagrange Multipliers Computed for Date Set**
**Given Sample of $D = 0.044$ m and $\bar{y} = 0.022$ m**

| $m$ Value | $\lambda_0$ | $\lambda_1$ |
|---|---|---|
| $m = 1/2$ | 2.374 | −0.176 |
| $m = 2/3$ | 1.967 | 0.331 |
| $m = 3/4$ | 2.176 | 0.441 |
| $m = 2$ | 1.783 | 0.123 |
| $m = 3$ | 0.528 | −0.316 |
| $m = 4$ | 0.332 | 0.0004 |

Upon integrating by parts, the solution of Equation 5.53 is found to be

$$\frac{1}{(2m-1)\lambda_1}\{[1-(m-1)\lambda_0]^{(2m-1)/(m-1)}-[1-(m-1)(\lambda_0+\lambda_1 D)]^{(2m-1)/(m-1)}\}$$

$$-D[1-(m-1)(\lambda_0+\lambda_1 D)]^{(2m-1)/(m-1)} = m^{m/(m-1)}\lambda_1\bar{y} \qquad (5.54)$$

Equations 5.51 and 5.53 are solved simultaneously to determine the Lagrange multipliers $\lambda_0$ and $\lambda_1$. Their analytical solution is not tractable but the numerical solution is relatively straightforward.

It is seen that for given $D$ and $\bar{y}$, the values of Lagrange multipliers depend on the value of $m$ used, as shown for sample $D = 0.044$ m and $\bar{y} = 0.022$ m in Table 5.5. It can be seen from the table that $\lambda_1$ becomes small and less effective when $m$ becomes large. With increasing $m$, $\lambda_0$ also becomes smaller. These values suggest that it is important that an appropriate value of $m$ is selected. It may now be interesting to examine the sensitivity of $f(y)$ to $m$, as shown in Figure 5.9. It is observed from the figure that $f(y)$ is not highly sensitive to $m$ between $0.5 \leq m \leq 0.75$, but outside of this interval, it becomes highly sensitive. A plot of the entropy of travel distance as a function of the zeroth Lagrange multiplier for various values of $\lambda_1$ in Figure 5.10 shows that Tsallis entropy increases as $\lambda_0$ increases. This discussion shows that the Lagrange parameters play a fundamental role in the entropy-based derivations. It may, therefore, be desirable to derive the physical meaning of the Lagrange multipliers with the use of the random walk hypothesis.

### 5.3.2 DETERMINATION OF DISTRIBUTION PARAMETERS
####       FROM THE RANDOM WALK HYPOTHESIS

If the movement of sediment particles follows a "random walk," then it can be shown that the PDF, $f(x, y, z)$, describing the random walk at time $t$ after the release of particles follows a parabolic diffusion equation in which the gradient of the diffusion coefficient varies in the vertical direction (Chiu, 1967, 1987). If it is assumed that the

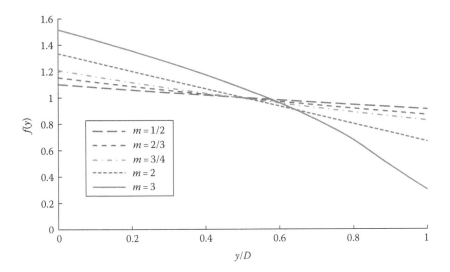

**FIGURE 5.9** Plot of $f(y)$ for various $m$ values.

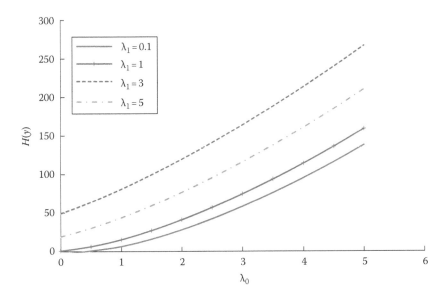

**FIGURE 5.10** Plot of Tsallis entropy versus zeroth Lagrange multiplier for various values of $\lambda_1$ for $m = 3$.

momentum transfer coefficient in the vertical direction, $\varepsilon_y$, is constant and is equal to the depth-averaged value denoted as $\overline{\varepsilon_y}$ and the vertical component of the fluid velocity $u_y$ is approximately equal to the negative of the depth-averaged settling velocity $v_s$, then the PDF of sediment particle movement can be shown to follow the Euler ordinary differential equation that is linear. However, empirical data show that the PDF of $y$ is nonlinear. Therefore, it is hypothesized that the sediment particle movement follows a nonlinear ordinary differential equation as

$$\varepsilon_s \frac{df(y)}{dy} + (\varepsilon_0 + v_s)[f(y)]^n = 0 \tag{5.55}$$

where $\varepsilon_0$ is a constant, perhaps equal to a constant value of the diffusion coefficient gradient, $\varepsilon_s$ is the depth-averaged value of the diffusion coefficient $\varepsilon_y$, and $n > 0$. If $n = 1$, Equation 5.55 reduces to the Euler equation.

The solution of Equation 5.55, using the initial condition: at $y = 0$, $f(y) = f_0$, follows

$$f(y) = \left[ f_0^{-n+1} - \frac{\varepsilon_0 v_s}{\varepsilon_s}(1-n)y \right]^{1/(1-n)} \tag{5.56}$$

The initial value $f_0$ needs to be specified. Let $a = (1-n)\frac{(\varepsilon_0 + v_s)}{\varepsilon_s}$. Then, Equation 5.56 can be written as

$$f(y) = [f_0^{-n+1} - ay]^{1/(1-n)} \tag{5.57}$$

The Tsallis entropy of Equation 5.57 can be written as

$$H(y) = \frac{1}{m-1} + \frac{(1-n)}{a(m+1-n)} \left\{ \left[ f_0^{-n+1} - aD \right]^{(m+1-n)/(1-n)} - [f_0]^{(m+1-n)/(1-n)} \right\} \tag{5.58}$$

Now, comparing Equation 5.49 with Equation 5.57, it is seen that

$$m - 1 = -n + 1 \tag{5.59}$$

$$\lambda_0 = \frac{1}{m-1} \left[ 1 - m f_0^{1-n} \right] \tag{5.60}$$

$$\lambda_1 = \frac{m}{m-1} a \tag{5.61}$$

Equations 5.58 through 5.60 are physically based, and therefore, the Lagrange multipliers of Equation 5.49 have physical meaning but these relationships need to be verified, however. The relations between the Lagrange multipliers and $f_0$

**TABLE 5.6**
**Lagrange Multipliers Relating to $f_0$ and $a$**

| $m$ Value | $\lambda_0$ | $\lambda_1$ | $N$ | $f_0$ | $A$ |
|-----------|-------------|-------------|------|--------|--------|
| $m = 1/2$ | 2.374 | −0.176 | 1.50 | 0.547 | −0.375 |
| $m = 2/3$ | 1.967 | 0.331 | 1.33 | 0.491 | −0.296 |
| $m = 3/4$ | 2.176 | 0.441 | 1.25 | 0.489 | −0.234 |
| $m = 2$ | 1.783 | 0.123 | 0.00 | −0.392 | 0 |
| $m = 3$ | 0.528 | −0.316 | −1.00 | −0.032 | −6 |
| $m = 4$ | 0.332 | 0.0004 | −2.00 | 0.003 | −24 |

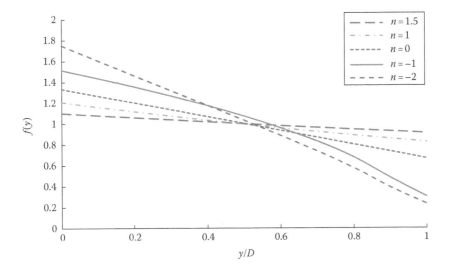

**FIGURE 5.11**   Plot of $f(y)$ versus $y$ for different values of $n$ using the relations from Equation 5.58 to 5.60.

and $a$ are computed in Table 5.6, based on the relations stated in Equations 5.58 through 5.60. Using parameters computed in this manner, $f(y)$ is plotted against $y/D$ in Figure 5.11.

### 5.3.3   DETERMINATION OF INITIAL VALUE OF PDF, $f_0$

Now we derive $f_0$ and $\bar{y}$ in terms of physically measurable quantities. Substituting Equation 5.57 into Equation 5.46, one obtains

$$\int_0^D \left[ f_0^{-n+1} - ay \right]^{1/(-n+1)} dy = 1 \tag{5.62}$$

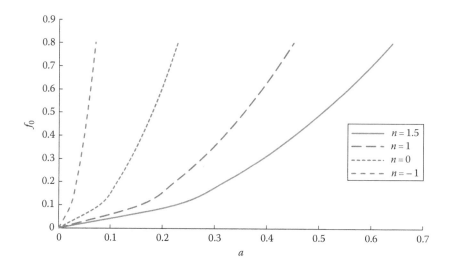

**FIGURE 5.12**   Plot of $f_0$ versus $a$ for various values of $n [D = 0.044$ m and $\bar{y} = 0.022$ m].

Integration of Equation 5.62 yields

$$f_0^{2-n} - \left[ f_0^{1-n} - aD \right]^{(2-n)/(-n+1)} = a\frac{(2-n)}{-n+1} \tag{5.63}$$

Equation 5.63 defines $f_0$ in terms of $a$ and $n$. A plot of $f_0$ versus $a$ for various values of $n$ in Figure 5.12 shows that $f_0$ increases with increasing $a$. Fundamental to determining parameter $a$ is the determination of diffusion coefficient.

Now the average value of $y$ is defined. Substituting Equation 5.57 in Equation 5.47, one gets

$$\int_0^D y \left[ f_0^{-n+1} - ay \right]^{1/(-n+1)} dy = \bar{y} \tag{5.64}$$

Solution of Equation 5.64 by parts leads to

$$\frac{1}{3-2n} \left[ f_0^{1-n} - aD \right]^{(3-n)/n} - \frac{1}{3-2n} f_0^{3-n} - \frac{D}{2-n} \left[ f_0^{1-n} - aD \right]^{(2-n)/(1-n)} = (1-n)a\bar{y} \tag{5.65}$$

This expresses $\bar{y}$ in terms of physically measurable quantities.

### 5.3.4   DETERMINATION OF DIFFUSION COEFFICIENT

The diffusion coefficient $\varepsilon_s$ is related to the average $(\bar{\varepsilon}_y)$ of the values of the momentum transfer coefficient in the $y$-direction, $\varepsilon_y$, as

$$\varepsilon_s = \beta\overline{\varepsilon_y} \tag{5.66}$$

where

$\beta$ is a parameter

$\overline{\varepsilon_y}$ is the depth-averaged value of $\varepsilon_y$, which is obtained from

$$\tau = \rho\varepsilon_y \frac{du}{dy} \tag{5.67}$$

The use of Equation 5.67 requires the knowledge of velocity derivative. Singh and Hao (2011) and Cui and Singh (2012) obtained the velocity distribution using the Tsallis entropy, which can be written as

$$u(y) = \frac{k}{\alpha_1}\left[\left(\frac{\alpha_v}{k}\right)^k + \alpha_1 \frac{y}{D}\right]^{1/k} - \frac{\alpha_v}{\alpha_1}, \ k = \frac{q}{q-1}, \ \alpha_v = \frac{q}{q-1}u_D^{1-q} = \alpha_0 + \frac{1}{q-1} \tag{5.68}$$

where

$q$ is the entropy index that may be different from the entropy index $m$ of Equation 5.45

$\alpha_0$ and $\alpha_1$ are the Lagrange multipliers

The derivative of velocity is now obtained as

$$\frac{du}{dy} = \frac{1}{D}\left[\left(\frac{\alpha_v}{k}\right)^k + \alpha_1 \frac{y}{D}\right]^{(1-k)/k} \tag{5.69}$$

The shear stress $\tau$ is derived as follows. Consider a wide channel of depth $D$ having a slope $S$. The shear stress is maximum at the channel bed ($y = 0$) and equals $\tau_0 = \rho gDS$ and decreases monotonically with increasing $y$ and becomes zero at the water surface. If $\tau$ is assumed to be a random variable, then it has a PDF $f(\tau)$. Now, let $\tau(y)$ be the shear stress at $y$. At any distance greater than $y$, the shear stress is less than $\tau$. It can be intuitively stated that the probability of shear stress less than or equal to $\tau$ can be expressed as $(D - y)/D$ or the cumulative distribution of $\tau$ in terms of $y$ is

$$F(\tau) = \frac{D-y}{D} \tag{5.70}$$

Then, the PDF of $\tau$ can be expressed as

$$f(\tau) = -\left(D\frac{d\tau}{dy}\right)^{-1} \tag{5.71}$$

Note that $f(\tau)$ is yet to be determined. By definition,

$$\int_0^{\tau_0} f(\tau)d\tau = 1 \tag{5.72}$$

The PDF $f(\tau)$ can be determined by maximizing the Tsallis entropy

$$H(\tau) = \frac{1}{r-1} \int_0^{\tau_0} \{f(\tau) - [f(\tau)]^{r-1}\} d\tau \tag{5.73}$$

where $r$ is the Tsallis entropy index, which may be different from entropy indices $m$ and $q$ defined earlier. Using the method of Lagrange multipliers, Equation 5.73 is maximized, subject to Equation 5.72, as

$$L = \frac{1}{r-1} \int_0^{\tau_0} \{f(\tau) - [f(\tau)]^{r-1}\} d\tau - \gamma_0 \left[ \int_0^{\tau_0} f(\tau)d\tau - 1 \right] \tag{5.74}$$

where $\gamma_0$ is the Lagrange multiplier. Differentiating Equation 5.74 with respect to $f(\tau)$ and equating the derivative yield

$$f(\tau) = \left[ \frac{1}{r-1} - \gamma_0 \right]^{1/(r-1)} \tag{5.75}$$

Substitution of Equation 5.75 in Equation 5.72 yields

$$f(\tau) = \frac{1}{\tau_0} \tag{5.76}$$

which is a uniform distribution.

Substitution of Equation 5.76 in Equation 5.71 leads to

$$\frac{1}{\tau_0} = -\frac{1}{D} \frac{dy}{d\tau} \tag{5.77}$$

Equation 5.77 results in

$$\tau(y) = \tau_0 \left( 1 - \frac{y}{D} \right) \tag{5.78}$$

If the flow is assumed steady uniform and one-dimensional, then Equation 5.78 can be obtained from the use of the momentum conservation equation or equation of motion in deterministic hydraulics.

### 5.3.5  Sediment Concentration Distribution

Now we derive the sediment concentration distribution. Let $C(y)$ be the sediment concentration at distance $y$ from the bed, and let there be $N$ sediment particles of a given size with settling velocity $v_s$ between $y = 0$ and $y = D$. The quantity $f(y)dy$ denotes the probability that a sediment particle is between positions $y$ and $y + dy$. Then, $Nf(y)dy$ denotes the number of particles between $y$ and $y + dy$. If $N$ is taken to represent the volume (or weight) of sediment particles of the specified size then $Nf(y)$ will denote the sediment concentration by volume (or weight). Depending on the way $f(y)$ is specified, the vertical distribution of sediment concentration can be expressed in two ways.

### 5.3.6  Tsallis Entropy–Based Sediment Concentration Distribution

Using Equation 5.49, $C(y)$ can be expressed as

$$C(y) = Nf(y) = Nm^{1/(1-m)}[1-(m-1)(\lambda_0 + \lambda_1 y)]^{1/(m-1)} \qquad (5.79)$$

From Equation 5.79, the initial sediment concentration can be expressed as

$$C_0 = C(y=0) = Nf(y=0) = Nm^{1/(1-m)}[1-(m-1)\lambda_0]^{1/(m-1)} \qquad (5.80)$$

However, the PDF of $C$ and its entropy still need to be determined.

Sediment concentration monotonically decreases from the channel bed to the water surface. If $C$ is assumed to be a random variable then the probability of sediment concentration less than or equal to $C$ can be expressed as

$$F(C) = \frac{D-y}{D} = \int_y^D \frac{1}{D}dy = \int_{C_D}^C -\frac{1}{D}\frac{dy}{dC}dC \qquad (5.81)$$

where $C_D$ is the sediment concentration at the water surface $y = D$. Then the PDF of $C$ can be inferred from Equation 5.81 as

$$f(C) = -\left(D\frac{dC}{dy}\right)^{-1} \qquad (5.82)$$

From Equation 5.79, $dC/dy$ can be expressed as

$$\frac{dC}{dy} = -\frac{m-1}{m}N^{m-1}\lambda_1 C^{2-m} \qquad (5.83)$$

Inserting Equation 5.83 in Equation 5.82, one gets

$$f(C) = \frac{1}{D\lambda_1}N^{1-m}\frac{m}{m-1}C^{m-2} = \psi\, C^{m-2} \qquad (5.84)$$

where

$$\psi = \frac{N^{1-m}}{D\lambda_1} \frac{m}{m-1}$$

(5.85)

It may be advantageous to define $f(C)$ at $y = 0$. To that end, from Equation 5.79,

$$\frac{dC}{dy} = -Nm^{1/(1-m)}\lambda_1[1-(m-1)(\lambda_0 +\lambda_1 y)]^{(2-m)/(m-1)}$$

(5.86)

Inserting Equation 5.86 in Equation 5.82, one obtains

$$f(C) = \frac{1}{DN\lambda_1} m^{1/(m-1)}[1-(m-1)(\lambda_0 + \lambda_1 y)]^{(m-2)/(m-1)}$$

(5.87)

The value of $f(C)$ at $y = 0$ is obtained from Equation 5.74 as

$$f(C)\big|_{y=0} = \frac{1}{DN\lambda_1} m^{1/(m-1)}[1-(m-1)\lambda_0]^{(m-2)/(m-1)}$$

(5.88)

Using Equation 5.84, the Tsallis entropy of the sediment concentration distribution can be written as

$$H(C) = \frac{1}{m-1}\left\{ C_0 - C_D - \frac{\psi}{D}\left[C_0^{m(2-m)+1} - C_D^{m(2-m)+1}\right]\right\}$$

(5.89)

With the use of Equation 5.84, the values of $C_0$ and $C_D$ can be obtained by using the normalizing constraint and the mean constraint. Using the normalizing constraint,

$$\int_{C_D}^{C_0} f(C)dC = \int_{C_D}^{C_0} \psi C^{m-2}dC = \frac{\psi}{m-1}\left[C_0^{m-1} - C_D^{m-1}\right] = 1$$

(5.90)

Equation 5.90 yields

$$C_0^{m-1} - C_D^{m-1} = \frac{m-1}{\psi}$$

(5.91)

Since $C_0$ is much larger than $C_D$, Equation 5.91 can be approximated as

$$C_0^{m-1} \approx \frac{m-1}{\psi}$$

(5.92)

Now using the mean constraint,

$$\int_{C_D}^{C_0} Cf(C)dC = \int_{C_D}^{C_0} \psi C^{m-1}dC = \frac{\psi}{m}\left[C_0^m - C_D^m\right] = \overline{C} \tag{5.93}$$

Equation 5.93 can be utilized to determine the mean constraint value or if this is known then $C_D$ can be determined. Approximately,

$$\frac{\psi}{m}C_0^{\,m} = \overline{C} \tag{5.94}$$

To estimate the sediment concentration, integration of Equation 5.86, which should satisfy the assumption of Equation 5.81, leads to

$$F(C) = \int_{C_D}^{C} f(C)dC = \int_{C_D}^{C} \psi C^{m-2}dC = \psi\frac{1}{m-1}C^{m-1} = 1 - \frac{y}{D} \tag{5.95}$$

Thus, the sediment concentration distribution is obtained as

$$C = \left[\frac{(m-1)}{\psi}\left(1 - \frac{y}{D}\right)\right]^{1/(m-1)} \tag{5.96}$$

It may be interesting to compare this distribution with the one that can be derived using the random walk model.

### 5.3.7 RANDOM WALK MODEL-BASED DISTRIBUTION

The procedure here is analogous to that for the Tsallis entropy-based method discussed earlier. In a manner of Equation 5.79, one can write Equation 5.57 as

$$C(y) = N\left[f_0^{1-n} - ay\right]^{1/(1-n)} \tag{5.97}$$

Differentiating Equation 5.97 with respect to $y$ yields

$$\frac{dC}{dy} = -aN^{1-n}C^n \tag{5.98}$$

Substituting Equation 5.98 in Equation 5.83, the PDF of $C$ is obtained as

$$f(C) = \frac{N^{n-1}}{aD}C^{-n} \tag{5.99}$$

Now the Tsallis entropy of Equation 5.99 can be expressed as

$$H(C) = \frac{1}{s-1}\left[ C_0 - C_D - \frac{1}{1-n}\left(\frac{N^{1-n}}{aD}\right)^s \left(C_0^{1-n} - C_D^{1-n}\right)\right] \qquad (5.100)$$

where $s$ denotes the Tsallis entropy index whose value can be different from the value of $m$. Equation 5.99 can be utilized to determine $C_0$ and $C_D$ using the normalizing constraint and the mean constraint. Using Equation 5.99 in the normalizing constraint,

$$\int_{C_D}^{C_0} \frac{N^{n-1}}{aD} C^{-n} dC \Rightarrow C_0^{1-n} - C_D^{1-n} = \frac{a(1-n)D}{N^{n-1}} \qquad (5.101)$$

Since $C_0 \gg C_D$, Equation 5.101 can be approximated as

$$C_0^{1-n} \approx \frac{a(1-n)D}{N^{n-1}} \qquad (5.102)$$

Using Equation 5.99 in the mean constraint,

$$\int_{C_D}^{C_0} \frac{N^{n-1}}{aD} C^{-n} C dC \Rightarrow C_0^{2-n} - C_D^{2-n} = \frac{a(2-n)D}{N^{n-1}} \overline{C} \qquad (5.103)$$

Equations 5.102 and 5.103 give $C_0$ and $C_D$. Equation 5.103 can be approximated as

$$C_0^{2-n} = \frac{a(2-n)D}{N^{n-1}} \overline{C} \qquad (5.104)$$

To estimate the sediment concentration, one integrates Equation 5.99 that should satisfy the assumption of Equation 5.84,

$$F(C) = \int_{C_D}^{C} f(C) dC = \int_{C_D}^{C} \frac{N^{n-1}}{aD} C^{-n} dC = \frac{N^{n-1}}{aD} \frac{1}{1-n} C^{1-n} = 1 - \frac{y}{D} \qquad (5.105)$$

Thus,

$$C = \left[(1-n)aDN^{1-n}\left(1 - \frac{y}{D}\right)\right]^{\frac{1}{1-n}} \qquad (5.106)$$

Comparing Equation 5.96 with Equation 5.106, one gets

$$m = 2 - n \tag{5.107}$$

$$aDN^{1-m} = \frac{1}{\psi} \tag{5.108}$$

### Example 5.9

Experimental data on sediment concentration observed by Einstein and Chien (1955) for experiment S-16 are given in Table 5.7. For Einstein and Chien's data, for example, experiment S-16, both $C_0 = 618$ g/L and $\bar{C} = 265.44$ g/L are given. Determine parameters $m$ and $\psi$ by solving Equations 5.90 and 5.91 and parameters $n$ and $a$ by solving Equations 5.98 and 5.100.

#### Solution

Parameters $m$ and $\psi$ are computed by solving Equations 5.90 and 5.91 with given $C_0 = 265.44$ g/L and $\bar{C} = 618$ g/L that yield $m = 1.877$ and $\psi = 0.023$. Similarly, parameters $a$ and $n$ are computed by solving Equations 5.98 and 5.100 that yield $n = 0.123$ and $a = 0.375$. It is found that Equation 5.106 ($m = 2 - n = 0.023$) holds in this case.

### Example 5.10

Field data on sediment concentration collected on the Atrisco Feeder Canal near Bernalillo, New Mexico (S5 Rio Grande), are given in Table 5.8. The Atrisco Feeder Canal is about 17.7 m (58 ft) wide, and the average depth is approximately 0.52 m (1.70 ft). The collected sediment concentration is tabulated in Table 5.8, where $C_0 = 51$ g/L and $\bar{C} = 17.02$ g/L. Determine parameters $m$ and $\psi$ by solving Equations 5.91 and 5.92 and parameters $n$ and $a$ by solving Equations 5.107 and 5.108.

**TABLE 5.7**
**Experimental Data for Experiment S-16**

| y (m) | C (g/L) | y (m) | C (g/L) |
|-------|---------|-------|---------|
| 0.004 | 618     | 0.018 | 146.7   |
| 0.005 | 621     | 0.023 | 90.6    |
| 0.006 | 563     | 0.027 | 58.5    |
| 0.007 | 464.5   | 0.032 | 38.4    |
| 0.009 | 358     | 0.038 | 24.8    |
| 0.012 | 260     | 0.044 | 16.48   |
| 0.015 | 190.7   |       |         |

**TABLE 5.8**

**Field Data Collected from Atrisco Feeder Canal**

| y (m) | C (g/L) | y (m) | C (g/L) |
|-------|---------|-------|---------|
| 0.553 | 3.75 | 0.256 | 18.375 |
| 0.490 | 4.5625 | 0.165 | 20 |
| 0.433 | 5.625 | 0.108 | 25.5625 |
| 0.370 | 7.25 | 0.046 | 51 |

**Solution**

Parameters $\psi$ and $m$ are computed by solving Equations 5.94 and 5.95 that yield $m = 1.501$ and $\psi = 0.0699$, while parameters $a$ and $n$ are computed by solving Equations 5.102 and 5.104 that yield $n = 0.499$ and $a = 0.449$. The relationship between parameters stated in Equation 5.106 also holds in this case.

**Example 5.11**

With parameters computed in Example 5.9 for experimental data (S-16), compute sediment concentration using both the Tsallis entropy and random walk methods and plot it.

**Solution**

Figure 5.13 plots the sediment concentration determined for both methods for a sample data of experiment S-16. Both methods satisfactorily agree with observations. The difference between two methods is less than 0.005 g/L, which is not even visible in the figure. Results show that the sediment concentration estimated from the Tsallis entropy is found to be equivalent to that from the random walk model.

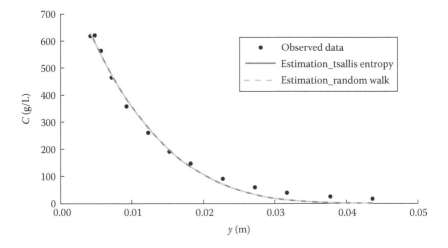

**FIGURE 5.13** Sediment concentration determined for both methods for data of experiment S-16.

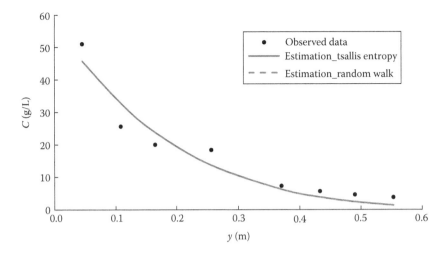

**FIGURE 5.14**   Sediment concentration determined for both methods for field data collected from Atrisco Feeder Canal.

### Example 5.12

With parameters computed in Example 5.10 for field data, compute sediment concentration using both the Tsallis entropy and random walk methods and plot it.

#### Solution

Figure 5.14 plots the sediment concentration estimated for both methods for field data S5. Both methods satisfactorily agree with observations. The difference between the two methods is less than 0.001 g/L, which is not visible in the figure. Results show that the sediment concentration estimated from Tsallis entropy is found to be equivalent to that from the random walk model.

## REFERENCES

Ackers, P. and White, W.R. (1973). Sediment transport: New approach and analysis. *Journal of Hydraulic Engineering*, 99(11), 2041–2060.

Bagnold, R.A. (1966). An approach to the sediment transport problem form general physics. U. S. Geological Survey Professional Paper 422-I, Washington, DC.

Chang, F.M., Simons, D.B., and Richardson, E.V. (1967). Total bed-material discharge in alluvial channels. *Proceedings, XII IAHR Congress*, Fort Collins, CO.

Chiu, C.L. (1967). Stochastic model of motion of solid particles. *Journal of the Hydraulics Division*, 95(5), 203–218.

Chiu, C.L. (1987). Entropy and probability concepts in hydraulics. *Journal of Hydraulic Engineering*, 113(5), 583–600.

Chiu, C.L. (1988). Entropy and 2-D velocity distribution in open channels. *Journal of Hydraulic Engineering*, 114(7), 738–756.

Chiu, C.L. (1989). Velocity distribution in open channel flow. *Journal of Hydraulic Engineering*, 115(5), 576–594.

Chiu, C.L., Jin, W., and Chen, Y.C. (2000). Mathematical models of distribution of sediment concentration. *Journal of Hydraulic Engineering*, 126(1), 16–23.

Choo, T.H. (2000). An efficient method of the suspended sediment-discharge measurement using entropy concept. *Water Engineering Research*, 1(2), 95–105.

Chow, V.T. (1959). Open Channel Hydraulics. 680 p., McGraw-Hill Book Publishing Company, New York.

Christiansen, J.E. (1935). Distribution of silt in open channels. *Transactions, American Geophysical Union*, 14, 480–481.

Coleman, N.L. (1981). Velocity profiles with suspended sediment. *Journal of Hydraulic Research*, 19(3), 211–229.

Cui, H. and Singh, V.P. (2012). On the cumulative distribution function for entropy-based hydrologic modeling. *Transactions of the ASABE*, 55(2), 429–438.

Cui, H. and Singh, V.P. (2013). One dimensional velocity distribution in open channels using Tsallis entropy. *Journal of Hydrologic Engineering*, 18(3), 331–339.

Cui, H. and Singh, V.P. (2014). Two dimensional velocity distribution in open channels using Tsallis entropy. *Journal of Hydrologic Engineering*, 19(2), 290–298.

Einstein, H.A. (1950). The bed load function for sediment transportation in open channels. Technical Bulletin 1026, USDA, Soil Conservation Service, Washington, DC.

Einstein, H.A. and Chien, N. (1955). Effects of heavy sediment concentration near the bed on velocity and sediment distribution. MRD Series Report No. 8, University of California at Berkeley and Missouri River Division, U.S. Army Corps of Engineers, Omaha, NE.

Jaynes, E.T. (1957a). Information theory and statistical mechanics 1. *Physical Review*, 106(4), 620–630.

Jaynes, E.T. (1957b). Information theory and statistical mechanics 2. *Physical Review*, 108(2), 171–190.

Lane, E.W. and Kalinske, A.A. (1941). Engineering calculations of suspended sediment. *Transactions, American Geophysical Union*, 20(3), 603–607.

Luo, H. and Singh, V.P. (2011). Entropy theory for two-dimensional velocity distribution. *Journal of Hydrologic Engineering*, 16(4), 303–315.

Molinas, A. and Wu, B. (2001). Transport of sediment in large sand-bed rivers. *Journal of Hydraulic Research*, 30(2), 135–146.

O'Brien, M.P. (1933). Review of the theory of turbulent flow and its relation to sediment transport. *Transaction, American Geophysical Union*, 12, 487–491.

Rouse, H. (1937). Modern conceptions of the mechanics of turbulence. *Transactions of ASCE*, 102, 463–543.

Rubey, W.W. (1933). Settling velocities of gravel, sand, and silt particles. *American Journal of Science*, 5th series, 25(148), 38.

Simons, D.B. and Senturk, F. (1992). *Sediment Transport Technology*. Water Resources Publications, Littleton, CO.

Singh, V.P. (1996). *Kinematic Wave Modeling in Water Resources: Environmental Hydrology*. John Wiley, New York.

Singh, V.P. and Hao, L. (2011). *Journal of Hydrologic Engineering*, 16(9), 725–735.

Singh, V.P. (2011). Hydrologic synthesis using entropy theory: Review. *Journal of Hydrologic Engineering*, 16(5), 421–433.

Stewart, B.A., Woolhiser, D.A., Wischmeier, W.H., Caro, J.H., and Frere, M.H. (1976a). Control of water pollution from cropland, Vol. 1. Report No. EPA-600/2–75–026a, Environmental Protection Agency, Office of Research and Development, Washington, DC.

Stewart, B.A., Woolhiser, D.A., Wischmeier, W.H., Caro, J.H., and Frere, M.H. (1976b). Control of water pollution from cropland, Vol. 2. Report No. EPA-600/2–75–026b, Environmental Protection Agency, Office of Research and Development, Washington, DC.

Stokes, G.G. (1851). On the effect of the internal friction of fluids on the motion of pendulums. *Transactions of Cambridge Philosophical Society*, 9, 8–106.

Sturm, T.W. (2010). *Open Channel Hydraulics*. McGraw-Hill, New York.

Tsallis, C. (1988). Possible generalization of Boltzmann-Gibbs statistics. *Journal of Statistical Physics*, 52(1–2), 479–487.

van Rijn, L.C. (1984). Sediment transport, Part 2: Suspended load transport. *Journal of Hydraulic Engineering*, 110(11), 1733–1754.

Yang, C.T. (1973). Incipient motion and sediment transport. *Journal of Hydraulic Engineering*, 99(10), 1679–1704.

Zanke, U. (1977). *Berechnung der Sinkgeschwindigkeiten von sedimenten*, Mitt. des Franzius instituts fuer Wasserbau, Heft 46, Seite 243. Technical University, Hannover, Germany.

# 6 Suspended Sediment Discharge

Sediment is one of the largest carriers of contaminants and is one of the most important issues in environmental hydraulics. It is, therefore, not surprising that sediment transport has received considerable attention for over a century. Watershed management, river training works, reservoir sedimentation, design of a variety of hydraulic works, and pollutant transport require calculation of sediment discharge or yield. Sediment discharge is computed by employing different combinations of different methods of computing channel cross-sectional velocity and suspended sediment concentration distribution. The velocity may be computed empirically or using the entropy theory, and the same is true of the sediment concentration. The objective of this chapter is to discuss the computation of suspended sediment discharge using different combinations of velocity and sediment concentration distributions.

## 6.1 PRELIMINARIES

Flow in natural channels often contains sediment, and some rivers carry huge quantities of sediment. Yellow River in China and Kosi River in India are two of the largest sediment carrying rivers in the world. There are 13 large rivers in the world that transport sediment more than $10^8$ tons per year (Chien and Wan, 2003). Rivers with high sediment content complicate flood control and aggravate reservoir sedimentation.

Einstein (1950) divided the total sediment discharge into bed load discharge and suspended sediment discharge based on the position and characteristics of particle movement and summed the two parts to estimate the total sediment discharge. It is widely known that the majority of rivers throughout the world transport more suspended sediment than bed load. Too much suspended sediment may lead to reservoir deposition, scouring, and siltation in the downstream channel, which may upset the balance between flow and sediment concentration. Therefore, the estimation of suspended sediment transport is vital for the design of hydraulic structures influencing or controlling the sediment discharge regime and for the estimation of average rate of erosion in a basin.

The suspended sediment discharge can be obtained from flow discharge and suspended sediment concentration, while flow discharge can be determined with the use of velocity distribution. There are many different methods of determining the velocity distribution and sediment concentration, such as entropy based, hydraulics based, and empirical. Different methods for velocity distribution have been reviewed by

Luo and Singh (2011) and compared by Cui and Singh (2013), and different methods for sediment concentration have been discussed by Cui (2011).

## 6.2    SEDIMENT DISCHARGE

The suspended sediment discharge can be computed by integrating sediment concentration and velocity over the cross section as

$$Q_s = \int cu dA \qquad (6.1)$$

where
  $A$ is the cross-sectional area (which is a function of flow depth)
  $dA$ is the elemental cross-sectional area
  $Q_s$ is the sediment discharge
  $c$ is the sediment concentration at depth $y$
  $u$ is the velocity at depth $y$

Equation 6.1 can be simplified by using mean sediment concentration and mean velocity as

$$Q_s = \overline{cu}A \qquad (6.2)$$

where
  $\overline{c}$ is the mean sediment concentration over the flow cross-sectional area
  $\overline{u}$ is the mean velocity of the channel cross section

In Equation 6.1, the velocity distribution $u(y)$ and sediment concentration distribution $c(y)$ can be obtained empirically as well as using entropy theory. That is suspended sediment discharge can be derived using different combinations of entropy-based and empirical methods of velocity distribution and sediment concentration distribution. Hydraulics-based methods of velocity distribution will not be considered in this chapter.

One empirical velocity distribution and one entropy-based velocity distribution are considered, that are the Prandtl–von Karman universal law (von Karman, 1935; Chiu, 1987, 1989) and Cui and Singh's (2013, 2014) Tsallis (1988) entropy-based velocity distribution. One empirical sediment concentration distribution and one entropy-based method are considered to compute the suspended sediment discharge. These include the empirical (partly) Rouse equation (Rouse, 1937) and Tsallis entropy–based (Cui, 2011; Cui and Singh, 2012) sediment concentration equation.

### 6.2.1    PRANDTL–VON KARMAN VELOCITY DISTRIBUTION

The Prandtl–von Karman universal velocity distribution for open-channel flow can be expressed as

$$u = \frac{u_*}{\kappa} \ln \frac{y}{y_0} \tag{6.3}$$

where
   $u$ is the velocity at a vertical depth $y$ above the channel bed
   $u_*$ represents the shear velocity
   $\kappa$ is the von Karman universal constant
   $y_0$ is the depth of the shear velocity

Using the Shannon entropy, Singh (2011) derived an expression for $y_0$ as

$$y_0 \exp\left(\frac{k u_D}{u_*}\right) \approx D \tag{6.4}$$

where $u_D$ is the velocity at depth $y = D$. The shear velocity $u_*$ can be computed with known channel characteristics as

$$u_* = \sqrt{gDS} \tag{6.5}$$

where
   $g$ is the acceleration due to gravity
   $D$ is the flow depth
   $S$ is the friction slope that is approximated by the channel slope for uniform flow

The von Karman universal constant $\kappa$ has a value of 0.4 for clear water and a value as low as 0.2 for heavily sediment-laden water.

## 6.2.2 TSALLIS ENTROPY–BASED VELOCITY DISTRIBUTION

Using the Tsallis entropy and the principle of maximum entropy, Cui and Singh (2013) derived a 2-D velocity distribution as

$$\frac{u}{u_{max}} = 1 - \frac{1}{G}\left(1 - \left[(1-G)^{m/(m-1)} + (1-(1-G)^{m/(m-1)})\left(\frac{y}{D}\right)^{\eta}\right]^{(m-1)/m}\right) \tag{6.6}$$

where
   $\eta$ is the exponent in the cumulative distribution function defined by Cui and Singh (2012)
   $m$ is the exponent defined in the Tsallis entropy definition (or entropy parameter) and was recommended as 3 by Cui and Singh (2013)
   $G$ is a dimensionless parameter expressed as

$$G = \frac{\lambda_1 u_{max}}{\lambda_1 u_{max} + \lambda_*} \tag{6.7}$$

Similar to the entropy parameter $m$, $G$ can be used as an index of the uniformity of the velocity distribution and is related to the relationship between the mean and maximum velocity as

$$\frac{\bar{u}}{u_{max}} = \Psi(G) = \frac{1}{A} \int \left[ 1 - \frac{1}{G} \left\{ 1 - (1-G)^{m/(m-1)} + \{1 - (1-G)^{m/(m-1)}\} \left(\frac{y}{D}\right)^a \right\}^{(m-1)/m} \right] dA$$

(6.8)

Cui and Singh (2013) simplified the ratio of the mean and maximum velocity using a polynomial regression as

$$\frac{\bar{u}}{u_{max}} = 0.554G^2 - 0.077G + 0.568$$

(6.9)

which showed good agreement with observed values. With records of mean and maximum velocity for a given cross section, the dimensionless parameter $G$ can be easily determined without solving nonlinear equations for Lagrange multipliers $\lambda_1$ and $\lambda_*$. Thus, the velocity distribution (Equation 6.6) for fixed $m = 3$ can be reduced to

$$\frac{u}{u_{max}} = 1 - \frac{1}{G} \left( 1 - \left( (1 + 0.5\ln G)\left(\frac{y}{D}\right)^a - 0.5\ln G \right)^{2/3} \right)$$

(6.10)

Equation 6.10 is used to compute the sediment discharge with $G$ computed using Equation 6.9. This method is discussed in more detail in Chapter 4.

### 6.2.3 ROUSE EQUATION

The Rouse equation (Rouse, 1937), derived from the Prandtl–von Karman logarithmic velocity equation and linear shear stress distribution, can be written as

$$\frac{c}{c_a} = \left( \frac{(D-y)}{y} \frac{a}{(D-a)} \right)^{R_0}$$

(6.11)

where
  $c_a$ is the reference concentration of sediment with settling velocity $\omega_s$ at a distance $y = a$, which is arbitrarily taken as $0.05y_0$ above the bed, $y_0$ is the depth of the shear velocity defined in the Prandtl–von Karman velocity distribution
  $D$ is the flow depth
  $R_0$ is referred to as the Rouse number, which is defined as $R_0 = \omega_s/(\beta\kappa u_*)$, where $\omega_s$ is the settling velocity, $\beta$ is the coefficient of proportionality, $\kappa$ is the von Karman constant, and $u_*$ is the shear velocity

### 6.2.4 TSALLIS ENTROPY–BASED SEDIMENT CONCENTRATION

Applying the Tsallis entropy, Cui (2011) derived a sediment concentration equation as

$$\frac{c}{c_0} = 1 - \frac{1}{N}(1 - ((1 + 0.5\ln N)F(c) - 0.5\ln N)^{2/3}) \tag{6.12}$$

where
$F(c)$ is the cumulative distribution function
$N$ is a dimensionless parameter defined as

$$N = \frac{\lambda_1 c_0}{\lambda_1 c_0 + \lambda_*} \tag{6.13}$$

where $\lambda_1$ and $\lambda_*$ are the Lagrange multipliers in the probability density function of sediment concentration. As shown by Cui (2011), $N$ can be computed from the relation between the mean and maximum sediment concentration values as

$$\frac{\bar{c}}{c_0} = 0.554N^2 - 0.077N + 0.568 \tag{6.14}$$

With $N$ computed from Equation 6.14 instead of Equation 6.13 involving $\lambda_1$ and $\lambda_*$, the suspended sediment concentration can be computed using Equation 6.12.

## 6.3 SUSPENDED SEDIMENT DISCHARGE

The suspended sediment discharge can be computed by using different combinations of velocity and sediment concentration distributions. These include three different combinations: (1) Tsallis entropy–based velocity and concentration distributions, (2) Tsallis entropy–based sediment concentration distribution and Prandtl–von Karman velocity distribution, and (3) Tsallis entropy–based velocity distribution and Rouse sediment concentration equation.

### 6.3.1 TSALLIS ENTROPY–BASED VELOCITY DISTRIBUTION AND SEDIMENT CONCENTRATION DISTRIBUTION

The Tsallis entropy–based velocity distribution (Cui and Singh, 2013) given by Equation 6.10 and sediment concentration (Cui, 2011) given by Equation 6.12 are substituted into Equation 6.1 to obtain the suspended sediment discharge per unit width (designated as 1) as

$$q_s^1 = u_{\max}c_0 \int_0^D \left[ 1 - \frac{1}{N}(1 - ((1 + 0.5\ln N)F(c) - 0.5\ln N)^{2/3}) \right]$$

$$\left[ 1 - \frac{1}{G}(1 - ((1 + 0.5\ln G)F(u) - 0.5\ln G)^{2/3}) \right] dy \tag{6.15}$$

where $q_s^1$ is $Q_s$ per unit width for case 1. To get an explicit solution of Equation 6.15 is difficult; however, it can be simplified with the use of mean values. The first term in the integration can be replaced by mean sediment concentration given by Equation 6.14, and the second term can be replaced by mean velocity given by Equation 6.9, such that Equation 6.15 reduces to

$$q_s^1 = Du_{max}c_0(0.554G^2 - 0.777G + 0.568)(0.554N^2 - 0.777N + 0.568) \quad (6.16)$$

Equation 6.16 provides a simple method to compute sediment discharge and is seen to be equivalent to Equation 6.2 with mean velocity and mean sediment concentration represented by Equations 6.9 and 6.14. Entropy-based parameters $G$ and $N$ are fixed for each channel cross section (Cui, 2011). Thus, once these entropy parameters have been obtained for some known cross section, with observed maximum velocity and sediment concentration, the sediment discharge can be obtained with ease.

### Example 6.1

Compute sediment discharge for experimental data set S1 observed by Einstein and Chien (1955), using the Tsallis entropy–based velocity distribution and sediment concentration distribution.

#### Solution

Given $u_{max}$ and $u_m$, $G$ is obtained as 0.525 following the example from Chapter 2. Given $c_0$ and $c_m$, $N$ is obtained as 0.05 from Chapter 5. Thus, the sediment discharge (g/L) can be computed for the first case from Equation 6.16 as

$$q_s^1 = Du_{max}c_0(0.554G^2 - 0.777G + 0.568)(0.554N^2 - 0.777N + 0.568)$$

$$= 0.17\,(\text{ft}) \times 6.416\,(\text{ft/s}) \times 58\,(\text{g/L})(0.554 \times 0.525^2 - 0.777 \times 0.525 + 0.568)$$

$$\times (0.554 \times 0.05^2 - 0.777 \times 0.05 + 0.568)$$

$$= 1.828$$

### 6.3.2   PRANDTL–VON KARMAN VELOCITY DISTRIBUTION AND TSALLIS ENTROPY–BASED SEDIMENT CONCENTRATION DISTRIBUTION

The sediment discharge was derived from the Tsallis entropy–based sediment concentration derived by Cui (2011) in conjunction with the Prandtl–von Karman velocity distribution. Substituting Equations 6.3 and 6.12 into Equation 6.1, one obtains the sediment discharge per unit width as

$$q_s^2 = \frac{u_* c_0}{\kappa} \int_0^D \left[1 - \frac{1}{N}(1 - ((1 + 0.5\ln N)F(c) - 0.5\ln N)^{2/3}\right]\ln\frac{y}{y_0}\,dy \quad (6.17)$$

The sediment discharge equation (Equation 6.17) can be simplified by replacing the first term in the integration with the mean sediment concentration equation (Equation 6.14) as

$$q_s^2 = \frac{u_* c_0}{\kappa}(0.554N^2 - 0.077N + 0.568)\int_0^D \ln\frac{y}{y_0}\,dy \tag{6.18}$$

Integration over $y$ leads to

$$q_s^2 = \frac{u_* c_0}{\kappa}(0.554N^2 - 0.077N + 0.568)[D\ln D - \ln y_0] \tag{6.19}$$

With entropy parameter $N$ obtained and the universal constant $\kappa$, sediment discharge can be computed when shear velocity and maximum sediment concentration are given.

### Example 6.2

Compute sediment discharge using Prandtl–von Karman velocity distribution and Tsallis entropy–based sediment concentration distribution for the data given in Table 6.1.

### Solution

For additional information on velocity given in Table 6.1, $\kappa = 0.305$, $y_0 = 0.001$ ft, $D = 0.17$ ft, $c_0 = 58$ g/m³, and $u_* = 0.378$ ft/s, is given. Using $N = 0.05$ from Example 6.1, sediment discharge (g/s) is computed as

$$q_s^2 = \frac{u_* c_0}{\kappa}(0.554N^2 - 0.077N + 0.568)[D\ln D - \ln y_0]$$

$$= \frac{0.378 \times 58}{0.305}(0.554 \times 0.05^2 - 0.777 \times 0.05 + 0.568)[0.17\ln 0.17 - \ln 0.001]$$

$$= 2.0\,g/s$$

---

**TABLE 6.1**

**Statistics for Velocity and Sediment Concentration for Experimental Run S1**

| Variable | Value |
|---|---|
| $u_*$ | 0.378 ft/s |
| $u_{max}$ | 6.416 ft/s |
| $u_m$ | 4.377 ft/s |
| $c_0$ | 58 g/L |
| $c_m$ | 31.05 g/L |
| $Q$ | 1.713 g/s |

### 6.3.3 Tsallis Entropy–Based Velocity Equation and Rouse Sediment Concentration Equation

Substituting the Tsallis entropy–based velocity equation (Equation 6.10) and the Rouse equation (Equation 6.11) into Equation 6.1, the specific sediment discharge (discharge per unit width) is obtained as

$$q_s^3 = u_{max}C_0 \int_0^D \left(\frac{y_0 - y}{y}\right)^{\omega_s/(\beta\kappa u*)} \left[1 - \frac{1}{G}((1 + 0.5\ln G)F(u) - 0.5\ln G)^{2/3}\right]dy \quad (6.20)$$

Replacing the second part in integral by the mean velocity given by Equation 6.9, Equation 6.20 reduces to

$$q_s^3 = u_{max}C_0(0.554G^2 - 0.777G + 0.568)\int_0^D \left(\frac{y_0 - y}{y}\right)^{\omega_s/(\beta\kappa u*)} dy \quad (6.21)$$

The integration in the second part of Equation 6.21 can be computed numerically.

### Example 6.3

Compute sediment discharge using the Tsallis entropy–based velocity equation and the Rouse sediment concentration equation for the data used in Example 6.1.

#### Solution

For this example, the solution is not explicit, and numerical integration is needed. The sediment discharge (g/s) is computed as

$$q_s^3 = u_{max}C_0(0.554G^2 - 0.777G + 0.568)\int_0^D \left(\frac{y_0 - y}{y}\right)^{\omega_s/(\beta\kappa u*)} dy$$

$$= 6.416 \times 58(0.554 \times 0.525^2 - 0.777 \times 0.525 + 0.568)\int_0^D \left(\frac{0.001 - y}{y}\right)^{9.423} dy$$

$$= 1.266\,g/s$$

### 6.3.4 Comparison

For 15 sets of experimental sediment concentration measurements collected in a laboratory by Einstein and Chien (1955) from the sediment-laden flow, sediment discharge is computed for each combination model, as plotted in Figure 6.1. It can be seen from the figure that sediment discharge estimated by both the Tsallis entropy methods leads to the smallest errors. The accuracy of the second method using the empirical method for the velocity part is higher than that of the third one using empirical methods for sediment concentration. This shows that the sediment

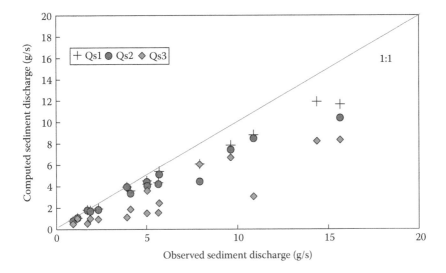

**FIGURE 6.1** Comparison of sediment discharge computed by three methods with observed sediment discharge.

concentration computed using the Rouse equation has more impact than the velocity distribution computed by the Prandtl–von Karman equation. However, a correction factor may be needed for better estimation since all the methods underestimate sediment discharge.

## 6.4   MODIFICATION FOR SEDIMENT DISCHARGE

A correction factor ω can be introduced to reduce the bias in computed sediment discharge. The correction factor can be computed by the least square method and should be consistent for each cross section and is summarized in Table 6.2. A higher correction factor implies less accuracy of the computed values.

**TABLE 6.2**
**Correction Factor for Two Methods**

| Qs1 | 1.16 |
|-----|------|
| Qs2 | 1.24 |
| Qs3 | 2.29 |

*Source:*   Einstein, H.A. and Chien, N., Effects of heavy sediment concentration near the bed on velocity and sediment distribution, MRD Series Report No.8, University of California at Berkeley and Missouri River Division, U.S. Army Corps of Engineers, Omaha, NE, 1955.

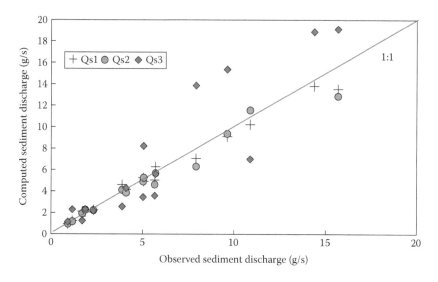

**FIGURE 6.2** Computed sediment discharge with correction factor.

The computed sediment discharge values, corrected with the use of the correction factor, become closer to the observed values, and the error is reduced by more than 60%. Figure 6.2 shows that the computed sediment discharge after correction is much closer to the 1:1 line than earlier, and the points are distributed both above and below the line for every method. Comparing with other methods, the fully entropy-based sediment discharge after correction is the closest to the observed values.

## REFERENCES

Chien, N. and Wan, Z. (2003). *Mechanics of Sediment Transport*. Science Press, Beijing, China.

Chiu, C.L. (1987). Entropy and probability concepts in hydraulics. *Journal of Hydraulic Engineering*, 113(5), 583–600.

Chiu, C.L. (1989). Velocity distribution in open channel flow. *Journal of Hydraulic Engineering*, 115(5), 576–594.

Cui, H. (2011). Estimation of velocity distribution and suspended sediment discharge in open channels using entropy. MS thesis, Texas A&M University, College Station, TX.

Cui, H. and Singh, V.P. (2012). On the cumulative distribution function for entropy-based hydrologic modeling. *Transactions of the ASABE*, 55(2), 429–438.

Cui, H. and Singh, V.P. (2013). Two dimensional velocity distribution in open channels using Tsallis entropy. *Journal of Hydrologic Engineering*, 18(2), 290–298.

Cui, H. and Singh, V.P. (2014). Computation of suspended sediment discharge in open channels by combining Tsallis entropy-based methods and empirical formulas. *Journal of Hydrologic Engineering*, 19(1), 18–25.

Einstein, H.A. (1950). The bed load function for sediment transportation in open channels. *Technical Bulletin 1026*. USDA, Soil Conservation Service, Washington, DC.

Einstein, H.A. and Chien, N. (1955). Effects of heavy sediment concentration near the bed on velocity and sediment distribution. MRD Series Report No. 8. University of California at Berkeley and Missouri River Division, U.S. Army Corps of Engineers, Omaha, NE.

Luo, H. and Singh, V.P. (2011). Entropy theory for two-dimensional velocity distribution. *Journal of Hydrologic Engineering*, 16(4), 303–317.

Rouse, H. (1937). Modern conceptions of the mechanics of turbulence. *Transactions of ASCE*, 102, 463–543.

Singh, V.P. (2011). Derivation of power law and logarithmic velocity distributions using the Shannon entropy. *Journal of Hydrologic Engineering*, 16(5), 478–483.

Tsallis, C. (1988). Possible generalization of Boltzmann-Gibbs statistics. *Journal of Statistical Physics*, 52(1–2), 479–487.

von Karman, T. (July 1935). Some aspects of the turbulence problem. *Mechanical Engineering*, 57, 407–412.

# 7 Sediment Concentration in Debris Flow

A debris flow commonly comprises more than 50% sediment by volume and the sediment particles may range in size from clay to boulders several meters in diameter (Major and Pierson, 1992). Thus, it is a dense, poorly sorted, solid–fluid mixture. In mountainous regions, prolonged heavy rainfall occurring over saturated hillslopes, earthquakes, and human activities are the main causes of debris flow. In debris flow, debris or rocks concentrate at the head of flow and move downslope with high concentration and strong destructive force. As they continue to flow downhill and through channels, with the addition of water, sand, mud, trees, boulders, and other material, they increase in volume and sediment concentration. Debris flows have a strong erosive force, grow during the movement of gathering debris eroded from the torrent bed or banks, and are capable of transporting huge volumes of sediment. Because of their power, these flows destroy whatever comes in their way; that is, they kill people and animals; decimate roads, bridges, railway tracks, homes and other property; and fill reservoirs.

Debris flow is caused in three ways: (1) the gully bed material is mobilized or the sediment particles from the gully bed are entrained by water runoff; (2) a natural dam formed by a landslide fails; and (3) a landside block is liquefied. The discussion in this chapter is restricted to the first way. Sediment concentration and sediment particle size are two fundamental factors that govern debris flow and its characteristics (Egashira et al., 2001; Takahashi, 1978). Of typical concern associated with debris flow are the formation, movement, and deposition of debris; delineation of flood zones; damage assessment; amongst others. To address these concerns, the equilibrium sediment concentration and its vertical distribution are needed. Debris flows, involving mixtures of debris and water, exhibit characteristics that are different from those of water flow or sediment-laden river flow. Although a physically based hydraulic model can be constructed for modeling debris flow, uncertainties in debris flow variables and parameters of such a model may, however, limit its potential. Lien and Tsai (2003) employed the Shannon entropy (Shannon, 1948) for debris flow modeling. This chapter employs the Tsallis entropy for the determination of sediment concentration distribution in debris flow.

## 7.1 NOTATION AND DEFINITION

Consider a debris flow over an erodible bed, as shown in Figure 7.1. It is assumed that the flow is steady and uniform, where the depth of flow is $h_0$, and the sediment concentration decreases monotonically from a maximum value of $c_m$ at the channel bottom to an arbitrary value of $c_h$ at the water surface. Let $c(y)$ be the sediment concentration at a vertical distance $y$ ($0 \leq y \leq h_0$) from the channel bed. The sediment concentration is defined as the volume of sediment divided by the volume of fluid–sediment mixture and is dimensionless. Thus, it is expressed as a fraction or in

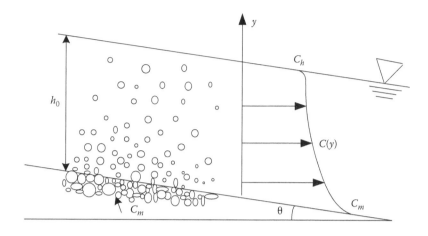

**FIGURE 7.1**   Steady uniform debris flow.

percent by volume. For application of the Tsallis entropy, it is assumed that the time-averaged sediment concentration $C$ is a random variable.

## 7.2   METHODOLOGY FOR THE DETERMINATION OF DEBRIS FLOW CONCENTRATION

Determination of debris flow concentration using the entropy theory (Singh and Cui, 2015) entails (1) definition of the Tsallis entropy, (2) specification of constraints, (3) maximization of entropy, (4) determination of the Lagrange multipliers, (5) determination of probability density function (PDF) and maximum entropy, (6) cumulative probability distribution function (CDF) hypothesis, and (7) sediment concentration distribution. Each of these components is now discussed.

### 7.2.1   DEFINITION OF TSALLIS ENTROPY

Let the concentration $C$ in debris flow be the random variable with PDF, $f(c)$. Then, the Tsallis entropy (Tsallis, 1988) of $C$, $H(C)$, can be expressed as

$$H(C) = \frac{1}{m-1}\left\{1 - \int_{c_h}^{c_m}[f(c)]^m dc\right\} = \frac{1}{m-1}\int_{c_h}^{c_m} f(c)\{1-[f(c)]^{m-1}\}dc \qquad (7.1)$$

where
   $c,\ c_h \le c \le c_m$, is the value of random variable $C$
   $c_m$ is the maximum value of $C$ or concentration at the bed
   $c_h$ is the concentration at the water surface
   $m$ is the entropy index
   $H$ is the entropy of $f(c)$ or $C$

Equation 7.1 is a measure of the uncertainty of variable $C$. The quantity $f(c)dc$ defines the probability of sediment concentration occurring between $c$ and $c + dc$. The objective is to derive $f(c)$, which is accomplished by maximizing $H$, subject to specified constraints, in accordance with the principle of maximum entropy (POME) (Jaynes, 1957).

### 7.2.2 SPECIFICATION OF CONSTRAINTS

Since $f(c)$ is a PDF, it must satisfy

$$\int_{c_h}^{c_m} f(c)dc = 1 \tag{7.2}$$

which is a statement of the total probability theorem. One of the simplest constraints is the mean or equilibrium sediment concentration by volume, denoted as $\overline{c}$ or $c_D$. The mean value may be known or obtained from observations and can be expressed as

$$\int_{c_h}^{c_m} cf(c)dc = E[c] = \overline{c} = c_D \tag{7.3}$$

In order to keep the algebra simple, additional constraints are not employed.

### 7.2.3 MAXIMIZATION OF ENTROPY

The entropy $H$ of $C$, given by Equation 7.1, can be maximized, subject to Equations 7.2 and 7.3, in accordance with POME, by employing the method of Lagrange multipliers. To that end, the Lagrangian function $L$ is expressed as

$$L = \int_{c_h}^{c_m} \frac{f(c)}{m-1}\{1-[f(c)]^{m-1}\}dc - \lambda_0\left[\int_{c_h}^{c_m} f(c)dc - 1\right] - \lambda_1\left[\int_{c_h}^{c_m} cf(c)dc - \overline{c}\right] \tag{7.4}$$

where $\lambda_0$ and $\lambda_1$ are the Lagrange multipliers. Differentiating Equation 7.4 with respect to $f$, recalling the Euler–Lagrange calculus of variation, while noting $f$ as variable and $C$ as parameter, and equating the derivative to zero, one obtains

$$\frac{\partial L}{\partial f} = 0 \Rightarrow \frac{1}{m-1}[1 - mf(c)^{m-1}] - \lambda_0 - \lambda_1 c = 0 \tag{7.5}$$

Equation 7.5 leads to

$$f(c) = \left[\frac{m-1}{m}\left(\frac{1}{m-1} - \lambda_0 - \lambda_1 c\right)\right]^{1/(m-1)} \tag{7.6}$$

Equation 7.6 is the POME-based least-biased PDF of sediment or debris flow concentration $C$.

### 7.2.4 DETERMINATION OF LAGRANGE MULTIPLIERS

Equation 7.6 has unknown $\lambda_0$ and $\lambda_1$ that can be determined with the use of Equations 7.2 and 7.3. The Lagrange multiplier $\lambda_1$ is associated with mean concentration and $\lambda_0$ with the total probability. These multipliers have opposite signs, with $\lambda_1$ being positive and $\lambda_0$ being negative. Substitution of Equation 7.6 in Equation 7.2 yields

$$\int_{c_h}^{c_m} \left[\frac{m-1}{m}\left(\frac{1}{m-1} - \lambda_0 - \lambda_1 c\right)\right]^{1/(m-1)} dc = 1 \tag{7.7}$$

Integration of Equation 7.7 yields

$$\frac{1}{\lambda_1}\left(\frac{m-1}{m}\right)^{m/(m-1)}\left[\left(\frac{1}{m-1} - \lambda_0 - \lambda_1 c_m\right)^{m/(m-1)} - \left(\frac{1}{m-1} - \lambda_0 - \lambda_1 c_h\right)^{m/(m-1)}\right] = 1 \tag{7.8}$$

Likewise, substitution of Equation 7.6 in Equation 7.3 yields

$$\int_{c_h}^{c_m} c\left[\frac{m-1}{m}(\lambda_* + \lambda_1 c)\right]^{1/(m-1)} dc = \bar{c} = c_D \tag{7.9}$$

Equation 7.9 can be integrated by parts as

$$-\lambda_1 \bar{c}\left(\frac{m}{m-1}\right)^{m/(m-1)} = c_m\left(\frac{1}{m-1} - \lambda_0 - \lambda_1 c_m\right)^{m/(m-1)} - c_h\left(\frac{1}{m-1} - \lambda_0 - \lambda_1 c_h\right)^{m/(m-1)}$$

$$+ \frac{m-1}{2m-1}\frac{1}{\lambda_1}\left[\left(\frac{1}{m-1} - \lambda_0 - \lambda_1 c_m\right)^{(2m-1)/(m-1)}\right.$$

$$\left. -\left(\frac{1}{m-1} - \lambda_0 - \lambda_1 c_h\right)^{(2m-1)/(m-1)}\right] \tag{7.10}$$

Equations 7.8 and 7.10 can be solved numerically for $\lambda_0$ and $\lambda_1$ for specified values of $\bar{c}$, $c_m$, $c_h$, and $m$.

### 7.2.5  DETERMINATION OF PROBABILITY DENSITY FUNCTION
### AND MAXIMUM ENTROPY

Integrating Equation 7.6 from $c_h$ to $c$, we obtain the CDF of C, $F(c)$, as

$$F(c) = \left(\frac{m-1}{m}\right)^{m/(m-1)} \frac{1}{\lambda_1}\left[\left(\frac{1}{m-1} - \lambda_0 - \lambda_1 c_h\right)^{m/(m-1)} - \left(\frac{1}{m-1} - \lambda_0 - \lambda_1 c\right)^{m/(m-1)}\right]$$

(7.11)

If the debris flow at the water surface is negligible, that is, $c_h = 0$, then Equation 7.11 becomes

$$F(c) = \left(\frac{m-1}{m}\right)^{m/(m-1)} \frac{1}{\lambda_1}\left[\left(\frac{1}{m-1} - \lambda_0\right)^{m/(m-1)} - \left(\frac{1}{m-1} - \lambda_0 - \lambda_1 c\right)^{m/(m-1)}\right]$$

(7.12)

Now the maximum entropy of C is obtained by inserting Equation 7.6 in Equation 7.1:

$$H(c) = \frac{1}{m-1}\left\{(c_m - c_h) + \left(\frac{m-1}{m}\right)^{m/(m-1)} \frac{1}{(2m-1)\lambda_1}\left[\left(\frac{1}{m-1} - \lambda_0 - \lambda_1 c_m\right)^{(2m-1)/(m-1)} - \left(\frac{1}{m-1} - \lambda_0 - \lambda_1 c_h\right)^{(2m-1)/(m-1)}\right]\right\}$$

(7.13)

Equation 7.13 is expressed in terms of the Lagrange multiplier $\lambda_1$, the lower limit of concentration $c_h$, and the upper limit of concentration $c_m$.

### Example 7.1

Compute and plot $f(c)$ as a function of $\lambda_1$ and $m$. Take $m = 2$ and $\lambda_1 = 0.1, 1.0, 5.0$, and 10.0. Assume $c_h = 0$ and $c_m = 1$.

### Solution

For different values of $\lambda_1$, $f(c)$ is computed using Equation 7.6 as given in Table 7.1 and is shown in Figure 7.2. It is seen that when $\lambda_1 = 0.1$, $f(c)$ tends to be uniform with $f(c = 0) = 1.0$. When $\lambda_1 = 10$, $f(c)$ approaches 0 quickly. Figure 7.2 plots the PDF of C.

Then, for different values of $m$ (=1/2, 2/3, 3/4, 2, and 3), $f(c)$ is computed using Equation 7.6 with $\lambda_1 = 0.5$ as given in Table 7.2 and shown in Figure 7.3. It is seen that when $m$ is between $0.5 \leq m \leq 0.75$, $f(c)$ is not highly sensitive to $m$.

**TABLE 7.1**

**Values of $f(c)$ for Different Values of $\lambda_1$**

| | Values of $f(c)$ with $m = 2$ | | | |
|---|---|---|---|---|
| $c$ | $\lambda_1 = 0.1$ | $\lambda_1 = 1$ | $\lambda_1 = 5$ | $\lambda_1 = 10$ |
| 0 | 1.000 | 1.212 | 1.327 | 1.375 |
| 0.1 | 1.000 | 1.171 | 1.265 | 1.308 |
| 0.2 | 1.000 | 1.129 | 1.200 | 1.237 |
| 0.3 | 1.000 | 1.084 | 1.131 | 1.162 |
| 0.4 | 1.000 | 1.038 | 1.058 | 1.082 |
| 0.5 | 1.000 | 0.990 | 0.980 | 0.995 |
| 0.6 | 1.000 | 0.939 | 0.894 | 0.900 |
| 0.7 | 1.000 | 0.885 | 0.800 | 0.794 |
| 0.8 | 1.000 | 0.828 | 0.693 | 0.671 |
| 0.9 | 1.000 | 0.767 | 0.566 | 0.520 |
| 1 | 1.000 | 0.700 | 0.400 | 0.300 |

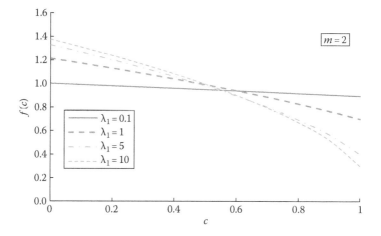

**FIGURE 7.2** Plot of $f(c)$ as a function of $\lambda_1$ and $m = 2$.

## Example 7.2

Compute and plot $f(c)$ as a function of $\lambda_1$. Take $m = 2$ and $\lambda_1 = 0.1$, 1.0, 5.0, and 10.0. Assume $c_h = 0$ and $c_m = 1$.

### Solution

For different values of $\lambda_1$, $f(c)$ is computed using Equation 7.12 with $m = 2$, as shown in Table 7.2 and Figure 7.3. For $\lambda_1 = 0.1$, $F(c)$ becomes linear (Table 7.3).

**TABLE 7.2**

**Values of f(c) for Different Values of m**

Values of $m = 2$ with $\lambda_1 = 0.5$

| c | m = 1/2 | m = 2/3 | m = 3/4 | m = 2 | m = 3 |
|---|---------|---------|---------|-------|-------|
| 0 | 1.100 | 1.152 | 1.206 | 1.333 | 1.514 |
| 0.1 | 1.078 | 1.118 | 1.159 | 1.267 | 1.436 |
| 0.2 | 1.057 | 1.086 | 1.114 | 1.200 | 1.354 |
| 0.3 | 1.037 | 1.054 | 1.072 | 1.133 | 1.267 |
| 0.4 | 1.017 | 1.024 | 1.031 | 1.067 | 1.173 |
| 0.5 | 0.998 | 0.995 | 0.992 | 1.000 | 1.071 |
| 0.6 | 0.979 | 0.967 | 0.955 | 0.933 | 0.958 |
| 0.7 | 0.961 | 0.941 | 0.920 | 0.867 | 0.829 |
| 0.8 | 0.943 | 0.915 | 0.887 | 0.800 | 0.677 |
| 0.9 | 0.926 | 0.890 | 0.854 | 0.733 | 0.479 |
| 1 | 0.909 | 0.866 | 0.824 | 0.667 | 0.000 |

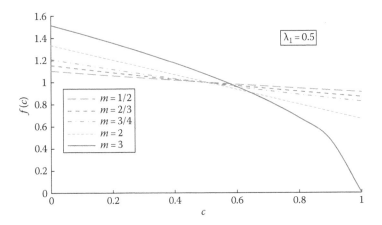

**FIGURE 7.3**   Plot of $f(c)$ as a function of $m$.

## Example 7.3

Compute and plot $F(c)$ for different values of $m = 1/2$, 2/3, 3/4, 2, and 3. Assume $\lambda_1 = 0.5$, $c_h = 0$, and $c_m = 1$.

### Solution

For different values of $m$, $F(c)$ is computed using Equation 7.12, as given in Table 7.4 and as shown in Figure 7.5. It is seen that when $m$ is between $0.5 \leq m \leq 0.75$, $F(c)$ is not sensitive to $m$.

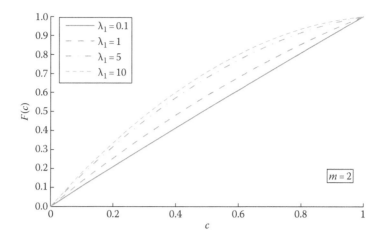

**FIGURE 7.4**  Plot of $F(c)$ as a function of $\lambda_1$ and $m = 2$.

**TABLE 7.3**
**Values of $F(c)$ for Different Values of $\lambda_1$**

| | Values of $F(c)$ with $m = 2$ | | | |
|---|---|---|---|---|
| $c$ | $\lambda_1 = 0.1$ | $\lambda_1 = 1$ | $\lambda_1 = 5$ | $\lambda_1 = 10$ |
| 0 | 0.000 | 0.000 | 0.000 | 0.000 |
| 0.1 | 0.105 | 0.130 | 0.164 | 0.175 |
| 0.2 | 0.208 | 0.253 | 0.314 | 0.333 |
| 0.3 | 0.311 | 0.370 | 0.450 | 0.475 |
| 0.4 | 0.413 | 0.480 | 0.571 | 0.600 |
| 0.5 | 0.513 | 0.583 | 0.679 | 0.708 |
| 0.6 | 0.613 | 0.680 | 0.771 | 0.800 |
| 0.7 | 0.711 | 0.770 | 0.850 | 0.875 |
| 0.8 | 0.808 | 0.853 | 0.914 | 0.933 |
| 0.9 | 0.905 | 0.930 | 0.964 | 0.975 |
| 1.0 | 1.000 | 1.000 | 1.000 | 1.000 |

## Example 7.4

Compute and plot $\lambda_0$ as a function of $\lambda_1$ and for various values of $m$. Assume $c_h = 0$ and $c_m = 1$. Take $m = 1/2, 2/3, 3/4, 2$, and 3.

### Solution

The values of $\lambda_0$ are computed using Equation 7.8 as

$$\frac{1}{\lambda_1}\left(\frac{m-1}{m}\right)^{m/(m-1)}\left[\left(\frac{1}{m-1}-\lambda_0-\lambda_1\right)^{m/(m-1)}-\left(\frac{1}{m-1}-\lambda_0\right)^{m/(m-1)}\right]=1$$

**TABLE 7.4**

**Values of F(c) for Different Values of m**

| | | Values of $F(c)$ with $\lambda_1 = 0.5$ | | | |
|---|---|---|---|---|---|
| $c$ | $m = 1/2$ | $m = 2/3$ | $m = 3/4$ | $m = 2$ | $m = 3$ |
| 0 | 0.000 | 0.000 | 0.000 | 0.000 | 0.000 |
| 0.1 | 0.122 | 0.123 | 0.123 | 0.130 | 0.146 |
| 0.2 | 0.238 | 0.239 | 0.240 | 0.253 | 0.284 |
| 0.3 | 0.349 | 0.350 | 0.351 | 0.370 | 0.414 |
| 0.4 | 0.455 | 0.456 | 0.458 | 0.480 | 0.535 |
| 0.5 | 0.556 | 0.557 | 0.559 | 0.583 | 0.646 |
| 0.6 | 0.652 | 0.654 | 0.656 | 0.680 | 0.747 |
| 0.7 | 0.745 | 0.746 | 0.748 | 0.770 | 0.836 |
| 0.8 | 0.833 | 0.834 | 0.836 | 0.853 | 0.911 |
| 0.9 | 0.918 | 0.919 | 0.920 | 0.930 | 0.968 |
| 1.0 | 1.000 | 1.000 | 1.001 | 1.000 | 1.000 |

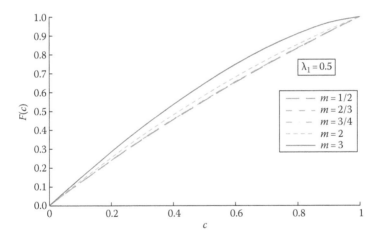

**FIGURE 7.5** Plot of $F(c)$ as a function of $m$.

This equation is solved by trial and error for various values of $m = 1/2, 2/3, 3/4, 2,$ and 3, and the values $\lambda_0$ so obtained are given in Table 7.4 and shown in Figure 7.6.

## Example 7.5

Compute and plot $H(C)$ as a function of $\lambda_0$ and for various $\lambda_1$ values. Assume $c_h = 0$ and $c_m = 1$. Take $m = 2$ and $\lambda_1 = 0.1, 1.0, 3.0,$ and $5.0$.

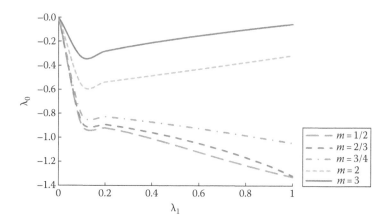

**FIGURE 7.6** Plot of $\lambda_0$ as a function of $\lambda_1$ for various values of $m$.

**TABLE 7.5**
**Values of $\lambda_0$ for Different Values of $m$**

| | Values of $\lambda_0$ | | | | |
|---|---|---|---|---|---|
| $\lambda_1$ | $m = 1/2$ | $m = 2/3$ | $m = 3/4$ | $m = 2$ | $m = 3$ |
| 1 | 0.000 | 0.000 | 0.000 | 0.000 | 0.000 |
| 2 | −0.898 | −0.871 | −0.812 | −0.566 | −0.333 |
| 3 | −0.923 | −0.894 | −0.829 | −0.541 | −0.285 |
| 4 | −0.962 | −0.925 | −0.851 | −0.514 | −0.246 |
| 5 | −1.011 | −0.961 | −0.874 | −0.488 | −0.213 |
| 6 | −1.067 | −1.003 | −0.900 | −0.461 | −0.183 |
| 7 | −1.124 | −1.051 | −0.927 | −0.434 | −0.156 |
| 8 | −1.181 | −1.105 | −0.956 | −0.406 | −0.130 |
| 9 | −1.236 | −1.167 | −0.986 | −0.378 | −0.106 |
| 10 | −1.287 | −1.239 | −1.017 | −0.350 | −0.082 |

**Solution**

For different values of $\lambda_1$, $H(C)$ is computed using Equation 7.13, as given in Table 7.6 and shown in Figure 7.7. The Tsallis entropy value increases with $\lambda_0$ (Table 7.6).

### 7.2.6 CUMULATIVE PROBABILITY DISTRIBUTION HYPOTHESIS

It is hypothesized that the probability of debris flow concentration being less than or equal to a given value $c$ can be expressed as $(h_0 - y)/h_0$. Then the cumulative distribution function of $C$, $F(c)$, in terms of flow depth can be written as

$$F(c) = \frac{h_0 - y}{h_0} = 1 - \frac{y}{h_0} \tag{7.14}$$

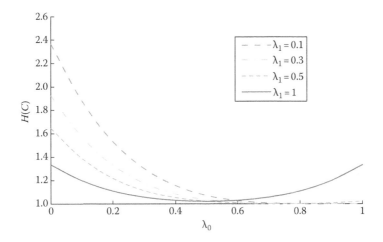

**FIGURE 7.7**   Plot of $H(C)$ as a function of $\lambda_0$ for various values of $\lambda_1$.

**TABLE 7.6**
**Values of $H(C)$ for Different Values of $\lambda_1$**

| | Values of $H(C)$ | | | |
| --- | --- | --- | --- | --- |
| $\lambda_0$ | $\lambda_1 = 0.1$ | $\lambda_1 = 0.3$ | $\lambda_1 = 0.5$ | $\lambda_1 = 1$ |
| 1 | 2.365 | 1.924 | 1.646 | 1.333 |
| 2 | 1.876 | 1.570 | 1.387 | 1.197 |
| 3 | 1.532 | 1.329 | 1.217 | 1.109 |
| 4 | 1.301 | 1.175 | 1.112 | 1.057 |
| 5 | 1.155 | 1.084 | 1.052 | 1.029 |
| 6 | 1.070 | 1.034 | 1.021 | 1.021 |
| 7 | 1.026 | 1.011 | 1.007 | 1.029 |
| 8 | 1.007 | 1.003 | 1.002 | 1.057 |
| 9 | 1.001 | 1.000 | 1.002 | 1.109 |
| 10 | 1.000 | 1.000 | 1.007 | 1.197 |

Equating Equation 7.14 to the CDF derived as Equation 7.11, we obtain

$$F(c) = \left(\frac{m-1}{m}\right)^{m/(m-1)} \frac{1}{\lambda_1} \left[ \left(\frac{1}{m-1} - \lambda_0 - \lambda_1 c_h\right)^{m/(m-1)} - \left(\frac{1}{m-1} - \lambda_0 - \lambda_1 c\right)^{m/(m-1)} \right]$$

$$= 1 - \frac{y}{h_0}$$

$$(7.15)$$

### 7.2.7   SEDIMENT CONCENTRATION DISTRIBUTION

Let $\lambda_* = (1/(m-1)) - \lambda_1$. Then, Equation 7.15 can be written as

$$c = \frac{\lambda_*}{\lambda_1} - \frac{1}{\lambda_1}\frac{m}{m-1}\left\{-\lambda_1\frac{m-1}{m}\left(1-\frac{y}{h_0}\right)+\left[\frac{m-1}{m}(\lambda_*-\lambda_1 c_h)\right]^{m/(m-1)}\right\}^{(m-1)/m} \tag{7.16}$$

If $c_h = 0$, Equation 7.16 reduces to

$$c = \frac{\lambda_*}{\lambda_1} - \frac{1}{\lambda_1}\frac{m}{m-1}\left\{-\lambda_1\frac{m-1}{m}\left(1-\frac{y}{h_0}\right)+\left(\frac{m-1}{m}\lambda_*\right)^{m/(m-1)}\right\}^{(m-1)/m} \tag{7.17}$$

Equation 7.17 is the debris flow concentration distribution defined in terms of flow depth.

## 7.3   REPARAMETERIZATION

The debris flow concentration distribution can be simplified by using a dimensionless entropy parameter defined as

$$\mu = \frac{\lambda_1 c_m}{\lambda_1 c_m - \lambda_*} \tag{7.18}$$

Dividing Equation 7.17 by $c_m$, we obtain

$$\frac{c}{c_m} = \frac{\lambda_*}{\lambda_1 c_m} - \frac{1}{\lambda_1 c_m}\frac{m}{m-1}\left\{-\lambda_1\frac{m-1}{m}\left(1-\frac{y}{h_0}\right)+\left[\frac{m-1}{m}\lambda_*\right]^{m/(m-1)}\right\}^{(m-1)/m} \tag{7.19}$$

Since $\lambda_*/\lambda_1 c_m = 1 - (1/\mu)$ from Equation 7.18, Equation 7.19 can be recast as

$$\frac{c}{c_m} = 1 - \frac{1}{\mu}\left(1-\left\{\left(\frac{m}{m-1}\right)^{m/(m-1)}\frac{\mu}{c_m}\left(-\frac{y}{h_0}\right)+1\right\}^{(m-1)/m}\right) \tag{7.20}$$

If $c_h = 0$ at $y = h_0$, Equation 7.20 reduces to

$$0 = 1 - \frac{1}{\mu}\left\{1-\left[1-\left(\frac{m}{m-1}\right)^{m/(m-1)}\frac{\mu}{c_m}\right]^{(m-1)/m}\right\} \tag{7.21}$$

Equation 7.21 suggests that

$$\left(\frac{m}{m-1}\right)^{m/(m-1)}\frac{\mu}{c_m} = 1 - (1-\mu)^{m/(m-1)} \tag{7.22}$$

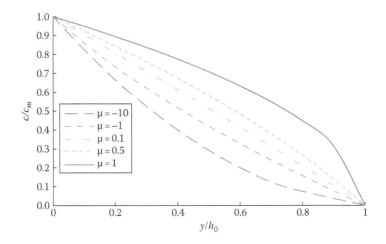

**FIGURE 7.8**   Plot of $y/h_0$ versus $c/c_m$ for various values of $\mu$.

Substituting Equation 7.22 into Equation 7.20, the dimensionless sediment concentration distribution with $c_h = 0$ becomes

$$\frac{c}{c_m} = 1 - \frac{1}{\mu}\left(1 - \left\{(1-\mu)^{m/(m-1)} + [1 - (1-\mu)^{m/(m-1)}]\left(1 - \frac{y}{h_0}\right)\right\}^{(m-1)/m}\right) \qquad (7.23)$$

Equation 7.23 expresses the debris flow concentration distribution as a function of vertical distance $y$.

### Example 7.6

Plot Equation 7.23 for $\mu = -10$ to $+1$.

#### Solution

The dimensionless debris flow concentration distribution $c/c_m$ is plotted as a function of $y/h_0$, as shown in Figure 7.8. The distribution is not sensitive to $\mu$ beyond the range $(-10, 1)$. As $\mu$ tends to zero, $c/c_m$ decreases linearly with $y/h_0$. This suggests that parameter $\mu$ can be regarded as a measure of the uniformity of sediment concentration distribution.

## 7.4   EQUILIBRIUM DEBRIS FLOW CONCENTRATION

Now the dimensionless equilibrium sediment concentration can be derived. Inserting Equation 7.23 in Equation 7.3 and integrating with the condition that $C = c_m$ at $y = 0$ and $C = c_h$ at $h = h_0$, we obtain

$$\frac{c_D}{c_m} = 1 - \frac{1}{\mu}\left[1 - \frac{m}{2m-1}\frac{[1-(1-\mu)^{(2m-1)/(m-1)}]}{[1-(1-\mu)^{m/(m-1)}]}\right] \qquad (7.24)$$

## Example 7.7

Compute and plot $c_D/c_m$ as a function of $\mu$ for different values of $m$.

### Solution

For $m = 1/2$, $2/3$, $3/4$, 2, and 3, $c_D/c_m$ is computed using Equation 7.24, as shown in Table 7.7 and Figure 7.9.

It can be seen from the figure that the value of $c_D/c_m$ monotonically increases with increasing $\mu$ and it is not sensitive to $\mu$ beyond the range $(-5, 1)$. When $m > 1$,

**TABLE 7.7**

**Values of $c_D/c_m$ for Different Values of $m$**

| $\mu$ | $m = 1/2$ | $m = 2/3$ | $m = 3/4$ | $m = 2$ | $m = 3$ |
|---|---|---|---|---|---|
| −5 | 0.193 | 0.143 | 0.093 | −0.046 | −0.377 |
| −4.5 | 0.206 | 0.154 | 0.102 | 0.018 | −0.247 |
| −4 | 0.220 | 0.167 | 0.113 | 0.085 | −0.120 |
| −3.5 | 0.237 | 0.182 | 0.126 | 0.156 | 0.004 |
| −3 | 0.257 | 0.200 | 0.143 | 0.230 | 0.125 |
| −2.5 | 0.280 | 0.222 | 0.164 | 0.308 | 0.243 |
| −2 | 0.308 | 0.250 | 0.192 | 0.389 | 0.358 |
| −1.5 | 0.341 | 0.286 | 0.231 | 0.474 | 0.470 |
| −1 | 0.381 | 0.333 | 0.286 | 0.562 | 0.579 |
| −0.5 | 0.432 | 0.400 | 0.368 | 0.654 | 0.685 |
| 0.01 | 0.502 | 0.503 | 0.503 | 0.749 | 0.788 |
| 0.5 | 0.619 | 0.667 | 0.714 | 0.848 | 0.888 |
| 1 | 0.985 | 0.990 | 0.995 | 0.950 | 0.985 |

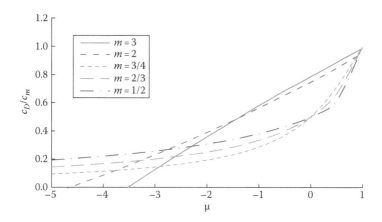

**FIGURE 7.9** Plot of $c_D/c_m$ as a function of $\mu$ for various values of $m$.

$c_D/c_m$ is highly sensitive to $\mu$ and increases from 0 to 1 rapidly. When $m < 1$, $c_D/c_m$ increases slowly from 0.1 to 0.5 until $\mu$ becomes equal to 0, and then it rapidly goes to 1.

Takahashi (1978) theoretically derived a relation for computing the equilibrium sediment concentration of debris flow occurring due to the mobilization of bed particles by water. His relation can be expressed as

$$c_D = \frac{\rho \tan\theta}{(\rho_s - \rho)(\tan\phi - \tan\theta)} \tag{7.25}$$

where
   $\theta$ is the angle of inclination of the channel bed from the horizontal
   $\phi$ is the angle of internal friction
   $\rho_s$ is the density of sediment
   $\rho$ is the density of water

Although Equation 7.25 has been widely used for calculating the equilibrium sediment concentration $c_D$ at the forefront part of debris flow, in a steady uniform state it yields unrealistic results in some cases.

## Example 7.8

Compute $c_D$ when $\tan\phi = 0.756$, $\tan\theta = 0.466$ ($\theta = 25°$), $\rho_s = 2.6$ g/cm$^3$, and $\rho = 1$ g/cm$^3$ (Lien and Tsai, 2003).

### Solution
Substituting the given values in Equation 7.25, one obtains

$$c_D = \frac{1 \times 0.466}{(2.6 - 1) \times (0.756 - 0.466)} = 1.004 \approx 1.0$$

This value implies that $c_D$ may be greater than unity that obviously is unrealistic, for flow cannot occur when the sediment concentration is that high.

## Example 7.9

Compute and plot $c_D/c_m$ as a function of $\tan\theta$ (%) [$\phi = 0.06$.] for the data observed by Takahashi (1978), as tabulated in Table 7.8. Take $c_m = 0.756$, $\rho_s = 2.6$ g/cm$^3$. Plot the $c_D/c_m$ as a function of $\mu$ in the same plot for comparison.

### Solution
The values of $c_D/c_m$ are computed using Equation 7.25, as given in Table 7.9 and shown in Figure 7.10. The value of $c_D/c_m$ computed from Equation 7.24 is plotted for $\mu$ changing from −4 to 2 as well in Figure 7.10.

It is seen that the estimation using the Tsallis entropy with $m = 3$ fits the observations with $r^2 = 0.873$, higher than that ($r^2 = 0.828$) of Equation 7.25. It suggests that entropy parameter $\mu$ is related to the inclination of the channel bed from the horizontal.

**TABLE 7.8**

**Observed $c_D/c_m$ with tan θ**

| tan θ (%) | $c_D/c_m$ | tan θ (%) | $c_D/c_m$ |
|-----------|-----------|-----------|-----------|
| 23.08 | 0.15 | 30.58 | 0.65 |
| 23.08 | 0.25 | 36.54 | 0.66 |
| 12.31 | 0.20 | 36.35 | 0.69 |
| 15.96 | 0.21 | 36.35 | 0.76 |
| 15.77 | 0.28 | 40.38 | 0.76 |
| 23.08 | 0.43 | 40.58 | 0.80 |
| 26.92 | 0.52 | 40.58 | 0.82 |
| 30.77 | 0.56 | 46.73 | 0.83 |
| 30.58 | 0.58 | 46.73 | 0.81 |
| 30.58 | 0.61 | 46.73 | 0.78 |

**TABLE 7.9**

**$c_D/c_m$ Is Computed from Equation 7.25**

| tan θ (%) | $c_D/c_m$ | tan θ (%) | $c_D/c_m$ |
|-----------|-----------|-----------|-----------|
| 5 | 0.059 | 35 | 0.689 |
| 10 | 0.114 | 40 | 0.864 |
| 15 | 0.189 | 45 | 1.059 |
| 20 | 0.284 | 50 | 1.274 |
| 25 | 0.399 | 55 | 1.509 |
| 30 | 0.534 | 60 | 1.764 |

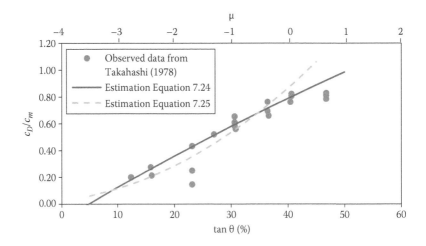

**FIGURE 7.10** Plot of $c_D/c_m$ as a function of tan θ (%) [φ = 0.06].

From flume experiments, Ou and Mizuyama (1994) developed an empirical relation for global sediment concentration using the channel bed slope as the principal factor:

$$c_{DT} = \frac{4.3(\tan\theta)^{1.5}}{1 + 4.3c_m(\tan\theta)^{1.5}} \tag{7.26}$$

where $c_{DT}$ is the average global sediment concentration of debris flow. This equation has been found to yield reasonable values of concentration even at higher channel bed slopes.

## Example 7.10

Compute and plot $c_{DT}$ as a function of $\tan\theta$ (%) [$\phi = 0.06$] for various values of $c_m$. Take $c_m = 0.5, 0.6, 0.7, 0.8$, and $0.9$.

### Solution

For various values of $c_m$, $c_{DT}$ is computed using Equation 7.26, as given in Table 7.10, and Figure 7.11 plots $c_{DT}$ as a function of $\tan\theta$ (%) [$\phi = 0.06$] for various values of $c_m$.

## Example 7.11

Compute and plot $c_D/c_m$ as a function of $\tan\theta$ (%) [$\phi = 0.06$] using Equation 7.26. Plot $c_D/c_m$ as a function of $\mu$ in the same plot for comparison.

### Solution

The values of $c_D/c_m$ are computed using Equation 7.26, as given in Table 7.11 and shown in Figure 7.12. The value of $c_D/c_m$ computed from Equation 7.24 is plotted for $\mu$ changing from −4 to 2 as shown in Figure 7.12.

Using the concept developed by Bagnold (1954) and experimental data of Takahashi (1978) and Ou and Mizuyama (1994), Lien and Tsai (2000) derived an

---

### TABLE 7.10
### Values of $c_{DT}$ as a Function of $\tan\theta$ (%) [$\phi = 0.06$] for Various Values of $c_m$

| $\tan\theta$ | $c_m = 0.5$ | $c_m = 0.6$ | $c_m = 0.7$ | $c_m = 0.8$ | $c_m = 0.9$ |
|---|---|---|---|---|---|
| 0 | 0.000 | 0.000 | 0.000 | 0.000 | 0.000 |
| 10 | 0.127 | 0.126 | 0.124 | 0.123 | 0.121 |
| 20 | 0.323 | 0.312 | 0.303 | 0.294 | 0.286 |
| 30 | 0.522 | 0.496 | 0.473 | 0.451 | 0.432 |
| 40 | 0.705 | 0.658 | 0.618 | 0.582 | 0.550 |
| 50 | 0.864 | 0.795 | 0.736 | 0.686 | 0.642 |
| 60 | 1.000 | 0.909 | 0.833 | 0.769 | 0.714 |

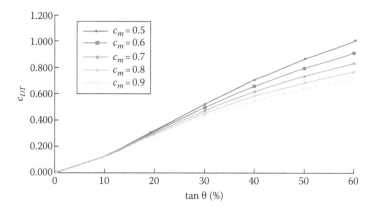

**FIGURE 7.11** Plot of $c_{DT}$ as a function of $\tan \theta$ (%) [$\phi = 0.06$].

---

**TABLE 7.11**

**$c_D/c_m$ Is Computed from Equation 7.26**

| tan θ (%) | $c_D/c_m$ | tan θ (%) | $c_D/c_m$ |
|---|---|---|---|
| 5 | 0.049 | 35 | 0.520 |
| 10 | 0.139 | 40 | 0.583 |
| 15 | 0.224 | 45 | 0.641 |
| 20 | 0.305 | 50 | 0.695 |
| 25 | 0.381 | 55 | 0.744 |
| 30 | 0.453 | 60 | 0.789 |

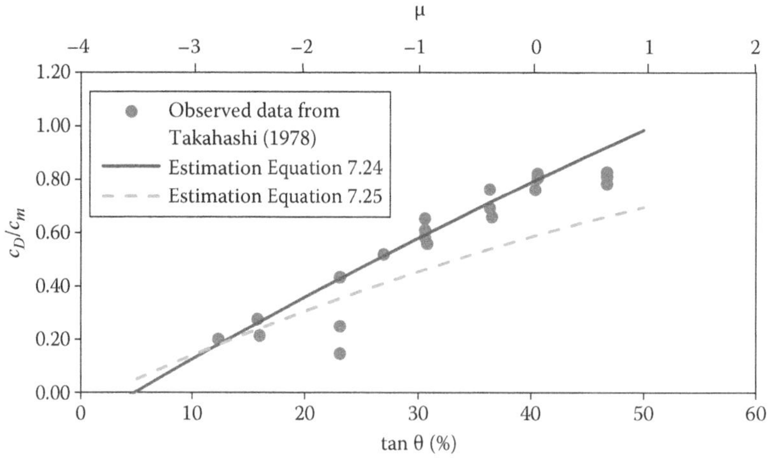

---

**FIGURE 7.12** Plot of $c_D/c_m$ as a function of $\tan \theta$ (%) [$\phi = 0.06$].

equilibrium sediment concentration equation for simulating the forefront part as well as the global average of debris flow. Lien and Tsai (2003) further modified this equation as follows.

Because of the balance in particle exchange between debris flow and channel bed at the dynamic equilibrium condition, the sediment concentration in debris flow tends to attain saturation. Under this condition, the effective shear stress in debris flow that acts on sediment particles resting on the channel bed is balanced by the critical shear stress of the particles. This means that sediment particles are in incipient motion. One can then express

$$T - F \tan\alpha = \tau_c \tag{7.27}$$

where

$$T = [(\rho_s - \rho)c_D + c]gh_0 \sin\theta \tag{7.28}$$

$$F = (\rho_s - \rho)gc_Dh_0 \cos\theta \tag{7.29}$$

where
  tan α is the dynamic friction coefficient varying from 0.32 to 0.75
  $T$ is the particle shear stress
  $F$ is the normal stress
  $\tau_c$ is the critical stress for the incipient motion of grains in the channel bed

## Example 7.12

Compute and plot $F$, $T$, and $\tau_c$ as functions of tan θ. Let $h_0 = 1$ m, $\rho_s = 2.6$ g/cm³.

### Solution

The average sediment concentration $c_D$ is computed using Equation 7.25, and substituting in Equations 7.28 and 7.29, $F$ and $T$ are computed. Then, $\tau_c$ is computed using Equation 7.27, as given in Table 7.12 and shown in Figure 7.13 that plots $F$, $T$, and $\tau_c$ as functions of θ.

For the incipient motion of uniformly sized bed material, Shields (1936) expressed the internal friction ψ as

$$\psi = \frac{\tau_c}{(\rho_s - \rho)gd_s} \tag{7.30}$$

where
  ψ is called the Shields parameter, representing the angle of inclination of channel bed from the horizontal
  $\tau_c$ is the shear stress
  $d_s$ is the particle diameter

**TABLE 7.12**

**Values of $F$, $T$, and $\tau_c$ as Functions of $\theta$**

| tan θ (%) | T | F | $\tau_c$ |
|---|---|---|---|
| 0 | 0.000 | 0.000 | 0.000 |
| 10 | 1.991 | 1.632 | 0.963 |
| 20 | 4.671 | 3.616 | 2.393 |
| 30 | 8.158 | 5.635 | 4.608 |
| 40 | 11.938 | 6.712 | 7.709 |
| 50 | 15.126 | 6.387 | 11.103 |
| 60 | 17.578 | 5.244 | 14.275 |

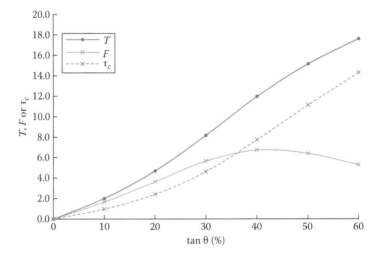

**FIGURE 7.13**    Plot of $F$, $T$, and $\tau_c$ as functions of $\theta$.

If the flow is fully developed, $\psi$ yielded by Equation 7.30 ranges from 0.04 to 0.06. Inserting Equations 7.28 and 7.29 in Equation 7.27, one obtains

$$\tau_c = [(\rho_s - \rho)c_D + c]gh_0 \sin\theta - (\rho_s - \rho)gc_D h_0 \cos\theta \tan\alpha \qquad (7.31)$$

Equating to Equation 7.30, one obtains

$$\tau_c = [(\rho_s - \rho)c_D + c]gh_0 \sin\theta - (\rho_s - \rho)gc_D h_0 \cos\theta \tan\alpha = \psi(\rho_s - \rho)gd_s \qquad (7.32)$$

By rearranging Equation 7.32, the equilibrium sediment concentration can be expressed as

$$\frac{c_D}{c_m} = \frac{1}{2}[(1 + \chi + \beta) \pm \sqrt{(1 + \chi + \beta)^2 - 4\alpha}] \qquad (7.33)$$

where

$$\chi = \frac{\rho \tan\theta}{c_m(\rho_s - \rho)(\tan\alpha - \tan\theta)} \tag{7.34}$$

and

$$\beta = \frac{\eta}{c_m \cos\theta(\tan\alpha - \tan\theta)} \tag{7.35}$$

where $\eta$ is a parameter obtained experimentally and $\tan\alpha$ is the dynamic friction coefficient shown in Equation 7.27. If $\tan\theta = \tan\alpha$, the equilibrium concentration is then given by

$$\frac{c_D}{c_m} = \frac{1}{1 + \eta[(\rho_s - \rho)/\rho](1/\sin\theta)} \tag{7.36}$$

## Example 7.12

Take $\eta = 0.04$. Compute and plot $c_D/c_m$ versus $\tan\theta$ (%) using Equation 7.33.

### Solution

$c_D/c_m$ is computed using Equation 7.33 and $\chi$ and $\beta$ are computed using Equations 7.34 and 7.35 for different values of $\tan\theta$ (%), as shown in Table 7.13 and Figure 7.14.

## Example 7.13

Compute and plot $c_D/c_m$ versus $\tan\theta$ (%) using Equation 7.36. Take $\eta = 0.04$.

### Solution

The values of $c_D/c_m$ are computed using Equation 7.36 for different values of $\tan\theta$ (%), as given in Table 7.14 and shown in Figure 7.15.

**TABLE 7.13**

$c_D/c_m$, $\chi$, and $\beta$ for Different Values of $\tan\theta$ (%)

| $\tan\theta$ | $\chi$ | $\beta$ | $c_D/c_m$ |
|---|---|---|---|
| 0 | 0.000 | 0.084 | 0.000 |
| 10 | 0.156 | 0.100 | 0.140 |
| 20 | 0.385 | 0.125 | 0.324 |
| 30 | 0.752 | 0.167 | 0.548 |
| 40 | 1.438 | 0.248 | 0.738 |
| 50 | 3.180 | 0.455 | 0.837 |
| 60 | 16.534 | 2.057 | 0.884 |

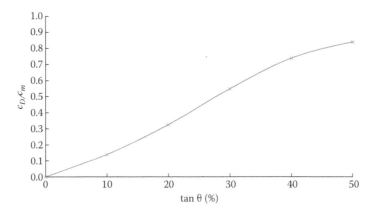

**FIGURE 7.14**  Plot of $c_D/c_m$ versus tan θ (%) using Equation 7.31.

**TABLE 7.14**
**Values of $c_D/c_m$ for Various Values of tan θ (%)**

| tan θ (%) | $c_D/c_m$ | tan θ (%) | $c_D/c_m$ |
|-----------|-----------|-----------|-----------|
| 5 | 0.064 | 35 | 0.617 |
| 10 | 0.138 | 40 | 0.699 |
| 15 | 0.225 | 45 | 0.763 |
| 20 | 0.323 | 50 | 0.806 |
| 25 | 0.424 | 55 | 0.820 |
| 30 | 0.524 | 60 | 0.803 |

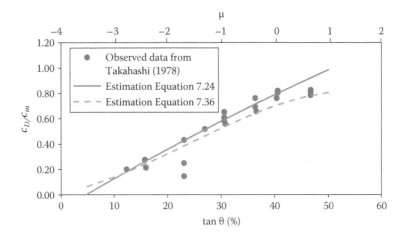

**FIGURE 7.15**  Plot of $c_D/c_m$ versus tan θ (%) using Equation 7.36.

It can be seen from Figure 7.14 that the estimated $c_D/c_m$ using Equation 7.36 is slightly lower than the observed value when $\theta$ is larger than 30°. However, the estimation using Equation 7.36 has an $r^2 = 0.880$, slightly larger than the entropy method (Equation 7.24 of $r^2 = 0.873$). There are two values observed at $\tan \theta = 23.08\%$ that fall far apart from the whole data trend in Figure 7.15. If these are considered as outliers, then $r^2$ increases to 0.949 for the entropy method and to 0.946 for Equation 7.36.

## REFERENCES

Bagnold, R.A. (1954). Experiments on gravity-free dispersion of large solid spheres in a Newtonian fluid under shear. *Proceedings*, Series A, Royal Society, London, U.K. Vol. 225, pp. 49–63.

Egashira, S., Itoh, T., and Takeuchi, H. (2001). Transition mechanism of debris flow over rigid bed to over erodible bed. *Physics and Chemistry of Earth*, 26(2), 169–174.

Jaynes, E.T. (1957). Information theory and statistical mechanics I. *Physical Review*, 106, 620–630.

Lien, H.P. and Tsai, F.W. (2000). Debris flow control by using slit dams. *International Journal of Sediment Research*, 15(4), 391–407.

Lien, H.P. and Tsai, F.W. (2003). Sediment concentration distribution of debris flow. *Journal of Hydraulic Engineering*, 129(12), 995–1000.

Major, J.J. and Pierson, T. (1992). Debris flow rheology: experimental analysis of fine-grained slurries. *Water Resources Research*, 28(5), 841–857.

Ou, G. and Mizuyama, T. (1994). Predicting the average sediment concentration of debris flow. *Journal of Japanese Erosion Control Engineering Society*, 47(4), 9–13 (Japanese).

Shannon, C.E. (1948). The mathematical theory of communications, I and II. *Bell System Technical Journal*, 27, 379–423.

Shields, A. (1936). Anwendung der ahnlichkeitsmechanik und der turbulenzforschung auf die geschiebebewegung, Mitt. Der Preuss. Versuchanstait der Wasserbau und Schiffbau, Berlin, Germany, 26, pp. 98–109 (in German).

Takahashi, T. (1979). Mechanical characteristics of debris flow. *Journal of Hydraulics Division*, ASCE, 104(8), 1153–1167.

## ADDITIONAL READING

Chen, C.L. (1988). General solutions for viscoelastic debris flow. *Journal of Hydraulic Engineering*, 113(5), 259–281.

Egashira, S., Miyamoto, K., sand Itoh, T. (1997). Constitutive equations of debris flow ad their applicability. *Proceedings of the First International Conference on Debris Flow Hazard Mitigation: Mechanics, Prediction and Assessment*, ASCE, San Francisco, California, pp. 340–347.

Jin, M. and Fread, D.L. (1999). Modeling of mud/debris unsteady flows. *Journal of Hydraulic Engineering*, 125(8), 827–834.

Singh, V.P. and Cui, H. (2015). Modeling sediment concentration in debris flow by Tsallis entropy. *Physica A*, 420, 49–58.

Takahashi, T. (1978). Mechanical characteristics of debris flow. *Journal of Hydraulic Engineering*, 104(8), 1153–1167.

Tsallis, C. (1988). Possible generalizations of Boltzmann-Gibbs statistics. *Journal of Statistical Physics*, 52(1/2), 479–487.

Tsubaki, T., Hashimoto, H., and Suetsugi, T. (1982). Grain stresses and flow properties of debris flow. *Proceedings of the Japanese Society of Civil Engineers*, San Francisco, California, Vol. 317, pp. 79–91 (in Japanese).

Wan, Z. and Wang, Z. (1994). *Hyperconcentrated Flow*, Monograph Series of IAHR. Balkema, Rotterdam, the Netherlands.

# 8 Stage–Discharge Relation

The stage–discharge relation, often called the rating curve, is employed for myriad purposes, including the determination of discharge for a prescribed stage in natural and engineered channels; calibration of physically based hydraulic and hydrological models; catchment routing; evaluation of flood inundation and floodplain mapping; constructing continuous records of discharge, sediment discharge, or sediment concentration; construction of pollutant graphs; estimation of storage variation; hydraulic design; and damage assessment (Singh, 1993). There are different types of rating curves, such as stage–discharge relation, sediment rating curve (Kazama et al., 2005), pollutant rating curve, and drainage basin rating curve. Since rating curves are of similar form from an algebraic viewpoint, fundamental to most rating curves is the estimation of discharge. The rating curves employed in practice are either of parabolic or of power form. Parameters of these curves are determined either graphically or using least-square, maximum likelihood, pseudo-maximum likelihood, or segmentation method (Petersen-Overlier and Reitan, 2005). The objective of this chapter is to present the derivation of rating curves using the Tsallis entropy. The entropy theory permits a probabilistic characterization of the rating curve and hence the probability density function (PDF) of discharge underlying the curve. It also permits a quantitative assessment of the uncertainty of discharge obtained from the rating curve.

## 8.1 METHODS FOR THE CONSTRUCTION OF RATING CURVES

There are several methods for deriving a rating curve that can be classified as graphical, hydraulic, artificial intelligence, and statistical.

### 8.1.1 GRAPHICAL METHOD

The graphical method involves plotting observed discharge and stage data on a graph paper and fitting an equation to the data collected, and it is commonly used to construct rating curves.

### 8.1.2 HYDRAULIC METHOD

The hydraulic method uses dimensional analysis or the mass and momentum conservation equations. Baiamonte and Ferro (2007) derived a stage–discharge relation from flume measurements on a sloping channel using dimensional analysis and the concept of self-similarity. Liao and Knight (2007) suggested three formulae for rating curves for prismatic channels. Petersen-Overlier (2004) used nonlinear regression and Jones formula to account for hysteresis due to unsteady flow. The U.S.

Geological Survey (USGS) used a simplified hydraulic approach for estimating peak discharge in the absence of direct measurements, such as during floods. Discharge is determined from a 1-D flow model based on Manning's roughness, measurements of channel geometry, and water surface elevation (Rantz, 1982). A similar method involves step-backwater surface models and Manning's $n$ for defining the shape of rating curves for stages where no measurements are made (Bailey and Ray, 1966). Indirect methods of discharge estimates entail extrapolation on estimated empirical roughness coefficients that can significantly vary (Jarrett, 1984).

Hydraulic models rely on empirical roughness parameterization for a specific flow condition, do not express roughness as a function of stage, and may, therefore, not accurately generate the complete rating curve. Kean and Smith (2005) developed a hydraulic method for generating curves for geomorphologically stable channels in which channel roughness is determined from field measurements of channel geometry; the physical roughness of the bed, banks and floodplains; and vegetation density on the banks and floodplain. They obtained accurate discharge estimates at two USGS gaging stations on White Water River, Kansas, United States, which provided.

### 8.1.3 Artificial Intelligence Methods

Artificial intelligence techniques have recently been employed for constructing rating curves. These include artificial neural network (ANN), genetic algorithm (GA), gene expression (GE), gene expression programming (GEP), and fuzzy logic. Bhattacharya and Solomatine (2000) used an ANN; Jain and Chalisgaonkar (2000) employed a three-layered forward ANN; Sudheer and Jain (2003) used an ANN with radial basis functions; Sahoo and Ray (2006) applied feed forward and back propagation and radial basis function ANNs; and Habib and Meselhe (2006) used ANNs and regression analysis to derive rating curves. Deka and Chandramouli (2003) compared an ANN, a modularized ANN, a conventional rating curve method, and a neuro-fuzzy method for deriving rating curves. Bhattacharya and Solomatine (2005) found ANNs and model tree 5 (M5) to be more accurate for constructing rating curves.

Guven and Aytek (2009) used GA for Schuylkill River at Berne, Pennsylvania. Lohani et al. (2006) employed the Takagi-Sugano (T5) fuzzy inference system for deriving rating curves for Narmada River in India. Ghimire and Reddy (2010) compared GA and M5 with GEP, multiple linear regression, and conventional stage–discharge relationship method. Azamathulla et al. (2011) compared GEP with GP, ANN, and two conventional methods. Sivapragasam and Muttil (2005) employed a support vector machine (SVM) for extrapolating rating curves that were tested at three gaging stations in Washington and found SVM to be better than the widely used logarithmic method, a higher-order polynomial, and ANN.

## 8.2 ERRORS IN RATING CURVES

A rating curve is often taken as a fixed curve, at least for a certain period. The rating curve may change with time and may not account for hysteresis in flow, and therefore, kinematic rating curves are not capable of representing looped conditions.

Hence, the field rating curve may not be accurate in determining streamflow when the stream bed profile and side slope characteristics change. It is prone to errors due to a number of factors: (1) errors in discharge measurements (Sauer and Meyer, 1992); (2) selection of a stable river cross-section; (3) maintenance of the stable cross-section; (4) abrupt changes in controls and submergence of controls causing irregularities in the slope of the stage–discharge relation; (5) variation in discharge for a given stage due to variations in slope, velocity, or channel conditions; (6) lack of permanent control; and (7) existence of more than one control for high and low flows (Yoo and Park, 2010). Using measurements containing errors and outliers, Sefe (1996) derived a single routing curve for Ukavaiigo River at Mehembo, Botswana. Hershey (1995) investigated errors in discharge due to errors in velocity and depth measurements.

## 8.3 FORMS OF RATING CURVES

A rating curve at a gaging station on a channel dominated by friction is normally expressed in power form (Kennedy, 1964) as

$$Q = a(y - y_0)^b + c \tag{8.1}$$

where
   $Q$ is the discharge ($L^3/T$, e.g. ft$^3$/s or m$^3$/s)
   $y$ is the stage or height of water surface (L, e.g. ft or m)
   $y_0$ is the height (L) when discharge is negligible and is usually taken as a constant value or is sometimes used as a fitting parameter
   $b$ is the exponent
   $a$ ($L^{3-b}/T$) and $c$ ($L^3/T$) are parameters; here L is the length dimension and T is the time dimension

Equation 8.1 specializes into three popular forms that are commonly employed (Corbett, 1962). The three forms have been popularly used and stem from river morphological characteristics (Singh, 1996).

*Type 1*: In this case, $y_0 = 0$ and $c = 0$. Equation 8.1 then becomes

$$Q = ay^b \tag{8.2}$$

or in logarithmic form

$$\log Q = \log a + b \log y \tag{8.3}$$

*Type 2*: In this case, $c = 0$. Equation 8.1 then becomes

$$Q = a(y - y_0)^b \tag{8.4}$$

or in logarithmic form

$$\log Q = \log a + b \log(y - y_0) \tag{8.5}$$

*Type 3*: In this case, $y_0 = 0$. Equation 8.1 then becomes

$$Q = ay^b + c \tag{8.6}$$

or in logarithmic form

$$\log(Q - c) = \log a + b \log y \tag{8.7}$$

It should be noted that the values of parameters $a$, $b$, and $c$ will vary from one relation to another. In the hydraulic literature, Equations 8.2 through 8.6 have been applied.

## 8.4   RANDOMNESS IN RATING CURVE

In order to account for the change in control from low flow to high flow, a segmentation method has been used to construct a rating curve (Overlier, 2006), suggesting an element of randomness in the stage–discharge curve. Hence, it may be reasonable to argue that temporally averaged discharge can be treated as a random variable. Although significant temporal variability in discharge has been recognized, adequate effort has not been made to account for its probabilistic characteristics when establishing rating curves and to quantify uncertainty in a rating curve. One way to accomplish the twin objectives of defining the probability distribution of discharge and the uncertainty of a rating curve is to use the entropy theory.

The entropy theory has advantages over other methods: (1) it accounts for the information available on the rating curve, such as moments (mean, variance, etc.) of discharge. These moments are more stable in time than individual measurements. (2) It quantifies the information or uncertainty associated with the curve. (3) It provides a tool for data sampling or to determine the number of measurements needed to determine a robust rating curve. (4) It obviates the need for estimating the rating curve parameters empirically or by curve fitting. (5) The parameters estimated by the entropy theory are expressed in terms of the specified constraints and have therefore physical meaning. Singh (2010a) derived the stage–discharge relation using the Shannon entropy (Shannon, 1948) and tested it with field data. Singh et al. (2014) employed the Tsallis entropy for deriving rating curves.

## 8.5   DERIVATION OF RATING CURVES

It is assumed that temporally averaged discharge $Q$ is a random variable with a PDF denoted as $f(Q)$. The procedure for deriving rating curves with the use of the Tsallis entropy comprises (1) definition of the Tsallis entropy, (2) specification of constraints, (3) maximization of entropy in concert with the principle of maximum entropy (POME), (4) derivation of the probability distribution of discharge, (5) determination of Lagrange multipliers, (6) maximum entropy, and (7) derivation of the rating curve. Each of these steps is discussed in what follows.

### 8.5.1 DEFINITION OF TSALLIS ENTROPY

The Tsallis entropy (Tsallis, 1988) of discharge, $Q$, or of $f(Q)$, denoted as $H(Q)$, can be expressed as

$$H(Q) = H[f(Q)] = \frac{1}{m-1} \int_{Q_0}^{Q_D} f(Q)\{1 - [f(Q)]^{m-1}\} dQ \qquad (8.8)$$

where
  $m$ is the entropy index
  $Q_0$ and $Q_D$ represent the lower and upper limits of discharge for integration

Equation 8.8 expresses a measure of uncertainty about $f(Q)$ measured by $\{1 - [f(Q)]^{m-1}/(m-1)\}$ or the average information content of sampled $Q$. Therefore, $f(Q)$ needs to be derived first, which involves maximizing $H(Q)$, subject to specified constraints. In order to determine the $f(Q)$ that is least biased toward what is not known and most biased toward what is known (with regard to discharge) the POME, developed by Jaynes (1957, 1982), is invoked. POME requires the specification of certain information on discharge, encoded in terms of what is termed constraints and leads to the most appropriate probability distribution that has the maximum entropy or uncertainty.

### 8.5.2 SPECIFICATION OF CONSTRAINTS

For deriving the stage–discharge relation, simple constraints can be specified, which are the total probability law written as

$$C_1 = \int_{Q_0}^{Q_D} f(Q) dQ = 1 \qquad (8.9)$$

and

$$C_2 = \int_{Q_0}^{Q_D} Q f(Q) dQ = \overline{Q} \qquad (8.10)$$

Equation 8.9 is the first constraint defining the total probability law, $C_1$, which must always be satisfied by the PDF of discharge, and Equation 8.10 is the second constraint $C_2$ that defines the mean discharge.

### 8.5.3 MAXIMIZATION OF ENTROPY

In order to obtain the least-biased PDF of $Q$, $f(Q)$, the Tsallis entropy, given by Equation 8.8, is maximized following POME, subject to Equations 8.9 and 8.10. To that end, the method of Lagrange multipliers is employed (Singh, 1998). The Lagrangian function, $L$, then becomes

$$L = \frac{1}{m-1} \int_{Q_0}^{Q_D} f(Q)\{1-[f(Q)]^{m-1}\}dQ - \lambda_0 \left[ \int_{Q_0}^{Q_D} f(Q)dQ - 1 \right]$$

$$- \lambda_1 \left[ \int_{Q_0}^{Q_D} Qf(Q)dQ - \bar{Q} \right] \tag{8.11}$$

Differentiating Equation 8.11 with respect to $f(Q)$ recalling the calculus of variation, one obtains

$$\frac{\partial L}{\partial f(Q)} = \frac{1}{m-1} \left( \int_{Q_0}^{Q_D} \{1-[f(Q)]^{m-1}\} dQ - \int_{Q_0}^{Q_D} (m-1)[f(Q)]^{m-1} dQ - \lambda_0 \left[ \int_{Q_0}^{Q_D} dQ \right] \right.$$

$$\left. - \lambda_1 \left[ \int_{Q_0}^{Q_D} QdQ \right] \right) \tag{8.12}$$

### 8.5.4 PROBABILITY DISTRIBUTION OF DISCHARGE

Equating the derivative in Equation 8.12 to 0, one obtains the entropy-based PDF of $Q$ as

$$f(Q) = \left[ \frac{m-1}{m} \left( \frac{1}{m-1} - \lambda_0 - \lambda_1 Q \right) \right]^{1/(m-1)} \tag{8.13}$$

It is interesting to note that at $Q = 0$, $f(Q)$ becomes

$$f(Q)\Big|_{Q=0} = \left[ \frac{m-1}{m} \left( \frac{1}{m-1} - \lambda_0 \right) \right]^{1/(m-1)} \tag{8.14}$$

For purposes of simplification, let

$$\lambda_* = \frac{1}{m-1} - \lambda_0 \tag{8.15}$$

With the use of Equation 8.15, Equation 8.13 can be cast as

$$f(Q) = \left[ \frac{m-1}{m} (\lambda_* - \lambda_1 Q) \right]^{1/(m-1)} \tag{8.16}$$

The cumulative probability distribution function (CDF) of $Q$ can be obtained by integrating Equation 8.16 from $Q_0$ to $Q$ as

$$F(Q) = \left( \frac{m-1}{m} \right)^{m/(m-1)} \frac{1}{\lambda_1} [(\lambda_* - \lambda_1 Q_0)^{m/(m-1)} - (\lambda_* - \lambda_1 Q)^{m/(m-1)}] \tag{8.17}$$

If $Q_0 = 0$, Equation 8.17 reduces to

$$F(Q) = \left( \frac{m-1}{m} \right)^{m/(m-1)} \frac{1}{\lambda_1} [\lambda_*^{m/(m-1)} - (\lambda_* - \lambda_1 Q)^{m/(m-1)}] \tag{8.18}$$

Equation 8.17 can also be written for $Q$ explicitly as

$$Q = \frac{\lambda_*}{\lambda_1} - \frac{1}{\lambda_1} \frac{m}{m-1} \left\{ -\lambda_1 \frac{m-1}{m} F(Q) + \left[ \frac{m-1}{m} (\lambda_* - \lambda_1 Q_0) \right]^{m/(m-1)} \right\}^{(m-1)/m} \tag{8.19}$$

and Equation 8.18 as

$$Q = \frac{\lambda_*}{\lambda_1} - \frac{1}{\lambda_1} \frac{m}{m-1} \left\{ -\lambda_1 \frac{m-1}{m} F(Q) + \left( \frac{m-1}{m} \lambda_* \right)^{m/(m-1)} \right\}^{(m-1)/m} \tag{8.20}$$

Equations 8.19 and 8.20 are quantile–probability relationships.

### 8.5.5 Determination of Lagrange Multipliers

The PDF, given by Equation 8.13, has two unknown Lagrange multipliers $\lambda_0$ and $\lambda_1$ that can be determined using Equations 8.9 and 8.10. Substituting Equation 8.13 in Equation 8.9, one obtains

$$\int_{Q_0}^{Q_D} \left[ \frac{m-1}{m} \left( \frac{1}{m-1} - \lambda_0 - \lambda_1 Q \right) \right]^{1/(m-1)} dQ = 1 \tag{8.21}$$

The solution of Equation 8.21 can be written as

$$\left( \frac{1}{m-1} - \lambda_0 - \lambda_1 Q_D \right)^{m/(m-1)} - \left( \frac{1}{m-1} - \lambda_0 - \lambda_1 Q_0 \right)^{m/(m-1)} = \left( \frac{m}{m-1} \right)^{m/(m-1)} \lambda_1 \tag{8.22}$$

Likewise, substitution of Equation 8.13 in Equation 8.10 yields

$$\int_{Q_0}^{Q_D} Q \left[ \frac{m-1}{m} \left( \frac{1}{m-1} - \lambda_0 - \lambda_1 Q \right) \right]^{1/(m-1)} dQ = \bar{Q} \tag{8.23}$$

Integration of Equation 8.23 yields

$$-\lambda_1 \bar{Q} \left( \frac{m}{m-1} \right)^{m/(m-1)} = Q_D \left( \frac{1}{m-1} - \lambda_0 - \lambda_1 Q_D \right)^{m/(m-1)} - Q_0 \left( \frac{1}{m-1} - \lambda_0 - \lambda_1 Q_0 \right)^{m/(m-1)}$$

$$+ \frac{m-1}{2m-1} \frac{1}{\lambda_1} \left[ \left( \frac{1}{m-1} - \lambda_0 - \lambda_1 Q_D \right)^{(2m-1)/(m-1)} \right.$$

$$\left. - \left( \frac{1}{m-1} - \lambda_0 - \lambda_1 Q_0 \right)^{(2m-1)/(m-1)} \right] \tag{8.24}$$

Equations 8.22 and 8.24 can, with the use of Equation 8.15, be recast, respectively, as

$$(\lambda_* - \lambda_1 Q_D)^{m/(m-1)} - (\lambda_* - \lambda_1 Q_0)^{m/(m-1)} = \left( \frac{m}{m-1} \right)^{m/(m-1)} \lambda_1 \tag{8.25}$$

$$-\lambda_1 \bar{Q} \left( \frac{m}{m-1} \right)^{m/(m-1)} = Q_D (\lambda_* - \lambda_1 Q_D)^{m/(m-1)} - Q_0 (\lambda_* - \lambda_1 Q_0)^{m/(m-1)}$$

$$+ \frac{m-1}{2m-1} \frac{1}{\lambda_1} \left[ (\lambda_* - \lambda_1 Q_D)^{(2m-1)/(m-1)} - (\lambda_* - \lambda_1 Q_0)^{(2m-1)/(m-1)} \right] \tag{8.26}$$

Equations 8.25 and 8.26 are implicit in the Lagrange multipliers $\lambda_*$ (or $\lambda_0$) and $\lambda_1$ but can be solved numerically.

## Example 8.1

The stage–discharge values for gaging station 08079600 on Brazos River, Texas, are given in Table 8.1. Determine the Lagrange multipliers $\lambda_0$, $\lambda_*$, and $\lambda_1$ for different values of $m$.

### Solution

From Table 8.1, the mean discharge $\bar{Q}$ is found to be 1.620 m³/s and maximum discharge is 13.188 m³/s. Substituting these values into Equations 8.25 and 8.26,

**TABLE 8.1**

**Observed Stage–Discharge Values for Gaging Station 08079600**

| Stage (m) | Discharge (m³/s) | Stage (m) | Discharge (m³/s) |
|---|---|---|---|
| 0.046 | 0.172 | 0.170 | 0.614 |
| 0.062 | 0.195 | 0.182 | 0.722 |
| 0.068 | 0.795 | 0.197 | 0.741 |
| 0.074 | 0.270 | 0.247 | 1.500 |
| 0.077 | 0.225 | 0.265 | 1.409 |
| 0.083 | 0.156 | 0.274 | 0.971 |
| 0.096 | 0.328 | 0.281 | 1.721 |
| 0.102 | 0.158 | 0.296 | 1.027 |
| 0.120 | 0.391 | 0.302 | 0.996 |
| 0.130 | 0.325 | 0.315 | 3.170 |
| 0.130 | 0.130 | 0.392 | 2.001 |
| 0.145 | 0.450 | 0.419 | 3.170 |
| 0.154 | 0.504 | 0.530 | 4.528 |
| 0.160 | 0.563 | 0.648 | 7.584 |
| 0.163 | 0.606 | 0.752 | 13.188 |

**TABLE 8.2**

**Computation of Lagrange Multipliers**

| $m$ Values | $\lambda_*$ | $\lambda_1$ |
|---|---|---|
| $m = 2/3$ | 0.228 | 0.103 |
| $m = 3/4$ | 0.138 | 0.033 |
| $m = 1.5$ | 0.231 | −0.010 |
| $m = 2$ | 0.006 | 0.002 |
| $m = 3$ | −0.232 | −0.001 |
| $m = 4$ | −0.223 | 0.001 |

the Lagrange multipliers can be solved for different $m$ values, which are given in Table 8.2. It is seen that $\lambda_*$ varies from −0.23 to 0.23, while $\lambda_1$ is less than 0.1 except for $m = 2/3$.

## 8.5.6 MAXIMUM ENTROPY

The maximum Tsallis entropy or uncertainty of discharge can be obtained by substituting Equation 8.13 in Equation 8.8:

$$H(Q) = \frac{1}{m-1}\left\{ (Q_D - Q_0) + \left(\frac{m-1}{m}\right)^{m/(m-1)} \frac{1}{(2m-1)\lambda_1}[(\lambda_* - \lambda_1 Q_D)^{(2m-1)/(m-1)} \right.$$

$$\left. -(\lambda_* - \lambda_1 Q_0)^{(2m-1)/(m-1)}]\right\} \tag{8.27}$$

### 8.5.7 HYPOTHESIS ON CUMULATIVE PROBABILITY DISTRIBUTION OF DISCHARGE

The goal is to express $Q$ in terms of flow depth, $y$. To that end, let the maximum stage (channel flow depth) be denoted as $D$. It is then assumed that all values of stage $y$ measured from the bed to any point between 0 and $D$ are equally likely. This assumption is not highly unlikely since at different times different values of stage do occur and this is also consistent with the Laplacian principle of insufficient reason. Then, it may be hypothesized that the cumulative probability distribution of discharge is the ratio of the stage to the point where discharge is to be considered and the stage up to the maximum water surface. The probability of discharge being equal to or less than a given value of $Q$ is $y/D$; at any stage (measured from the bed) less than a given value, $y$, the discharge is less than a given value, say $Q$; thus, the cumulative distribution function of discharge, $F(Q) = P$ (discharge $\leq$ a given value of $Q$), $P$ = probability, is expressed as

$$F(Q) = \frac{y}{D} \tag{8.28}$$

Equation 8.28 constitutes the fundamental hypothesis for deriving the stage–discharge relation using entropy. In Equation 8.28 on the left side the argument of function $F$ is variable $Q$, whereas on the right side the variable is $y$. The CDF of $Q$ is not linear in terms of $Q$, unless $Q$ and $y$ are linearly related. Of course, it is plausible that $F(Q)$ might have a different form. A similar hypothesis has also been employed when using the entropy theory for deriving infiltration equations by Singh (2010b,c), soil moisture profiles by Singh (2010d), and velocity distributions by Chiu (1987).

The PDF is obtained by differentiating Equation 8.25 with respect to $Q$:

$$f(Q) = \frac{dF(Q)}{dQ} = \frac{1}{D}\frac{dy}{dQ} \quad \text{or} \quad f(Q) = \left(D\frac{dQ}{dy}\right)^{-1} \tag{8.29}$$

The $f(Q)\, dQ = F(Q + dQ) - F(Q)$ term denotes the probability of discharge being between $Q$ and $Q + dQ$.

### 8.5.8 DERIVATION OF RATING CURVE

Substituting Equation 8.13 in Equation 8.29, one gets

$$\left[\frac{m-1}{m}(\lambda_* - \lambda_1 Q)\right]^{1/(m-1)} dQ = \frac{dy}{D} \tag{8.30}$$

Using the limits when $y = y_0$, $Q = Q_0$, integration of Equation 8.30 results in the rating curve:

$$Q = \frac{1}{\lambda_1}\left\{\lambda_* - \left[(\lambda_* - \lambda_1 Q_0)^{m/(m-1)} + \lambda_1\left(\frac{m-1}{m}\right)^{-m/(m-1)}(y - y_0)\right]^{(m-1)/m}\right\} \tag{8.31}$$

Equation 8.31 can be considered as a generalized rating curve.

If $Q_0 = 0$, then Equation 8.31 reduces to

$$Q = \frac{1}{\lambda_1}\left\{\lambda_* - \left[(\lambda_*)^{m/(m-1)} + \lambda_1\left(\frac{m-1}{m}\right)^{-m/(m-1)}(y)\right]^{(m-1)/m}\right\}\qquad(8.32)$$

Equation 8.32 can be considered in the form of

$$Q = c - a(y+d)^b \qquad (8.33)$$

where

$$a = \left(\frac{1}{\lambda_1}\right)^{1/m}\left(\frac{m}{m-1}\right)\qquad(8.34)$$

$$b = \frac{m-1}{m}\qquad(8.35)$$

$$c = \frac{\lambda_*}{\lambda_1}\qquad(8.36)$$

$$d = \frac{1}{\lambda_1}\left(\frac{m-1}{m}\right)^{m/(m-1)}(\lambda_* - \lambda_1 Q_0)^{m/(m-1)} - y_0 \qquad(8.37)$$

## Example 8.2

Construct the rating curve for different values of entropy index $m$ for the stage–discharge values given in Table 8.1 and determine an appropriate value of $m$. Choose the entropy index $m$ from the range of 0.5–4.

### Solution

To determine the entropy index $m$, the rating curve is calculated for different $m$ values for the gaging station, as plotted in Figure 8.1. The Lagrange multipliers are already computed in Example 8.1. It is seen from the figure that $m$ affects the rate of increase in the rating curve. Generally, for larger $m$ values, the rating curve becomes higher and increases faster. However, $m = 3$ fits the observations best. The rating curve estimated by $m = 3$ or 4 gives similar accuracy for stage less than 1 m, but as stage becomes higher, $m = 3$ is significantly better than $m = 4$. The root mean square (RMS) values are 13.22, 9.83, 6.45, 0.49, and 24.37 m³/s, for $m = 0.5$, 1.5, 2.5, 3, and 4, respectively. Thus, $m = 3$ seems appropriate.

## Example 8.3

Construct the rating curve with entropy index $m = 3.0$ for data given in Table 8.1. Also, construct the 95% confidence bands.

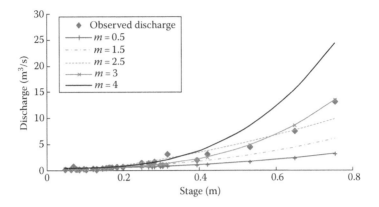

**FIGURE 8.1**    Rating curve of station 08079600 on Brazos River, Texas.

**Solution**

The Lagrange multipliers for entropy index $m = 3.0$ are solved for as in Example 8.1 and shown in Table 8.2, which are $\lambda_* = -0.232$ and $\lambda_* = -0.001$. Then, the rating curve is computed using Equation 8.32, as shown in Figure 8.2. By referring to the probability of a parameter being in an interval, the 95% confidence intervals of $Q$ in $[Q_{low}, Q_{up}]$ are computed from repeated simulations. Then, the confidence interval is calculated from the cumulative distribution of parameters conditioned on the observed data, so that $P(Q_{low} \leq Q \leq Q_{up}) = 0.95$.

## Example 8.4

Determine parameters of the rating curve with $m = 3$ for the stage–discharge data in Table 8.1.

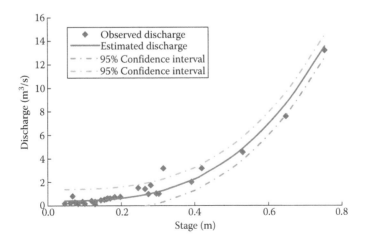

**FIGURE 8.2**    Rating curve of station 08079600 on Brazos River.

**Solution**

The Lagrange multipliers are computed by solving nonlinear equations (Equations 8.25 and 8.26) for given mean and maximum discharge values and the computed values are given in Table 8.2. With the computed Lagrange multipliers, rating curve parameters are computed from Equations 8.34 through 8.37. It can be seen from Equation 8.35 that the $b$ value is only determined by the $m$ value; thus, $b = 0.667$ is fixed as $m = 3$. The average values of rating curve parameters for gaging station 08079600 on Brazos River are $a = 0.448$, $b = 0.667$, $c = 4.634$, and $d = 33.32$, respectively.

## 8.6 REPARAMETERIZATION

It may be convenient to reparameterize by grouping the Lagrange multipliers so that the resulting rating curve has only one parameter. To that end, let

$$M = \frac{\lambda_1 Q_{max}}{\lambda_1 Q_{max} - \lambda_*} \quad \text{or} \quad Q_{max} = \frac{\lambda_* M}{\lambda_1 (M - 1)} \tag{8.38}$$

When $Q = Q_{max}$, $F(Q_{max}) = 1$ and Equation 8.19 with $Q_{min} = 0$ yields

$$Q_{max} = \frac{\lambda_*}{\lambda_1} - \frac{1}{\lambda_1} \frac{m}{m-1} \left[ -\lambda_1 \frac{m-1}{m} + \left( \frac{m-1}{m} \lambda_* \right)^{m/(m-1)} \right]^{(m-1)/m} \tag{8.39}$$

Dividing Equation 8.20 by $Q_{max}$, the result is

$$\frac{Q}{Q_{max}} = \frac{\lambda_*}{\lambda_1 Q_{max}} - \frac{1}{\lambda_1 Q_{max}} \frac{m}{m-1} \left\{ -\lambda_1 \frac{m-1}{m} F(Q) + \left[ \frac{m-1}{m} \lambda_* \right]^{m/(m-1)} \right\}^{(m-1)/m}$$

$$= \frac{\lambda_*}{\lambda_1 Q_{max}} \left( 1 - \frac{1}{\lambda_*} \left\{ -\lambda_1 \left( \frac{m-1}{m} \right)^{1/(m-1)} F(Q) + \lambda_*^{m/(m-1)} \right\}^{(m-1)/m} \right) \tag{8.40}$$

Noting from Equation 8.38 that $\lambda_*/\lambda_1 Q_{max} = 1 - (1/M)$, which when substituted in Equation 8.40 yields

$$\frac{Q}{Q_{max}} = 1 - \frac{1}{M} - \left( 1 - \frac{1}{M} \right) \left\{ - \left( \frac{1}{\lambda_*} \right)^{m/(m-1)} \lambda_1 \left( \frac{m}{m-1} \right)^{1/(m-1)} F(Q) + 1 \right\}^{(m-1)/m}$$

$$= 1 - \frac{1}{M} \left( 1 - \left\{ \left( \frac{m}{m-1} \frac{M}{1-M} Q_{max} \right)^{m/(m-1)} [F(Q) - 1] + 1 \right\}^{(m-1)/m} \right) \tag{8.41}$$

It is noted that when $Q = Q_{min} = 0$, $F(Q) = 0$. Thus, Equation 8.41 reduces to

$$0 = 1 - \frac{1}{M}\left\{1 - \left[1 - \left(\frac{m}{m-1}\frac{M}{1-M}Q_{max}\right)^{m/(m-1)}\right]^{(m-1)/m}\right\} \tag{8.42}$$

Rearranging Equation 8.42, one gets

$$\left(\frac{m}{m-1}\frac{M}{1-M}Q_{max}\right)^{m/(m-1)} = 1 - (1-M)^{m/(m-1)} \tag{8.43}$$

Interestingly, Equation 8.43 expresses $M$ as a function of $Q_{max}$ for a given entropy index $m$ and if $m$ is determined to be 3, it can be recast as

$$Q_{max} = \frac{2}{3}\left(\frac{1-M}{M}\right)[1-(1-M)^{3/2}]^{2/3} \tag{8.44}$$

Equation 8.44 shows that $M$ under the assumption that $Q_{min} = 0$ should be bounded by 1.0. However, $M$ computed from the Lagrange multipliers will be larger than 1 as long as $\lambda_*$ is positive. The values of $M$ computed from Equations 8.38 and 8.43 may be different. Equation 8.41 can be simplified by inserting the right side of Equation 8.43 as

$$\frac{Q}{Q_{max}} = 1 - \frac{1}{M}(1-\{(1-M)^{m/(m-1)} + [1-(1-M)^{m/(m-1)}]F(Q)\}^{(m-1)/m}) \tag{8.45}$$

Substituting Equation 8.28 in Equation 8.45, the result is

$$\frac{Q}{Q_{max}} = 1 - \frac{1}{M}\left(1 - \left\{(1-M)^{m/(m-1)} + \left[1-(1-M)^{m/(m-1)}\right]\frac{y}{D}\right\}^{(m-1)/m}\right) \tag{8.46}$$

In Equation 8.45, the Lagrange multipliers are replaced with entropy parameter $M$, and hence, the flow discharge can be determined from Equation 8.46 with only parameter, $M$. Parameter $M$ can be used as an index of the uniformity of the probability distribution of discharge, which is related to the maximum discharge.

Singh et al. (2014) plotted the $M$ value against the maximum discharge on a log–log paper, as shown in Figure 8.3. It can be seen from the figure that though the maximum discharge varies from 0 to 1600 m³/s, the $M$ value is bounded between 1 and 1.1, except for one outlier of 1.5. They found a linear relationship between $M$ and maximum flow, which using regression was written as

$$\log(M) = -0.047\log(Q_{max}) + 0.1294 \tag{8.47}$$

with a coefficient of determination of 0.715.

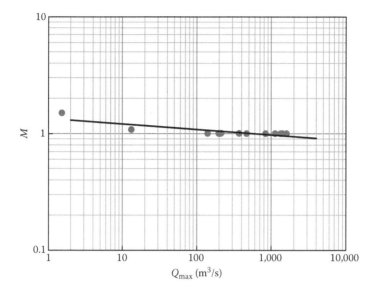

**FIGURE 8.3** Relationship between entropy index $M$ and maximum discharge.

## Example 8.5

For 13 USGS gaging stations, Singh et al. (2014) computed rating curves and provided values of $M$ computed using Equation 8.47 as well as from the Lagrange multipliers. The two sets of $M$ values are given in Table 8.3. Plot the two sets of the $M$ values against each other and comment.

### TABLE 8.3
### Computed $M$ from Two Methods

| Station | Location | $M$ | $M*$ |
|---------|----------|-----|------|
| 8079600 | Justiceburg | 1.0813 | 0.8503 |
| 8080500 | Aspermont | 1.0039 | 0.9932 |
| 8082000 | Salt-FK | 1.0036 | 0.9873 |
| 8082500 | Seymour | 1.0052 | 0.9820 |
| 8083100 | Clear-FK | 1.4992 | 0.4936 |
| 8088000 | South Bend | 1.0067 | 0.9882 |
| 8090800 | Dennis | 1.0010 | 0.9946 |
| 8089000 | Palo Pinto | 1.0014 | 0.9878 |
| 8096500 | Waco | 1.0013 | 0.9970 |
| 8098290 | Highbank | 1.0004 | 0.9978 |
| 8111500 | Hempstead | 1.0003 | 0.9984 |
| 8114000 | Richmond | 1.0002 | 0.9982 |
| 8116650 | Rosharon | 1.0002 | 0.9981 |
| Average | | 1.0465 | 0.9436 |

**Solution**

The $M$ values computed from the Lagrange multipliers are given in Table 8.3. The $M$ values, denoted as $M^*$, are obtained from the known values of $Q_{max}$ using Equation 8.47, as shown in Table 8.3. The values of $M$ and $M^*$ are compared in Figure 8.3. It can be seen that $M^*$ has a larger range than does $M$. The $M^*$ values are lower with an average value of 0.944 than the $M$ values with an average value of 1.047. The $M^*$ values are lower than 1, while the $M$ values are larger than 1. Either $M$ or $M^*$ is not uniformly distributed inside its range, but $M$ is clustered between 1 and 1.01 while $M^*$ between 0.8 and 1.

## Example 8.6

Validate the regression of $M$ on $Q_{max}$ obtained from Brazos River, whose parameters for Trinity River, Pearl River, Tennessee River, and Pee Dee river basins are shown in Table 8.3. The values of the Lagrange multipliers for the flow data of these rivers are given in Table 8.4.

**Solution**

Figure 8.4 shows the regression of $M$ on $Q_{max}$ obtained from Brazos River and other river basins. The value of $Q_{max}$ from the various rivers varies from 0.2 to 20,000 m³/s, but the $M$ value still remains within the range of 1 and 2.5, and the $M$ value for the Red River is the only one exceeding 2. It can be seen from Figure 8.4 that the regression equation (Equation 8.47) is generally valid. The observations from Trinity River, Pearl River, and Pee Dee River fall exactly on the regression line. Thus, it suggests that though the regression is obtained using the data from Brazos River, it is generally valid for other river basins also.

## Example 8.7

Construct the rating curves using entropy parameter $M$ for station 08082500 on the Brazos River and show how well it compares with observed flow values.

**TABLE 8.4**
**Lagrange Multipliers for Flow Data from Trinity River, Pearl River, Tennessee River, and Pee Dee River Basins**

| River | Station | $Q_{max}$ | $M$ | $\lambda_1$ | $\lambda_*$ |
|---|---|---|---|---|---|
| Pee Dee River | 02135000 | 155 | 1.05 | 89.9 | 699 |
| Pear River | 02489000 | 160 | 1.011 | 89 | 167 |
| Tennessee River | 03592718 | 8.03 | 1.748 | 5.71 | 34.4 |
| Mississippi River | 07010000 | 17,376 | 1.000 | 8311 | 24,040 |
| Red River | 07300000 | 0.184 | 2.39 | 0.123 | 7.248 |
| Trinity River | 08047000 | 26.6 | 1.225 | 19.2 | 115 |

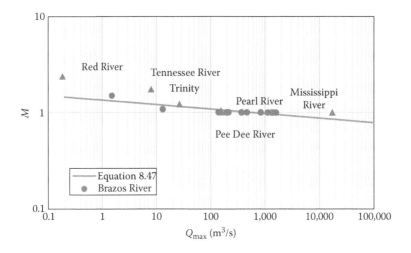

**FIGURE 8.4** Relation between $M$ and $Q_{max}$ for other rivers.

## Solution

The rating curve is determined either from Equation 8.31 with the known Lagrange multipliers or from Equation 8.33 with the known rating curve parameters. Figure 8.5 shows the agreement between predicted rating curves and observed curves for station 08080500. The predicted rating curve increases from 0 to exactly the observed maximum discharge value, and mimics the flow pattern of the station. The rating curve is also determined using entropy parameter $M$ defined by Equation 8.38. For station 08082500, with known values of $Q_{mean}$ and $Q_{max}$, $M$ is obtained as 1.005 from either solving for Lagrange multipliers or using relation to the ratio between $Q_{mean}$ and $Q_{max}$ defined by Equation 8.38. Then the rating curve is plotted against the observed values in Figure 8.6. The estimated rating

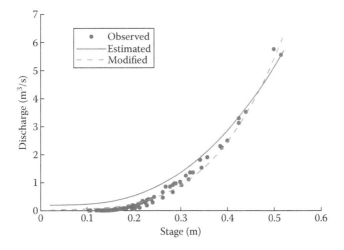

**FIGURE 8.5** Rating curve estimated using entropy parameter $M$ for station 08082500.

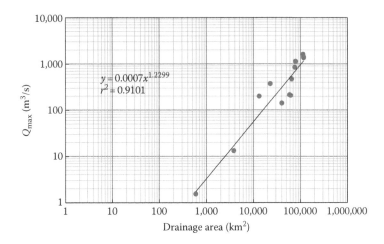

**FIGURE 8.6**  Relation between maximum discharge and drainage area.

curve is shifted from the observed values, due to the assumption of $Q_{min} = 0$ at $y = 0$ for deriving Equations 8.45 and 8.46). Therefore, correction may be needed to avoid the shift by manually fixing $Q_{min}$ at the observed stage rather than taking $y = 0$. After the correction, the modified rating curve fits observation well with an RMS value of 0.08 m³/s. This suggests that the use of entropy parameter $M$ is equivalent to the use of the Lagrange multiplier method.

## 8.7  RELATION BETWEEN MAXIMUM DISCHARGE AND DRAINAGE AREA

Thirteen stream gaging stations located on Brazos River, Texas, are selected from the USGS website. Table 8.5 tabulates the drainage area, maximum discharge, and mean discharge for these stations. The relationship between the drainage area and maximum discharge can be plotted, as shown in Figure 8.6, and can be expressed as

$$Q_{max} = 0.0007A^{1.229} \qquad (8.48)$$

where
$Q_{max}$ is in m³/s
$A$ is in km²

## 8.8  RELATION BETWEEN MEAN DISCHARGE AND DRAINAGE AREA

The relation between mean discharge and drainage area given in Table 8.5, as shown in Figure 8.7, can be expressed as

$$Q_{mean} = 9 \times 10^{-5} A^{1.234} \qquad (8.49)$$

**TABLE 8.5**

**Drainage Area and Maximum Discharge for Different Stations**

| Station | Location | Drainage Area (km²) | $Q_{max}$ (m³/s) | $Q_{mean}$ (m³/s) |
|---|---|---|---|---|
| 08079600 | Justiceburg | 3,797 | 13 | 2 |
| 08080500 | Aspermont | 22,782 | 371 | 32 |
| 08082000 | Salt-FK | 13,287 | 198 | 32 |
| 08082500 | Seymour | 40,243 | 140 | 26 |
| 08083100 | Clear-FK | 591 | 2 | 0.3 |
| 08088000 | South Bend | 58,723 | 213 | 17 |
| 08090800 | Dennis | 65,364 | 467 | 63 |
| 08089000 | Palo Pinto | 61,670 | 207 | 80 |
| 08096500 | Waco | 76,558 | 838 | 47 |
| 08098290 | Highbank | 78,829 | 1129 | 128 |
| 08111500 | Hempstead | 113,649 | 1605 | 293 |
| 08114000 | Richmond | 116,827 | 1407 | 306 |
| 08116650 | Rosharon | 117,428 | 1344 | 313 |

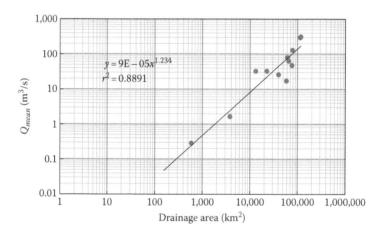

**FIGURE 8.7** Relation between mean discharge and drainage area.

## 8.9 RELATION BETWEEN ENTROPY PARAMETER AND DRAINAGE AREA

The relation between the entropy parameter $M$ given in Table 8.3 and the drainage area given in Table 8.5 is plotted in Figure 8.8. Thus, the relationship between $M$ and drainage area can be expressed as

$$M = 1.9716A^{-0.061} \tag{8.50}$$

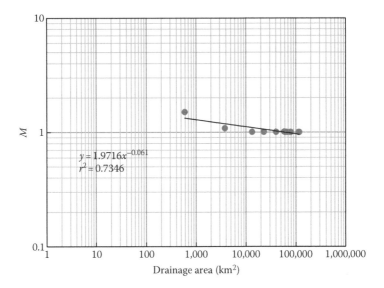

**FIGURE 8.8** Relation between entropy parameter $M$ and drainage area.

## Example 8.8

The stage–discharge data for a gaging station 0811150 where the drainage area is 113,649 km² is given in Table 8.5. Compute the rating curve using the relation given by Equations 8.48 and 8.49 as if the station is ungaged.

### Solution

Based on the power relationship between discharge and drainage area, the rating curve of station 08111500 having a drainage area of 113,649 km² is estimated. The mean and maximum discharge values are computed using regression equations (Equations 8.48 and 8.49) as 156 m³/s and 1456 m³/s, respectively, and the estimated values are slightly smaller than observed values shown in Table 8.5. However, the estimated differences are acceptable. With the estimated mean and maximum discharge values, rating curves are computed, as shown in Figure 8.9. It can be seen that the estimated rating curve is in close agreement with the observed curve and the RMS value is only 35.69 m³/s, which is slightly larger than the value computed when observed discharge values are used, which is 34.31 m³/s.

## Example 8.9

Determine the rating curve for station 08111500 with a drainage area of 113,649 km² with the use of entropy parameter $M$ determined from the drainage area.

### Solution

With the given drainage area, the value of $M$ is obtained as 0.969 from the power relationship shown in Figure 8.9. Then, the rating curve is determined using Equations 8.45 and 8.46, as plotted in Figure 8.10. The rating curve so determined has a shift from the observed curve, so a correction may be made by fixing $Q_{min}$ at

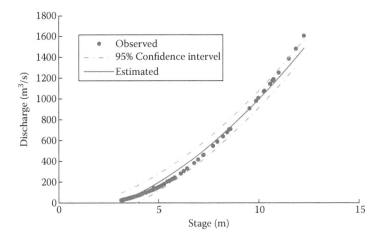

**FIGURE 8.9** Rating curve estimated for station 08111500.

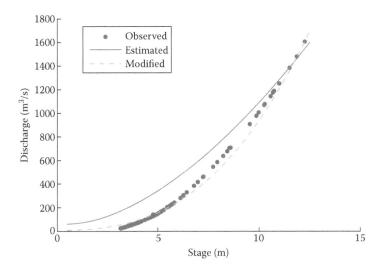

**FIGURE 8.10** Rating curve estimated using $M$ for station 08111500.

the observed minimum stage. It can be seen from the figure that the rating curve determined after the correction fits the observations well. The RMS decreases to 34.78 m³/s, which is closer to that when observed values are used for determining the rating curve.

## 8.10   EXTENSION OF RATING CURVE

Often, it is necessary to extend the rating curve for new observed discharges that are beyond the highest measured values (Torsten et al., 2002; Sivapragasam and Muttil, 2005). For a gaging station 08108250, the rating curve is extended from a stage of

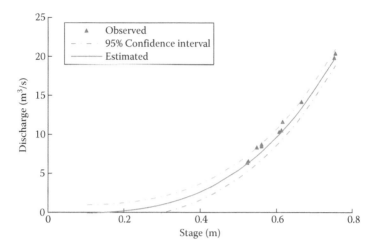

**FIGURE 8.11** Extended rating curve for gaged station 08082500.

0.5 m to a stage of 0.8 m, as shown in Figure 8.11, where the measured discharge is higher than 6 m³/s. The figure shows that the estimated discharges fit the extended rating curve well and all predicted discharge values fall within the 95% confidence interval. For ungaged station 08111500, where the rating curve is predicted using the relationship between the discharge and the drainage area, the rating curve is extended from a stage of 12 m to a stage of 15 m, as shown in Figure 8.12. The extended rating curve for this ungaged site is not as good as the one for the gaged site. Nevertheless, the observed values between 1500 and 2000 m³/s are within the upper 95% confidence interval, while the discharge exceeding 2000 m³/s falls outside the interval and produces a larger variance.

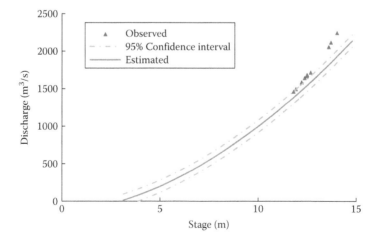

**FIGURE 8.12** Extended rating curve for ungaged station 08111500.

## REFERENCES

Azamathulla, H.M., Ghani, A.A., Leow, C.S., Chang, C.K., and Zakaria, N.A. (2011). Gene-expression programming for the development of a stage-discharge curve of the Pahang River. *Water Resources Management*, 25, 2901–2906, doi: 10.1007/s11269-011-9845-7.

Baiamonte, G. and Ferro, V. (2007). Simple flume for flow measurement in sloping open channel. *Journal of Irrigation and Drainage Engineering*, 133(1), 71–78.

Bailey, J.F. and Ray, H.A. (1966). Definition of stage-discharge relation in natural channels by step-backwater analysis. U.S. Geological Survey Water Supply Paper 1869-A, 34pp.

Bhattacharya, B. and Solomatine, D.P. (2000). Application of artificial neural network in stage-discharge relationship. *Proceedings of the Fourth International Conference on Hydroinformatics*, Iowa City, IA.

Bhattacharya, B. and Solomatine, D.P. (2005). Neural networks and M5 model trees in modeling water level-discharge relationship. *Neurocomputing*, 63, 381–396.

Chiu, C.L. 1987. Entropy and probability concepts in hydraulics. *Journal of Hydraulic Engineering*, 113(5), 583–600.

Corbett, D.M. (1962). Stream-gaging procedure-A manual describing methods and practices of the Geological Survey. Department of Interior, U.S. Geological Survey, Washington, D.C.

Deka, P. and Chandramouli, V. (2003). A fuzzy neural network model for deriving the river stage-discharge relationship. *Hydrological Sciences Journal*, 48(2), 197–209.

Ghimire, B.N.S. and Reddy, M.J. (2010). Development of stage-discharge rating curve in a river using genetic algorithms and model tree. *International Workshop on Advances in Statistical Hydrology*, Taormina, Italy.

Guven, A. and Aytek, A. (2009). A new approach for stage-discharge relationship: Gene-expression programming. *Journal of Hydrologic Engineering*, 14(8), 812–820.

Habib, E.H. and Meselhe, E.A. (2006). Stage-discharge relations for low-gradient tidal streams using data driven models. *Journal of Hydraulic Engineering*, 132(5), 482–492.

Hershey, R.W. 1995. *Streamflow Measurements*, 2nd edn., Chapman & Hall, England, U.K.

Jain, S.K. and Chalisgaonkar, D. (2000). Setting up stage-discharge relations using ANN. *Journal of Hydrologic Engineering*, 5(4), 428–433.

Jarrett, R.D. (1984). Hydraulics of high-gradient streams. *Journal of Hydraulic Engineering*, 110(11), 1519–1539.

Jaynes, E.T. (1957). Information theory and statistical mechanics, I. *Physical Review*, 106, 620–630.

Jaynes, E.T. (1982). On the rationale of maximum entropy methods. *Proceedings of the IEEE*, 70, 939–952.

Kazama, S., Suzuki, K., and Sawamoto, M. (2005). Estimation of rating-curve parameters for sedimentation using a physical model. *Hydrological Processes*, 19, 3863–3871.

Kean, J.W. and Smith, J.D. (2005). Generation and verification of theoretical curves in the Whitewater River basin Kansas. *Journal of Geophysical Research*, 110, F04012, doi: 1029/2004JF000250.

Kennedy, E.J. (1964). Discharge rating at gaging stations. Chapter A10, Book 3: *Applications of Hydraulics*, USG Geological Survey, 59pp., Washington, DC.

Liao, H. and Knight, D.W. (2007). Analytic stage-discharge formulas for flow in straight prismatic channels. *Journal of Hydraulic Engineering*, 133(10), 111–1122.

Lohani, A.K., Goel, N.K., and Bhatia, K.K.S. (2006). Takagi-Sugeno fuzzy inference system for modeling stage-discharge relationship. *Journal of Hydrology*, 331(1–2), 146–160.

Overlier, P. (2006). Modeling stage-discharge relationships affected by hysteresis using the Jones formula and nonlinear regression. *Hydrological Sciences*, 51(3), 365–388.

Petersen-Overleir, A. (2004). Accounting for heteroscedasticity in rating curve estimates. *Journal of Hydrology*, 292, 173–181.

Petersen-Overleir, A. and Reitan, T. (2005). Objective segmentation in compound rating curves. *Journal of Hydrology*, 311, 188–201.

Rantz, S.E. (1982). Measurement and computation of streamflow. Vol. 2, Computation of discharge. U.S. Geological Survey Water Supply Paper 2175, Washington, DC.

Sahoo, G.B. and Ray, C. (2006). Flow forecasting for a Hawaii stream using rating curves and neural networks. *Journal of Hydrology*, 317(1–2), 63–80.

Sauer, V.B. and Meyer, R.W. 1992. Determination of error in individual discharge measurements. U.S. Geological Survey Open-File Report, 92-144.

Sefe, F.T.K. (1996). A study of the stage-discharge relationship of the Ukavaiigo River at Mohembo, Botswana. *Hydrological Sciences Journal*, 41(1), 97–116.

Shannon, C.E. (1948). A mathematical theory of communications, I and II. *Bell System Technical Journal*, 27, 379–443.

Singh, V.P. (1993). *Elementary Hydrology*. Prentice Hall, Englewood Cliffs, NJ.

Singh, V.P. (1996). *Kinematic Wave Modeling in Water Resources: Surface Water Hydrology*. John Wiley & Sons, New York.

Singh, V.P. (1998). *Entropy-Based Parameter Estimation in Hydrology*. Kluwer Academic Publishers, Boston, MA.

Singh, V.P. (2010a). Derivation of rating curves using entropy theory. *Transactions of the ASABE*, 53(6), 1811–1821.

Singh, V.P. (2010b). Entropy theory for derivation of infiltration equations. *Water Resources Research*, 46, 1–20, W03527, doi: 10-1029/2009WR008193.

Singh, V.P. (2010c). Tsallis entropy theory for derivation of infiltration equations. *Transactions of the ASABE*, 53(2), 447–463.

Singh, V.P. (2010d). Entropy theory for movement of moisture in soils. *Water Resources Research*, 46, 1–12, W03516, doi: 10-1029/2009WR008288.

Singh, V.P., Cui, H., and Byrd, A.R. (2014). Derivation of rating curve by the Tsallis entropy. *Journal of Hydrology*, 513, 342–352.

Sivapragasam, C. and Muttil, N. (2005). Discharge curve extension-a new approach. *Water Resources Management*, 19, 505–520.

Sudheer, K.P. and Jain, S.K. (2003). Radial basis function neural network for modeling rating curves. *Journal of Hydrologic Engineering*, 8(3), 161–164.

Torsten, D., Gerd, M., and Torsten, S. (2002). Extrapolating stage-discharge relationships by numerical modeling. *Proceedings of International Conference on Hydraulic Engineering*, Warsaw, Poland.

Tsallis, C. (1988). Possible generalizations of Boltzmann-Gibbs statistics. *Journal of Statistical Physics*, Vol. 52(1/2), 479–487.

Yoo, C. and Park, J. (2010). A mixture-density-network based approach for finding rating curves: Facing multimodality and unbalanced data distribution. *KSCE Journal of Civil Engineering*, 14(2), 243–250.

# Section III

Hydrology

# 9 Precipitation Variability

Precipitation is the primary determinant for the assessment of potential availability of water resources in an area or country. Although precipitation is cyclic in nature, its distribution in both space and time is highly erratic, leading to unevenly distributed water resources. The uncertainty (or disorder) in the occurrence of precipitation, especially intensity, amount, and duration, in time and space is one of the primary constraints to the development and use of water resources. When developing a basin wide, regional, or nationwide strategy for water resources development as well as for meeting current and future water demands, the uncertainty in precipitation occurrence over a given area (e.g., basin, region, or country) can be the determining factor in making a decision on the priorities for area-wide development or demarcating the boundaries to establish the feasibility and necessity of development. Development and management of water resources require not only the aggregate precipitation but also its variability. The stability of water supply increases with decreasing spatial variability of precipitation. On the other hand, the less temporally variable the precipitation is, the more dependable the water supply is. When evaluating the availability of water resources in a watershed or investigating the relative availability of local or regional water resources, the temporal variability of precipitation becomes a major concern. The uncertainty or disorder of a precipitation variable can be calculated using entropy, provided the probability distribution function of the precipitation variable under consideration is known. This chapter discusses the use of Tsallis entropy for the evaluation of precipitation variability.

## 9.1 ENTROPY AS A MEASURE OF PRECIPITATION UNCERTAINTY

From the standpoint of evaluating the availability of water resources, precipitation characteristics of interest are intensity, amount, and duration of occurrence. Entropy can be employed to determine the uncertainty associated with any of these characteristics. If $p_i > 0$ is the probability of the $i$th precipitation event (e.g., amount, intensity, or duration), then the Tsallis entropy $S_m$ (Tsallis, 1988) can be expressed as

$$S_m = \frac{1}{m-1} \sum_{i=1}^{N} p_i \left( 1 - p_i^{m-1} \right) \tag{9.1}$$

where
  $N > 1$ is the number of events
  $m$ is the Tsallis entropy parameter

## 9.2   INTENSITY ENTROPY

Let the total amount of precipitation over a certain period of time (say day, week, or month) or simply precipitation intensity (amount per unit time) be a random variable. Then, its probability distribution can be derived from data and its entropy can be calculated. The entropy so obtained is referred to as "intensity entropy (*IE*)." *IE* can be evaluated as follows (Maruyama et al., 2005):

1. Obtain precipitation data available at a rain gauge. If data are available for *M* years and intensity is defined on a monthly basis, then the number of monthly intensity values would be $12 \times M = N$. All monthly precipitation data of the *M*-year record are considered as one data set without any consideration of sequence or chronology.
2. Split the whole range of precipitation values into *n* classes at an equal interval.
3. Count the number of values or frequency $M_i$ in each class $i$, $i = 1, 2, ..., n$.
4. Calculate the relative frequency $f_i = M_i/N$, $i = 1, 2, ..., n$, for each class *i*. The relative frequency is the probability mass associated with the class interval. This yields the probability mass function in discrete form. Dividing the relative frequency or probability mass by the width of the class interval one obtains the probability density function (PDF). This is done for each class interval, and thus, the PDF is obtained in discrete form for the precipitation intensity data.
5. Calculate the *IE* in terms of the relative frequencies obtained in the preceding step as

$$IE = \frac{1}{m-1} \sum_{i=1}^{n} \left( \frac{M_i}{N} \right) \left( 1 - \left( \frac{M_i}{N} \right)^{m-1} \right) \tag{9.2}$$

where
    *n* is the number of classes
    *N* is the number of values of all classes
    $f_i$ is the relative frequency for class *i*

The *IE* defined over a semi-infinite range of $0 \leq IE < \infty$, is a measure to decipher the disorder of precipitation intensity. Less disordered intensity is measured by smaller *IE*, pointing to a more skewed distribution of the frequency of precipitation. On the contrary, more highly disordered intensities result in larger *IE*, extending over a wider range of monthly precipitation intensity. It is, however, noted that the PDF of precipitation intensity is always defined over a positive abscissa, including its zero origin, due to the nonnegativity of precipitation. It then follows that an increase in *IE* results in an increase in the expected value of precipitation, flattening the graph of the function. This suggests that *IE* is positively correlated with the expected amount of precipitation and can, therefore, be an alternative to the aggregate amount of precipitation.

## TABLE 9.1
## Frequency Distribution Table for Climate Division 4 in Texas

| Class Interval (in.) | Frequency | Relative Frequency | Probability Density Function (Data) | Probability Density (Computed) |
|---|---|---|---|---|
| 0.00–1.48 | 152 | 0.107 | 0.072 | 0.099 |
| 1.48–2.96 | 413 | 0.289 | 0.195 | 0.212 |
| 2.96–4.45 | 412 | 0.289 | 0.195 | 0.278 |
| 4.45–5.93 | 234 | 0.164 | 0.110 | 0.225 |
| 5.93–7.42 | 123 | 0.086 | 0.058 | 0.112 |
| 7.42–8.90 | 65 | 0.046 | 0.031 | 0.034 |
| 8.90–10.38 | 14 | 0.010 | 0.007 | 0.006 |
| 10.38–11.86 | 10 | 0.007 | 0.005 | 0.001 |
| 11.86–13.34 | 3 | 0.002 | 0.001 | 0.000 |
| 13.34–14.80 | 1 | 0.001 | 0.000 | 0.000 |

## Example 9.1

Consider monthly precipitation data for climate division 4 in Texas and determine its probability distribution. Indicate the mean precipitation value for the distribution. Collect and employ precipitation series from the Full Network Estimated Precipitation FNEP (http://climatexas.tamu.edu/). Office of State Climatologist, Texas A&M University, College Station, Texas. The climate division 4 (East Texas) has data for a period of 118 years from 1895 to 2013.

### Solution

Monthly precipitation data are available for 119 years. The total number of monthly values is $12 \times 119 = 1428$. The empirical discrete frequencies are then computed, as shown in Table 9.1. Thus, the empirical distribution is obtained, as graphed in Figure 9.1. Also shown is the fitted PDF $p(x) = f(x)$, where $x$ is the monthly precipitation value. The fitted PDF is a gamma distribution whose parameters are $\alpha = 4.523$ and $\beta = 1.004$, given as

$$f(x) = \frac{1}{\Gamma(\alpha)\beta^{\alpha}} x^{k-1} e^{-x/\beta} = 0.082 x^{3.523} e^{-x/1.004}, \quad 0 < x < \infty$$

Mean precipitation value from the distribution $= (1/n)\sum_{i=1}^{n} x_i r_i = 3.803$ in., where $r_i$ is the relative frequency. The overall mean monthly precipitation is $\mu = 3.80$ in., with a standard deviation $\sigma = 2.08$ in. and a coefficient of variation $CV = \sigma/\mu = 0.55$.

## Example 9.2

Consider precipitation data for each month separately. Then, compute monthly precipitation intensity probability distributions. Indicate the mean precipitation value for the distribution. Also compute the entropy for each month.

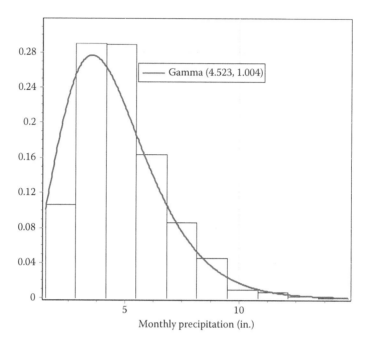

**FIGURE 9.1**    Empirical and fitted frequency distribution functions for monthly precipitation.

### Solutions

Frequency distributions for all 12 months are computed separately in the same way as in Example 9.1. For four sample months of January, April, August, and October, empirical and fitted distributions are shown in Figures 9.2 through 9.5. The gamma distribution is fitted to the empirical frequency distributions with parameters indicated in plots.

For each month, the mean values (in.) are given as follows:

|          | Jan  | Feb  | Mar  | Apr  | May  | Jun  | Jul  | Aug  | Sep  | Oct  | Nov  | Dec  |
|----------|------|------|------|------|------|------|------|------|------|------|------|------|
| Avg.     | 3.76 | 3.59 | 3.82 | 4.25 | 4.73 | 3.90 | 3.33 | 2.79 | 3.36 | 3.74 | 4.03 | 4.34 |
| St. Dev. | 1.96 | 1.59 | 1.67 | 2.10 | 2.31 | 2.17 | 1.72 | 1.65 | 1.85 | 2.70 | 2.35 | 1.98 |
| CV       | 0.52 | 0.44 | 0.44 | 0.49 | 0.49 | 0.56 | 0.52 | 0.59 | 0.55 | 0.72 | 0.58 | 0.46 |
| Skew     | 0.56 | 0.23 | 0.51 | 0.81 | 0.63 | 0.62 | 0.99 | 1.70 | 1.10 | 1.40 | 1.28 | 0.54 |

The *IE* computed for each month of the year and all months using monthly precipitation data of the climatic region 4 (East Texas) for the period 1895–2013 is given in the following:

| Jan  | Feb  | Mar  | Apr  | May  | Jun  | Jul  | Aug  | Sep  | Oct  | Nov  | Dec  | All  |
|------|------|------|------|------|------|------|------|------|------|------|------|------|
| 0.46 | 0.44 | 0.44 | 0.46 | 0.46 | 0.47 | 0.43 | 0.42 | 0.44 | 0.46 | 0.46 | 0.45 | 0.45 |

The monthly precipitation *IE* is plotted in Figure 9.6.

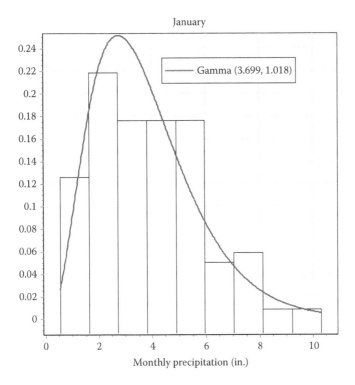

**FIGURE 9.2** Empirical and fitted frequency distribution functions of precipitation for the month of January.

### Example 9.3

Compute the *IE* of monthly precipitation for climate division 4 in Texas and plot it against skewness.

#### Solution

The monthly *IE* values computed in Example 9.2 show that a smaller *IE* value corresponds to a more skewed distribution and a higher *IE* value occurs for a less skewed distribution, as seen from Figure 9.7. It can also be seen that a positive correlation exists between *IE* and precipitation intensity. The value of *IE* is 0.54 for the climate division 4 (East Texas).

## 9.3 APPORTIONMENT ENTROPY OF TEMPORAL PRECIPITATION

The temporal distribution of precipitation or apportionment is important for assessing the availability of water resources. Since the temporally occurring values are random, the Tsallis entropy can be used to measure its uncertainty. For given precipitation data, frequencies of occurrence of discrete precipitation amounts spread over a given period can be determined and then can entropy (Maruyama and Kawachi, 1998).

Consider a historical precipitation time series $N$ years long. Let $r_{ij}$ be the aggregate precipitation during the $i$th time interval in the year. If the interval is 1 day, then

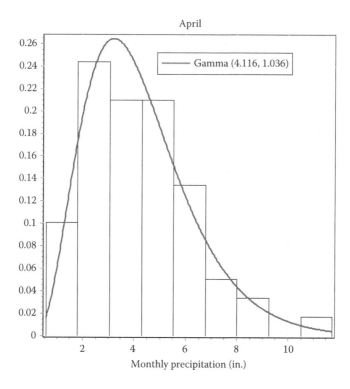

**FIGURE 9.3** Empirical and fitted frequency distribution functions of precipitation for the month of April.

$r_{ij}$ represents daily precipitation on the $i$th day in the $j$th year; here, $i = 1, 2, ..., 365$; $j = 1, 2, ..., N$. If the time interval is 1 month then $r_{ij}$ represents the precipitation amount in the $i$th month in the $j$th year, and $i = 1, 2, ..., 12$. To keep symbols simple, we can omit subscript $j$. For example daily precipitation values on January 1 and on December 31 for the same year can be expressed as $r_1$ and $r_{365}$, respectively. The aggregate precipitation during the year (annual precipitation), $R$, can then be expressed by the summation of $r_i$ from $i = 1$ to $i = 365$ as

$$R = \sum_{i=1}^{365} r_i \tag{9.3}$$

where the value of $r_i$ may be zero for some days and is finite for other days.

A precipitation series of $r_1, r_2, ..., r_n$ can thus be regarded as accumulated occurrences of unit rains for the $1, 2, ..., n$th days, respectively, and $r_i$ divided by the sample value of $R$ defines the probability $p_i$:

$$p_i = \frac{r_i}{R} \tag{9.4}$$

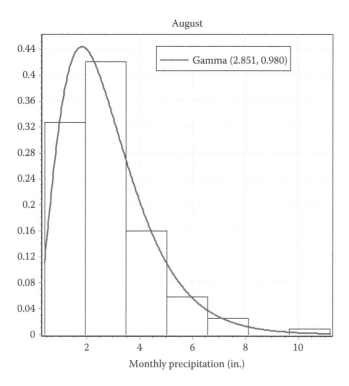

**FIGURE 9.4** Empirical and fitted frequency distribution functions of precipitation for the month of August.

This is the occurrence probability for the precipitation amount on the $i$th day, and therefore, its distribution represents the probabilistic characteristic of the over-a-year temporal apportionment of annual precipitation, that is of uncertainty of precipitation occurrence. In a similar manner, the probability distribution of precipitation can be defined for other time intervals, such as a week, 2 weeks, 1 month, or a season.

Substitution of Equation 9.4 into the Tsallis entropy equation (Equation 9.1) yields the value of entropy:

$$H = S_m = \frac{1}{m-1} \sum_{i=1}^{N} \frac{r_i}{R} \left[ 1 - \left( \frac{r_i}{R} \right)^{m-1} \right] \tag{9.5}$$

Equation 9.5 implies that the value of $H$ is independent of the sequential or chronological order of $r_i$ in the series. It is also seen that $H$ takes on a zero value when $R$ falls only on 1 day of the year and a maximum value when $R/n$ falls equally every day throughout the year, that is, the closer the entropy $H$ approaches its maximum value, the more uniform the precipitation apportionment is (i.e., the less temporally variable the precipitation is). $H$ can, thus, be regarded as a measure of precipitation

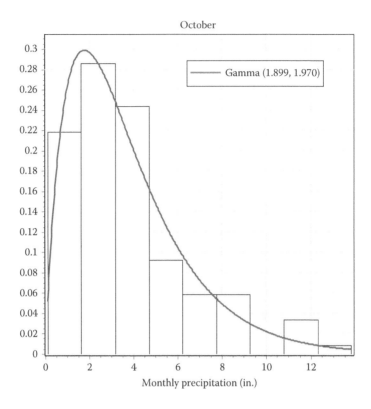

**FIGURE 9.5**  Empirical and fitted frequency distribution functions of precipitation for the month of October.

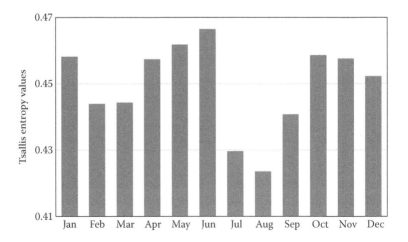

**FIGURE 9.6**  Month-wise precipitation intensity entropy. Computation is based on the precipitation data of climatic division 4 (East Texas) for the period of 1895–2013.

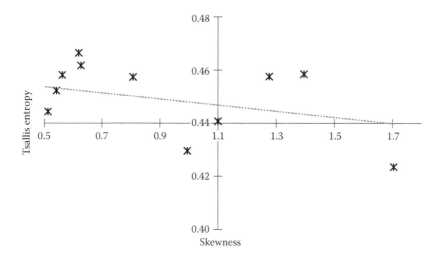

**FIGURE 9.7** Relation between intensity entropy and skewness of monthly precipitation PDF.

variability in a scalar sense. When yearly precipitation series for $M$ years is available at the same rain gauge, a better estimate of the annual entropy can be obtained by averaging the entropy values as

$$\overline{H} = \frac{1}{M} \sum_{j=1}^{M} H_j \qquad (9.6)$$

where

$\overline{H}$ is the average entropy of the $M$-year record
$H_j$ is the entropy of the $j$th year precipitation

## 9.3.1 APPORTIONMENT FROM EMPIRICAL DATA

Apportionment entropy ($AE$) can be computed in two ways. First, for precipitation data in a given year, $r_i$ is the precipitation for the $i$th day during the year. For example for January 1, $r_1$ is the sum of precipitation values that occurred on January 1 of that year. The same applies to other days. The frequency of precipitation occurrence on the $i$th day is expressed by Equation 9.4. In this manner, the temporal apportionment of precipitation within the year is computed. Then, yearly values are summed, and the average value is computed using Equation 9.6.

Second, if precipitation data are available for several years, then $r_i$ is the accumulated precipitation for the $i$th day for the whole period. Suppose 50 years of daily precipitation values are available then for January 1, $r_1$ is the sum of 50 precipitation values that occurred on January 1 of each year of the 50-year record. In that case, $R$ will be the sum of 50-year precipitation. The same applies to other days. The frequency of precipitation occurrence on the $i$th day is expressed by Equation 9.4. In this manner, the temporal apportionment of precipitation within the year is computed for the entire record.

### 9.3.2    APPORTIONMENT BY RANDOM EXPERIMENTATION

If yearly precipitation data are available but monthly or other short time interval data are not, then these data can be generated by random experimentation. In this manner, one can determine the throughout-the-year precipitation variability or disaggregate annual precipitation corresponding to a given time interval, such as daily. The random experiment, consisting of a number of trials, where each trial is considered in a probabilistic sense, can be conducted as follows.

If the annual precipitation for a given year is $R$, say $R = 1000$ mm, then the experiment may consist of 1000 trials. The value of $R$ is rounded off to the next nearest integer. Each trial contains 365 days, January 1 to December 31. From each trial one day is randomly selected, that means that any day of the year has the same probability of being selected. The selected day is assigned a value of precipitation of 1 mm. Let us suppose that January 3 happens to be selected during this trial, then it will be assigned a precipitation value of 1 mm. Similarly, the next trial is performed and a day is selected randomly. The selected day will again be assigned a precipitation value of 1 mm. In this manner, 1000 trials are made, and the total amount of precipitation associated with the selected days will be 1000 mm. If a particular day happens to be randomly selected 10 times, then it will be assigned a precipitation value of 10 mm (which is the sum of 10 selections). The days not selected will be assigned a zero value. When the precipitation values, thus generated, are plotted against days, the result will be the precipitation series for the year. The same can be done for other years. In this manner, the entire precipitation time series can be constructed by simply knowing yearly values.

### 9.3.3    CALCULATION OF PRECIPITATION APPORTIONMENT ENTROPY
####           AND ISOENTROPY LINES

The precipitation series with a 1-day resolution can be considered to describe the throughout-the-year precipitation distribution. The sequence of observed daily precipitation values in a year is described by a probability distribution of precipitation occurrence, and the Tsallis entropy value is obtained. The entropy is computed for all available yearly precipitation sequences. Then, an average of the entropies obtained over the years of interest is considered as the average annual entropy. The average annual entropy values, thus obtained, for the rain gauges scattered throughout the area are employed to construct an isoentropy map that delineates precipitation variability. Entropy can be considered as a measure of precipitation variability.

**Example 9.4**

> Consider precipitation records in the state of Texas, that is, long-term monthly precipitation data set from the Full Network Estimated Precipitation (FNEP), which has been created as an alternative to the National Climatic Data Center (NCDC) data set. The data set is developed statewide for individual climate divisions

(http://climatexas.tamu.edu/). There are 10 climate divisions in the state of Texas: 1 = High Plains, 2 = Low Rolling Plains, 3 = North Central, 4 = East Texas, 5 = Trans Pecos, 6 = Edwards Plateau, 7 = South Central, 8 = Upper Coast, 9 = Southern, 10 = Lower Valley. Data considered for each of the divisions are monthly precipitation values (in inches) over a period of 1895–2013. Compute yearly precipitation amounts for each climate division for the entire period of record. Also, compute the average yearly precipitation for the entire record. Using monthly precipitation values, compute the entropy of each climate division for each year. Compute the mean, standard deviation, and CV of monthly precipitation, considering only those months when it rained. Also, compute the standard deviation and CV of yearly entropy. Plot yearly precipitation as well as yearly entropy.

### Solution

Ten climate divisions, having records from 1895 to 2013, are selected. These climate divisions, along with their average annual precipitation amounts, are given in Table 9.2. The mean, standard deviation, and CV of monthly precipitation are computed, considering only those months when it rained. In this manner, these statistics are computed for each year and then they are averaged for the whole record, as shown in Table 9.3. Using monthly precipitation values, the Tsallis entropy is computed for each climate division for each year. To that end, the probability of precipitation in the $i$th month is computed by taking the ratio of monthly precipitation to yearly precipitation. This leads to the probability distribution of monthly precipitation apportionment. Using these probability values, the yearly entropy is computed, as shown in Table 9.4. Then, the average entropy for the period of record is computed using these yearly entropy values. Also, the standard deviation and CV of yearly entropy are computed, as shown in Table 9.4. Yearly precipitation as well as yearly entropy is plotted. Now, the standard deviation and CV of yearly entropy are computed, as shown in Table 9.4. Also, minimum and maximum entropy values are shown in Table 9.4 and Figure 9.8.

### TABLE 9.2
### Precipitation Amounts for Climate Divisions in the State of Texas

| Climate Divisions in Texas (TX) | | Average Yearly Precipitation (in.) | Std. Dev. (in.) | Coefficient of Variation |
|---|---|---|---|---|
| TX-01 | High Plains | 18.97 | 4.40 | 0.23 |
| TX-02 | Low Rolling Plains | 23.48 | 5.45 | 0.23 |
| TX-03 | North Central | 33.46 | 6.86 | 0.21 |
| TX-04 | East Texas | 45.63 | 8.30 | 0.18 |
| TX-05 | Trans Pecos | 12.33 | 3.76 | 0.30 |
| TX-06 | Edwards Plateau | 25.02 | 6.16 | 0.25 |
| TX-07 | South Central | 32.70 | 8.11 | 0.25 |
| TX-08 | Upper Coast | 46.13 | 10.38 | 0.23 |
| TX-09 | Southern | 23.24 | 6.20 | 0.27 |
| TX-10 | Lower Valley | 23.55 | 5.94 | 0.25 |

**TABLE 9.3**

**Mean Monthly Precipitation in Each Climate Division**

| Climate Division | | Average Monthly Precipitation (in.) | Std. Dev. Monthly Precipitation | CV |
|---|---|---|---|---|
| TX-01 | High Plains | 1.580 | 1.341 | 0.848 |
| TX-02 | Low Rolling Plains | 1.957 | 1.583 | 0.809 |
| TX-03 | North Central | 2.788 | 1.851 | 0.664 |
| TX-04 | East Texas | 3.803 | 2.081 | 0.547 |
| TX-05 | Trans Pecos | 1.027 | 1.003 | 0.977 |
| TX-06 | Edwards Plateau | 2.085 | 1.603 | 0.769 |
| TX-07 | South Central | 2.725 | 1.967 | 0.722 |
| TX-08 | Upper Coast | 3.844 | 2.465 | 0.641 |
| TX-09 | Southern | 1.936 | 1.641 | 0.848 |
| TX-10 | Lower Valley | 1.962 | 1.908 | 0.972 |

**TABLE 9.4**

**Average Yearly Entropy and Its Standard Deviation of Each Climate Division**

| Climate Division | | Average Entropy | Std. Dev. | Min Tsallis Entropy | Max Tsallis' Entropy |
|---|---|---|---|---|---|
| TX-01 | High Plains | 0.55 | 0.031 | 0.499 | 0.626 |
| TX-02 | Low Rolling Plains | 0.56 | 0.030 | 0.499 | 0.630 |
| TX-03 | North Central | 0.56 | 0.025 | 0.511 | 0.612 |
| TX-04 | East Texas | 0.54 | 0.023 | 0.499 | 0.599 |
| TX-05 | Trans Pecos | 0.53 | 0.028 | 0.475 | 0.614 |
| TX-06 | Edwards Plateau | 0.56 | 0.026 | 0.503 | 0.613 |
| TX-07 | South Central | 0.55 | 0.028 | 0.486 | 0.615 |
| TX-08 | Upper Coast | 0.54 | 0.026 | 0.490 | 0.602 |
| TX-09 | Southern | 0.55 | 0.031 | 0.454 | 0.614 |
| TX-10 | Lower Valley | 0.54 | 0.035 | 0.462 | 0.632 |

## Example 9.5

Conduct a random experiment for monthly precipitation apportionment and then do the same calculations as in Example 9.4. Now, compare the two cases and comment on the results. Place entropy values obtained in two ways on the Brazos River basin map. What do these two maps show?

### Solution

Now, we perform an experiment by generating random monthly precipitation values for each of the 10 climate divisions in the state of Texas over a 119-year period. The mean, standard deviation, and CV of monthly precipitation are computed.

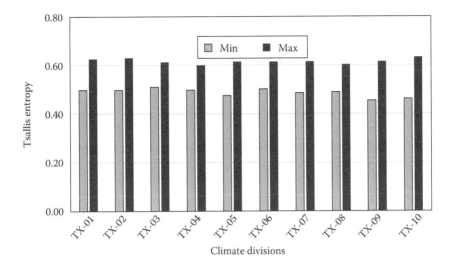

**FIGURE 9.8** Yearly minimum and maximum values of the Tsallis entropy (Min, Minimum; Max, Maximum).

In this manner, these statistics for each year are obtained and then they are averaged for the whole record, as given in Table 9.5 and Figure 9.9. Then, we calculate precipitation apportionment and compute entropy as in Example 9.4, using the Tsallis entropy. The average entropy of each climate division is determined, and the standard deviation and CV of yearly entropy are computed, and the minimum and maximum values of yearly entropy are obtained, as shown in Table 9.6.

**TABLE 9.5**

**Mean, Standard Deviation and Coefficient of Variation of Monthly Random Precipitation**

| Climate Division | | Average Monthly Precipitation (in.) | Std. Dev. Monthly Precipitation (in.) | CV |
|---|---|---|---|---|
| TX-01 | High Plains | 4.496 | 2.545 | 0.566 |
| TX-02 | Low Rolling Plains | 5.563 | 3.170 | 0.570 |
| TX-03 | North Central | 6.122 | 3.462 | 0.565 |
| TX-04 | East Texas | 6.973 | 4.020 | 0.577 |
| TX-05 | Trans Pecos | 2.704 | 1.595 | 0.590 |
| TX-06 | Edwards Plateau | 4.582 | 2.571 | 0.561 |
| TX-07 | South Central | 6.607 | 3.895 | 0.589 |
| TX-08 | Upper Coast | 7.495 | 4.318 | 0.576 |
| TX-09 | Southern | 6.232 | 4.079 | 0.654 |
| TX-10 | Lower Valley | 8.962 | 5.159 | 0.576 |

*Note:* Std. dev., Standard deviation.

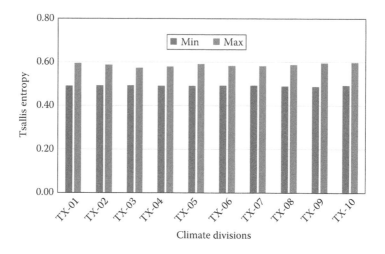

**FIGURE 9.9** Maximum and minimum yearly Tsallis entropy values for the climate divisions based on random precipitation data.

**TABLE 9.6**

**Average Entropy, Standard Deviation, and Values of Minimum and Maximum Entropy of Each Climate Division**

| Climate Division | | Average Tsallis Entropy | Std. Dev. Tsallis Entropy | Min Tsallis Entropy | Max Tsallis Entropy |
|---|---|---|---|---|---|
| TX-01 | High Plains | 0.536 | 0.022 | 0.490 | 0.595 |
| TX-02 | Low Rolling Plains | 0.536 | 0.023 | 0.493 | 0.587 |
| TX-03 | North Central | 0.533 | 0.022 | 0.493 | 0.573 |
| TX-04 | East Texas | 0.535 | 0.024 | 0.491 | 0.579 |
| TX-05 | Trans Pecos | 0.534 | 0.024 | 0.491 | 0.591 |
| TX-06 | Edwards Plateau | 0.539 | 0.023 | 0.493 | 0.583 |
| TX-07 | South Central | 0.531 | 0.023 | 0.492 | 0.582 |
| TX-08 | Upper Coast | 0.535 | 0.025 | 0.489 | 0.588 |
| TX-09 | Southern | 0.532 | 0.026 | 0.488 | 0.595 |
| TX-10 | Lower Valley | 0.535 | 0.025 | 0.493 | 0.597 |

Now, comparing the two cases, it is seen that the entropy calculated with the random experiment is larger than that for the first case using actual values. The reason is that in the second case the experiment is random, wherein the frequency of occurrence of precipitation tends to be uniform. This also explains the case of average entropy and yearly entropy. Due to random sampling, the standard derivation is less for the second case.

For precipitation in the case of random experiment, the mean is less and the standard derivation is also less than that in the first case. This is because random sampling will spread the distribution of the amount of precipitation uniformly during the record and thus will decrease the mean. For the same reason, the standard derivation is less which means the data are "smooth" or uniform during the record. In other words, random sampling of precipitation during the record can decrease the mean and standard derivation of precipitation, decrease the standard deviation of entropy, and increase the entropy (including average entropy and yearly entropy).

Now, entropy values obtained in two ways are mapped for the state of Texas, as shown in Figures 9.10 and 9.11.

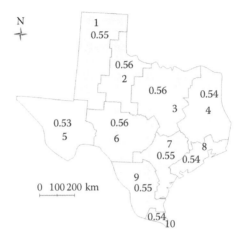

**FIGURE 9.10** Entropy of precipitation from observations for climate divisions in Texas. The climatic divisions in Texas are as follows: 1, High Plains; 2, Low Rolling Plains; 3, North Central; 4, East Texas; 5, Trans Pecos; 6, Edwards Plateau; 7, South Central; 8, Upper Coast; 9, Southern; 10, Lower Valley.

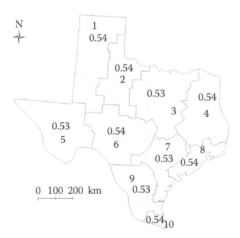

**FIGURE 9.11** Entropy of precipitation from random experimentation for the climate divisions in Texas. The climatic regions in Texas are as follows: 1, High Plains; 2, Low Rolling Plains; 3, North Central; 4, East Texas; 5, Trans Pecos; 6, Edwards Plateau; 7, South Central; 8, Upper Coast; 9, Southern; 10, Lower Valley.

## Example 9.6

Using the results from Example 9.4, construct a map of isoentropy lines for the climate division in Texas and comment on the information obtained from the map. Also, construct an isohyetal map. What does this map reflect? Also construct a map of the CV. What does this map reflect? Taken together, what do these three maps convey? Which map should be preferred and why?

### Solution

The isoentropy map is constructed, as shown in Figure 9.12, in which entropy indicates significant zones over the whole state of Texas and can delineate a plausible climatic map that qualitatively explains precipitation variability. The distribution of averaged annual precipitation, delineated by isohyetal lines, is given in Figure 9.13. This map indicates the spatial variability of mean precipitation. The distribution of

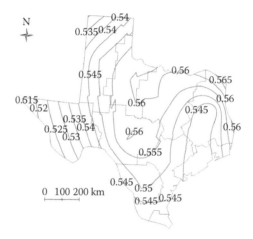

**FIGURE 9.12** Isoentropy lines for the state of Texas.

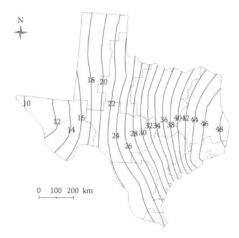

**FIGURE 9.13** Isohyets of annual rainfall (in.) for the state of Texas.

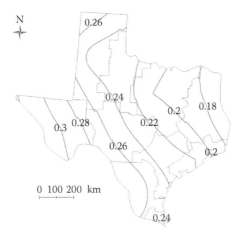

**FIGURE 9.14**   Coefficient of variation for the state of Texas.

averaged annual precipitation can be related to the zones delineated by isoentropy lines. The potential availability of water resources can be assessed or categorized by isoentropy lines. The spatial distribution of the CV of monthly precipitation is shown in Figure 9.14. The CV of monthly precipitation at each rain gauge is computed in the same way as entropy. The average monthly value of precipitation is obtained for each rain gauge for each year. Then, CV is computed for each year. Summing the CV values obtained for each year and dividing the sum by the number of years of record, the annually averaged CV value is obtained for each climate division. As compared with the entropy-based map, the values of CV show a similar pattern, as shown in Figure 9.12 and one can make a clear classification of the region. As a result, it does have a clear match with the climate division map. Taken together, the entropy can delineate a plausible climatic map that qualitatively explains precipitation variability. The isoentropy map is preferred, since it shows significant zones over the state of Texas.

## 9.4   ENTROPY SCALING

The Tsallis entropy can be computed for different timescales $t$, including $t \in \{1, 2, 7, 15, 30, 60, 120, 240, 360\,\text{days}\}$. Data of daily precipitation from the College Station Easterwood Airport for a period of 1960–2013 are used to compute the Tsallis entropy with $m = 3$. Results are presented in Table 9.7 and graphed in Figure 9.15. Then, entropy scaling is applied to the 10 climate divisions in Texas using the timescales of 1, 2, 3, and 6 months, as presented in Table 9.7 and Figure 9.16.

**TABLE 9.7**
**Entropy Scaling Using Tsallis Entropy**

| 1 D | 2 D | 7 D | 15 D | 30 D | 60 D | 90 D | 120 D | 240 D | 360 D |
|------|------|------|------|------|------|------|------|------|------|
| 0.15 | 0.23 | 0.41 | 0.47 | 0.49 | 0.49 | 0.49 | 0.49 | 0.49 | 0.49 |

D, days.

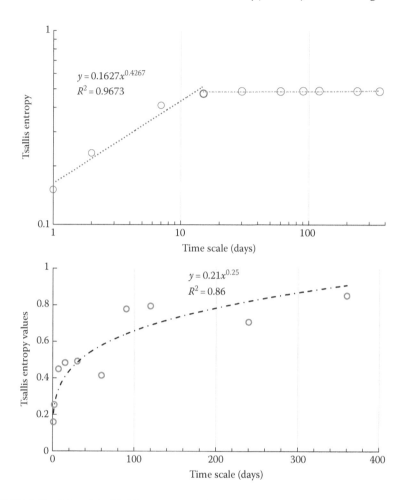

**FIGURE 9.15**   Plot of the Tsallis entropy scaling values with $m = 3$.

## 9.5   DETERMINATION OF WATER RESOURCES AVAILABILITY

Development and management of water resources require not only the knowledge of aggregate precipitation but also its space–time variability. For evaluating the availability of water resources in a watershed and development of a basin-wide strategy for water resources development, the temporal variability of precipitation is a major determinant. Further, the uncertainty associated with the temporal distribution of precipitation or apportionment is also important. Both precipitation $IE$ and $AE$ have been discussed, and now the objective is to use these entropies for assessing the availability of water resources in a given region.

### 9.5.1   PRECIPITATION DATA

For assessing the water resources availability in a region, the first step is to determine the availability of precipitation data. To that end, all rain gauges for the area under

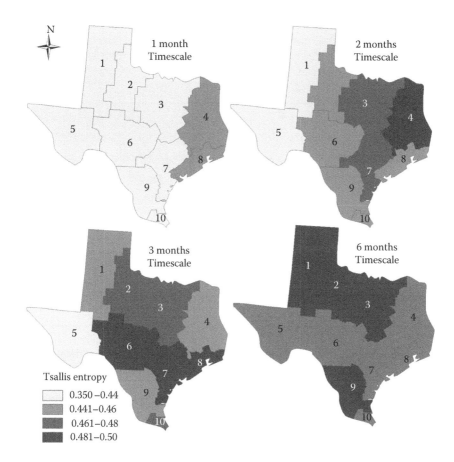

**FIGURE 9.16** Entropy scaling for the Texas climatic divisions. The climatic regions in Texas are as follows: 1, High Plains; 2, Low Rolling Plains; 3, North Central; 4, East Texas; 5, Trans Pecos; 6, Edwards Plateau; 7, South Central; 8, Upper Coast; 9, Southern; 10, Lower Valley.

consideration are selected. Daily precipitation data for these gauges are obtained. If some yearly precipitation values ($M$) measured at these gauges have missing values due to the failure of recording and/or some gauges have extremely short duration of data acquisition, then these precipitation series may be omitted from analysis. For a reliable estimation of entropy $H$ and mean entropy $\overline{H}$ at any rain gauge, precipitation series and rain gauges should be screened, based on two criteria: (a) Any yearly precipitation series to be selected must have a complete set of daily precipitation data, thus having 365 and 366 consecutive data for the common and leap years, respectively. (b) Any rain gauge to be selected must have at least 10 years or preferably more than 10 years of precipitation observations. (This may vary from one watershed or country to another.) The gauges must satisfy the criterion (a). Criterion (b) requiring $M \geq 10$ is based on the acceptance that the meteorological data consecutively observed over 10 years or more can be used for the description of quasi-averaged yearly meteorology (Maruyama and Kawachi, 1998).

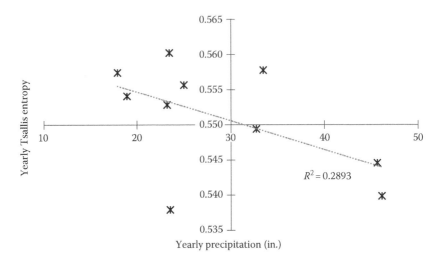

**FIGURE 9.17**    Relation between average yearly entropy and yearly precipitation.

### 9.5.2 CATEGORIZATION OF WATER RESOURCES AVAILABILITY

Of interest is the relation of entropy to precipitation. A diagram of averaged annual entropy on the ordinate and averaged annual precipitation on the abscissa shows that the average annual entropy and the average annual precipitation are less mutually related, with a small correlation coefficient (see Figure 9.17). This suggests that besides the aggregate precipitation, its temporal apportionment can be a significant aspect of precipitation data. When coupled, entropy and precipitation on a yearly basis become a measure of the throughout-the-year potential availability of water resources (Kawachi et al., 2001). To explain it qualitatively, the whole plotted area can be divided into, say, four parts, each delineated with two intersecting lines that pass through the means of the respective two variables. Then, in terms of water resources availability, the respective quadrants, I, II, III, and IV, can be comparatively categorized. This is explained using an example of the Texas climate divisions.

**Example 9.7**

Using the values obtained from Example 9.2, plot average annual entropy against average annual precipitation or precipitation on a rectangular paper. Then, divide the plot into four parts based on the lines issuing from the mean annual precipitation and the mean entropy. What do you conclude from each part? Should the watershed be divided into more than four parts and if so then what should those parts be?

**Solution**

The average annual entropy is plotted against average annual precipitation, as shown in Figure 9.17. The first and third quadrants seem less represented. However, from the second and fourth quadrants we can note a similar trend, as the average yearly Tsallis entropy tends to decrease, while the average yearly precipitation increases. This also indicates that regions with less precipitation have higher variability, suggesting that water needs to be managed properly. As a recommendation, we suggest that water needs to be stored for future use through a wise groundwater management strategy. On the other hand, one can infer that the available water resources are reliable in regions with high precipitation patterns.

### 9.5.3 Assessment of Water Resources Availability

Let the total amount of precipitation over a month or simply monthly precipitation (or intensity) be a random variable. Then, its probability distribution can be derived from data. The entropy is then obtained, and it is referred to as *IE*. The ratio of monthly precipitation to the sum of monthly precipitation values over a year (i.e., to annual precipitation) can also be considered as a random variable. These relative precipitation intensities over a year reflect the probabilistic characteristics of

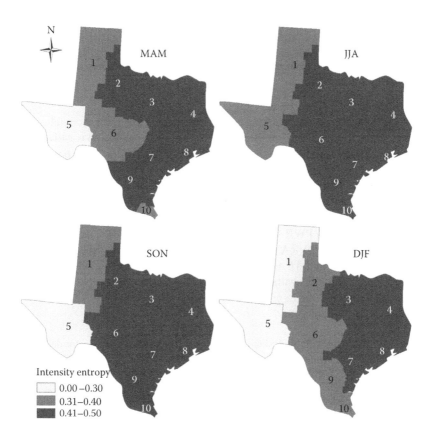

**FIGURE 9.18** Seasonal variability of intensity entropy across the Texas climatic regions. MAM, March–April–May; JJA, June–July–August; SON, September–October–November; DJF, December–January–February. The climatic regions in Texas are as follows: 1, High Plains; 2, Low Rolling Plains; 3, North Central; 4, East Texas; 5, Trans Pecos; 6, Edwards Plateau; 7, South Central; 8, Upper Coast; 9, Southern; 10, Lower Valley.

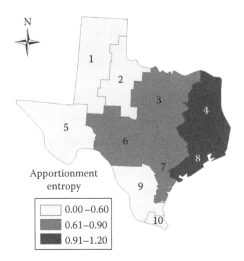

**FIGURE 9.19** Variability of the apportionment entropy across the Texas climatic regions. The climatic regions in Texas are as follows: 1, High Plains; 2, Low Rolling Plains; 3, North Central; 4, East Texas; 5, Trans Pecos; 6, Edwards Plateau; 7, South Central; 8, Upper Coast; 9, Southern; 10, Lower Valley.

precipitation occurrence in the year. Since these ratios comprise precipitation apportionment rates for all months in the year, the entropy so calculated is called the *AE*.

The potential water resources availability (PWRA) in an area can be assessed in terms of disorder in intensity and over-a-year apportionment of monthly precipitation. The disorder can be measured by the two entropies mentioned earlier, *IE* and *AE*. These entropies can be standardized and pairs of standardized *IE* and *AE* for different locations of rain gauges can be plotted. Then, simple clustering can be considered for delineating PWRA distributed over an area of interest and for classifying regional attributes of PWRA (Maruyama et al., 2005). Figures 9.18 and 9.19 depict the variability of *IE* and *AE* across different climate divisions in Texas. They reflect the uncertainty as to the availability of water resources.

## REFERENCES

Kawachi, T., Maruyama, T., and Singh, V.P. (2001). Precipitation entropy for delineation of water resources zones in Japan. *Journal of Hydrology*, 246, 36–44.

Maruyama, T. and Kawachi, T. (1998). Evaluation of precipitation characteristics using entropy. *Journal of Rainwater Catchment Systems*, 4(1), 7–10.

Maruyama, T., Kawachi, T., and Singh, V.P. (2005). Entropy-based assessment and clustering of potential water resources availability. *Journal of Hydrology*, 309, 104–113.

Tsallis, C. (1988). Possible generalizations of Boltzmann-Gibbs statistics. *Journal of Statistical Physics*, 52(1/2), 479–487.

# 10 Infiltration

Infiltration is fundamental to determining the runoff hydrograph, soil moisture and groundwater recharge, irrigation efficiency, life span of pavements, and leaching of nutrients. In hydrology, irrigation engineering, watershed management, and soil science, a number of infiltration equations have been developed, some of which are now commonly applied in hydrologic modeling and have been included in popular watershed hydrology models (Singh, 1989, 1995; Singh and Frevert, 2002a,b, 2006; Singh and Woolhiser, 2002). Some of the commonly used equations (Singh and Yu, 1990) are Green and Ampt (1911), Kostiakov (1932), Horton (1938), Philip two-term (Philip, 1957), Holtan (1961), and Overton (1964). These equations represent the potential or capacity rate of infiltration at a point. The objective of this chapter is to present the derivation of some popular infiltration equations using the Tsallis entropy theory.

## 10.1 PRELIMINARIES

In this chapter, infiltration rate will imply capacity or potential rate that is the maximum rate at which water enters the soil under no restriction on the supply of water. Clearly, the actual rate of infiltration is less than or equal to the potential rate, depending on the availability of water. It is known that infiltration is a function of antecedent soil moisture, soil characteristics, vegetation, land use, climatic characteristics, land slope, and supply of water or rainfall. Some of these factors vary in space, some vary in time, and some vary in both space and time. Soil characteristics vary significantly from one place to another, and antecedent soil moisture, which defines the initial infiltration, also significantly varies spatially. The infiltration parameters determined using point measurements are point values, or at best reflect average values. Although large spatial variability in infiltration is recognized, little effort has been made to account for its probabilistic characteristics, except for a few watershed models, as for example the BASINS (formerly Stanford Watershed Model) (Crawford and Linsley, 1966; Donigian and Imhoff, 2006). Crawford and Linsley (1966) were probably the first to consider spatial variations in infiltration capacity; from empirical data reported in the literature (Burgy and Luthin, 1956), they found large variations in infiltration capacity even in relatively homogeneous soils (uniform Yolo silt loam) and over small areas (40 ft × 20 ft). Considering infiltration capacity as a random variable, they expressed the cumulative probability distribution function of infiltration capacity as a function of area.

## 10.2 FORMULATION OF ENTROPY THEORY

Let the infiltration capacity (or infiltrability), as a function of time $t$, be defined as $I(t)$. It is assumed that the soil is dry, and water is applied to the dry soil with no limitation to the supply of water. At the beginning, infiltration will be maximum, and as time progresses, the infiltration capacity declines and may reach a steady or constant rate or even approach zero. The constant rate is often called the drainage rate. This capacity of infiltration is the potential rate and will be equal to or greater than the actual rate, depending on the supply of water. Since the infiltration capacity may significantly vary from one place to another, it is assumed that the spatially averaged infiltration capacity $I(t)$ is a random variable and would therefore have a probability density function (PDF). It is recognized that this assumption needs to be verified or may even be tenuous but even if it is weakly true it would not greatly mar the usefulness of the entropy theory.

Consider a discrete form of infiltration capacity $I$ with probability distribution $P$. The infiltration capacity can take on $N$ values with each value corresponding to a different time, $I = \{I_i, i = 1, 2,..., N\}$, occurring with probabilities $P = \{p_i, i = 1, 2,..., N\}$. The Tsallis entropy (Tsallis, 1988, 2002, 2004), denoted by $H$, can be written as

$$H = \frac{k}{m-1} \sum_{i=1}^{N} p_i \left(1 - p_i^{m-1}\right) = k \frac{1 - \sum_{i=1}^{N} p_i^m}{m-1} \tag{10.1}$$

where
    $k$ is a measure that keeps the units of $H$ consistent and is often taken as unity
    $p_i = p(I_i)$
    $m$ is any real number

Exponent $m$ influences the variability of $H$ of $I$ with the probability. The quantity $(1 - p_i^{m-1})/(m-1)$ is a measure of the uncertainty in $p_i$ of $I_i$.

If the infiltration capacity is defined as a continuous random variable with a PDF defined as $f(I)$, then the Tsallis entropy, $H(I)$, can be expressed as

$$H(I) = \frac{k}{m-1} \int_{I_L}^{I_U} f(I)\{1 - [f(I)]^{m-1}\}dI = \frac{k}{m-1} \left[1 - \int_{I_L}^{I_U} [f(I)]^m \, dI\right] \tag{10.2}$$

where
    $I_U$ and $I_L$ are, respectively, the upper and lower limits of integration for $I$
    $H$ describes the expected value of $\{1 - [f(I)]^{m-1}\}/(m-1)$

Considering $\{1 - [f(I)]^{m-1}\}/(m-1)$ as a measure of uncertainty, Equation 10.2 defines the average uncertainty associated with $f(I)$ and in turn with $I$. More uncertain $I$ is, more information will be needed to characterize it. The key here is to derive the least-biased $f(I)$.

## 10.3 METHODOLOGY FOR THE DERIVATION OF INFILTRATION EQUATIONS

The method for deriving infiltration equations using the entropy theory comprises the following parts (Singh, 2013): (1) application of principle of maximum entropy (POME), (2) specification of information on infiltration rate in terms of constraints, (3) maximization of entropy in accordance with POME, (4) derivation of the probability distribution of infiltration rate and its entropy, (5) formulation of continuity equation, (6) statement of cumulative distribution hypothesis, (7) relation between cumulative infiltration and infiltration rate, and (8) derivation of infiltration rate. The Tsallis entropy is already defined, and the remainder of these parts is outlined in what follows.

### 10.3.1 PRINCIPLE OF MAXIMUM ENTROPY

The principle of maximum entropy formulated by Jaynes (1957a,b, 1982) says that the least-biased probability distribution of $I$, $f(I)$, will be the one that will maximize $H(I)$ given by Equation 10.1 or 10.2, subject to the given information on $I$ expressed as constraints. In other words, if no information other than the given constraints is available, then the probability distribution should be selected such that it is least biased toward what is not known. Such a probability distribution is yielded by the maximization of the Tsallis entropy. Thus, one of the key points is to define constraints on $I$, for $f(I)$ depends on these constraints.

### 10.3.2 SPECIFICATION OF CONSTRAINTS

Information on $I(t)$ can be obtained using the knowledge of soil physics and experimental or field observations. For a given soil, one frequently measures infiltration and then characterizes the soil infiltration and more particularly the time capacity rate of infiltration or infiltration curve for the soil under the condition that water supply is not a limiting factor. If infiltration capacity rate observations are available, then information on the infiltration capacity rate can be expressed in terms of constraints, $C_r$, $r = 0, 1, 2, \ldots, n$, as

$$C_0 = \int_{I_L}^{I_U} f(I)dI = 1 \tag{10.3}$$

$$C_r = \int_{I_L}^{I_U} g_r(I)f(I)dI = \overline{g_r(I)}, \quad r = 1, 2, \ldots, n \tag{10.4}$$

where
$g_r(I)$, $r = 1, 2, \ldots, n$, represent some functions of $I$
$n$ denotes the number of constraints
$\overline{g_r(I)}$ is the expectation of $g_r(I)$

The constraints are analogous to moments. For example, if $r = 1$, and $g_1(I) = I$, then Equation 10.4 would correspond to the mean infiltration capacity rate; likewise, for $r = 2$, and $g_2(I) = (I - \bar{I})^2$, Equation 10.4 would denote the variance of $I$. For most infiltration equations used in hydrology, more than two constraints are not needed. The role of constraints cannot be overemphasized. The type of probability distribution that one obtains by maximizing the entropy depends on the type of constraints one defines. Thus, there is a one-to-one correspondence between the PDF and its constraints. In the case of deriving a specific infiltration equation, the problem becomes tricky, since its PDF is not known a priori. Hence, trial and error seems to be the only option in the beginning.

### 10.3.3 Maximization of Tsallis Entropy

In order to obtain the least-biased $f(I)$, the entropy given by Equation 10.2 is maximized, subject to Equations 10.3 and 10.4, and one simple way to achieve maximization is the use of the method of Lagrange multipliers. To that end, the Lagrangian function, $L$, can be expressed (Singh, 1998) as

$$L = \frac{1}{m-1} \int_{I_L}^{I_U} f(I)\{1 - [f(I)]^{m-1}\} dI + \lambda_0 \left[ \int_{I_L}^{I_U} f(I) dI - C_0 \right] + \sum_{r=1}^{n} \lambda_r \left[ \int_{I_L}^{I_U} f(I) g_r(I) dI - C_r \right]$$

(10.5)

where $\lambda_r$, $r = 0, 1, 2, …, n$, are the Lagrange multipliers. Recalling the Euler–Lagrange equation of calculus of variation, the least-biased $f(I)$ is obtained by maximizing $L$, noting that $f$ is variable and $I$ is parameter. Thus, differentiating Equation 10.5 and equating the derivative to zero, one gets

$$\frac{\partial L}{\partial f(I)} = 0 \Rightarrow \frac{1}{m-1}\{1 - [f(I)]^{m-1}\} - (m-1)[f(I)]^{m-1} + \lambda_0 + \sum_{r=1}^{n} \lambda_r g_r(I) = 0 \quad (10.6)$$

### 10.3.4 Derivation of Probability Distribution and Maximum Entropy

Solution of Equation 10.6 leads to the PDF of $I$ in terms of the given constraints:

$$f(I) = \left\{ \frac{1}{m} + \frac{(m-1)}{m} \left[ \lambda_0 + \sum_{r=1}^{n} \lambda_r g_r(I) \right] \right\}^{1/(m-1)}$$

(10.7)

The Lagrange multipliers, $\lambda_r$s, can be determined with the use of Equations 10.3 and 10.4. Equation 10.7 is the entropy-based PDF of infiltration rate of power type. The cumulative probability distribution function of $I$, $F(I)$, can be written as

$$F(I) = \int_{I_L}^{I} \left\{ \frac{1}{m} + \frac{m-1}{m} \left[ \lambda_0 + \sum_{r=1}^{m} \lambda_r g_r(I) \right] \right\}^{1/(m-1)} dI \qquad (10.8)$$

Substituting Equation 10.7 in Equation 10.2, one obtains the maximum entropy of $f(I)$ or $I$:

$$H(I) = \frac{1}{m-1} \int_{I_L}^{I_U} \left\{ \frac{1}{m} + \frac{(m-1)}{m} \left[ \lambda_0 + \sum_{r=1}^{n} \alpha_r g_r(I) \right] \right\}^{1/(m-1)}$$

$$\times \left[ 1 - \left\{ \frac{1}{m} + \frac{(m-1)}{m} \left[ \lambda_0 + \sum_{r=1}^{n} \lambda_r g_r(I) \right] \right\} \right] dI \qquad (10.9)$$

Equations 10.2 through 10.4, 10.7, and 10.9 constitute the building blocks of the entropy theory of infiltration, which is now illustrated by deriving six popular infiltration equations as examples (Singh, 2010).

### 10.3.5 FORMULATION OF CONTINUITY EQUATION

Consider a dry soil element, as shown in Figure 10.1, to which water is supplied without any limitation. The water infiltrates the soil element at a capacity rate of $I(t)$ and exits it at a rate of $I_c(t)$. The soil will have a maximum soil moisture retention capacity denoted by $S$. For a dry soil $S$ will be equal to the soil porosity multiplied by the soil elemental volume minus the volume of pore spaces occupied by roots, earthworms, or other objects. The soil elemental volume is computed by choosing an appropriate length of the element that depends on the soil type under consideration.

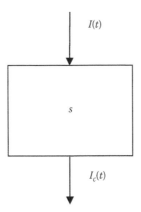

**FIGURE 10.1** Soil element with infiltration. $I(t)$, rate of infiltration; $I_c(t)$, rate of infiltration exiting the element; $S$, soil moisture retention capacity.

In general it is taken as the crop root zone depth that may be about 100 cm or about 3 ft. In a dry soil with no macropores, the maximum amount of water retained will be the same as the cumulative infiltration $J$; that is, $0 \leq J \leq S$. If $W$ is the amount of pore space available for infiltration of water at any time, then $W + J = S$.

The continuity equation for a soil element can now be expressed as follows (Singh and Yu, 1990):

$$\frac{dJ}{dt} = I(t) - I_c(t) \tag{10.10}$$

One can also express the continuity equation (Equation 10.10) as

$$J(t) = \int_0^t [I(t) - I_c(t)]dt \tag{10.11}$$

Strictly speaking $I_c$ varies in time but for the discussion in this chapter, it is assumed constant for two reasons, that is $I_c(t) = I_c$. First, the infiltration equations considered here assume a constant value of $I_c$. Second, the measurements of $I_c$ varying in time are usually not available.

## 10.3.6 Cumulative Probability Distribution Hypothesis

It is hypothesized that the cumulative probability distribution of infiltration $F(I)$ can be defined as the ratio of soil moisture potential $(W)$ to the maximum soil moisture retention $(S)$:

$$F(I) = \frac{W}{S} \tag{10.12}$$

Term $W$ defines the volume of pore space available for infiltration, that is $S$ minus the volume of water infiltrated $J$. Thus, $F(I)$ can also be defined as one minus the ratio of the cumulative infiltration to the maximum potential cumulative infiltration or maximum soil moisture retention, $S$:

$$F(I) = 1 - \frac{J}{S} \tag{10.13}$$

The hypothesis expressed by Equation 10.13 needs to be validated using field data or experimental observations. Differentiation of Equation 10.13 yields

$$dF(I)dI = -\frac{dJ}{S}; \quad dF(I) = f(I) = -\frac{1}{S}\frac{dJ}{dI} \tag{10.14}$$

where $f(I)$ is the PDF of $I(t)$, which is determined using the entropy theory.

### 10.3.7 Relation between Cumulative Infiltration and Infiltration Rate

Substitution of Equation 10.8 in Equation 10.14 and then integration result in

$$J = S \int_{I_U}^{I} \left\{ \frac{1}{m} + \frac{m-1}{m} \left[ \lambda_0 + \sum_{r=1}^{n} \lambda_r g_r(I) \right] \right\}^{1/(m-1)} dI \qquad (10.15)$$

Equation 10.15 expresses the relation between cumulative infiltration and infiltration capacity rate, and can be integrated. In a way, this equation describes what can be considered as infiltration rating curve. The explicit form of this relation depends on the form of $g_r(I)$, $r = 1, 2, ..., n$.

### 10.3.8 Derivation of Infiltration Equation

Noting $dJ(t)/dt$, differentiation of $J(t)$ from Equation 10.15 will lead to a general expression for $I(t)$ which is what is desired. This suggests that the key to deriving an infiltration equation is to derive its associated PDF whose derivation depends on the constraints specific to that infiltration equation. Application of the entropy theory is illustrated by deriving six popular infiltration equations, including the Horton, Kostiakov, Philip, Green–Ampt, Overton, and Holtan equations.

## 10.4 HORTON EQUATION

In the Horton equation, the initial infiltration capacity rate is denoted as $I_0$ and the steady or constant rate denoted as $I_c$. Thus, $I(t)$ will vary from $I_c$ to $I_0$. The simplest constraint that $f(I)$ must satisfy is as follows:

$$\int_{I_c}^{I_0} f(I)dI = 1 \qquad (10.16)$$

Applying POME and using the method of Lagrange multipliers (Singh, 1998), one obtains the Lagrangian function $L$ as

$$L = \frac{1}{m-1} \left\{ \int_{I_c}^{I_0} f(I)(1 - [f(I)]^{m-1}) dI + \lambda_0 \left[ \int_{I_c}^{I_0} f(I)dI - 1 \right] \right\} \qquad (10.17)$$

where $\lambda_0$ is the zeroth Lagrange multiplier. Differentiating Equation 10.17 with respect to $f$ and keeping in mind that $I$ is a parameter here, not a variable, and equating the derivative to zero, one gets

$$\frac{\partial L}{\partial f} \Rightarrow 0 = \frac{1}{m-1} \left\{ \int_{I_c}^{I_0} [1 - [f(I)]^{m-1} - (m-1)[f(I)]^{m-1}] dI \right\} + \lambda_0 \left[ \int_{I_c}^{I_0} dI \right] \qquad (10.18)$$

Equation 10.18 yields

$$f(I) = \left\{ \frac{m-1}{m} \left[ \frac{1}{m-1} + \lambda_0 \right] \right\}^{1/(m-1)} \tag{10.19}$$

Equation 10.19 is the Tsallis entropy-based PDF and contains one unknown parameter: the zeroth Lagrange multiplier.

For simplicity, let $\lambda_* = \lambda_0 + (1/(m-1))$ and $A = [((m-1)/m)\lambda_*]^{1/(m-1)}$. Equation 10.19 can be expressed as

$$f(I) = \left\{ \frac{m-1}{m}[\lambda_*] \right\}^{1/(m-1)} = A \tag{10.20}$$

Substituting Equation 10.20 in Equation 10.16, one obtains

$$\int_{I_c}^{I_0} f(I)dI = 1 = \int_{I_c}^{I_0} dI = \frac{1}{A} \tag{10.21}$$

Equation 10.21 gives the Lagrange multiplier $\lambda_0$ as

$$\lambda_0 = \frac{m}{m-1} \left( \frac{1}{I_0 - I_c} \right)^{m-1} - \frac{1}{m-1}; \quad \lambda_* = \lambda_0 + \frac{1}{m-1} = \frac{m}{m-1} \left( \frac{1}{I_0 - I_c} \right)^{m-1} \tag{10.22}$$

Substitution of Equation 10.22 in Equation 10.20 yields

$$f(I) = \frac{1}{I_0 - I_c} \tag{10.23}$$

Equation 10.23 is the PDF of infiltration rate from the Horton equation, which is uniform and depends only on the initial and steady infiltration capacity rates. The cumulative distribution function of $I$ would be linear, expressed as

$$F(I) = \int_{I_c}^{I} f(I)dI = \int_{I_c}^{I} \frac{1}{I_0 - I_c} dI = \frac{I - I_c}{I_0 - I_c} \tag{10.24}$$

Combining Equations 10.14 and 10.23, one obtains

$$\frac{1}{I_0 - I_c} dI = -\frac{1}{S} dJ \tag{10.25}$$

Integrating Equation 10.25, one obtains

$$\frac{I - I_c}{I_0 - I_c} = 1 - \frac{J}{S} \tag{10.26}$$

Equation 10.26 can be recast as

$$\frac{dJ}{dt} + \frac{J}{k} = I_0 - I_c, \quad \frac{S}{(I_0 - I_c)} = \frac{I_0 - I_c}{S} \tag{10.27}$$

Solution of Equation 10.27 yields the cumulative infiltration as

$$J = (I_0 - I_c)k - (I_0 - I_c)\exp\left(\frac{-t}{k}\right) \tag{10.28}$$

Differentiating Equation 10.28 with respect to $t$ and recalling the continuity equation (Equation 10.10), one obtains the infiltration rate as

$$I(t) = I_c + (I_0 - I_c)\exp\left(\frac{-t}{k}\right) \tag{10.29}$$

which is the Horton equation. Recall that

$$k = \frac{S}{(I_0 - I_c)} \tag{10.30}$$

Derivation of Equation 10.29 shows that the Horton equation requires no constraint other than the total probability theorem, which is not a constraint in a true sense, for all probability distributions must satisfy it. Parameter $k$ is expressed as the ratio of the maximum soil moisture retention and the initial infiltration capacity rate minus the steady-state infiltration rate. It has the dimension of time and indicates the time that it takes for the infiltrated water to fill the maximum moisture retention space, if the capacity rate of infiltration were the initial infiltration rate (i.e., the maximum infiltration rate) minus the steady rate, or the initial excess infiltration capacity rate. Infiltration observations, under the condition of no limit on water supply, provide initial and steady infiltration capacity rates and for a given soil with the knowledge of its porosity and its column height the value of $S$ (the maximum soil moisture retention) can be obtained. Thus, parameter $k$ can be computed using Equation 10.30 without calibration. This also provides a physical interpretation of parameter $k$, which can be interpreted as average travel time.

The entropy of the probability distribution underlying the Horton equation or the infiltration rate can be expressed as

$$H(I) = \frac{1}{m - 1}\int_{I_c}^{I_0} \{1 - [f(I)]^m\}dI = \frac{1}{m - 1}[(I_0 - I_c) - (I_0 - I_c)^{1-m}] \tag{10.31}$$

For $m = 2$, Equation 10.31 becomes

$$H(I) = (I_0 - I_c) - \frac{1}{(I_0 - I_c)} \tag{10.32}$$

Equation 10.31 states that the uncertainty of $f(I)$, or for that matter $I$, depends on the initial value of $I$, $I_0$, and steady rate $I_c$. This equation consists of two parts: $(I_0 - I_c)$ and $(I_0 - I_c)^{1-m}$. An important implication is that for a given soil the uncertainty of the Horton equation $m > 1$ is maximum when it is dry because that is when the initial infiltration will be maximum and as a result the first part will be much greater than the second part and hence the difference between these two parts will be greater, translating into greater entropy. This difference and hence entropy reduces as soil becomes wetter. This means that when sampling infiltration, greater care should be exercised in the beginning of infiltration and less toward the tail. This also means that infiltration observations should be more closely spaced temporally in the beginning but the time interval between observations can be increased with the progress of infiltration.

## Example 10.1

Data on field experiments on infiltration in Troupe sand in the Georgia Coastal Plain have been reported by Rawls et al. (1976) in a report published by the Agriculture Research Service of the U.S. Department of Agriculture. Characteristics of infiltration observations are as follows (Table 10.1): $D$ = the duration of the experiment = 123 min; $t_c$ = the time to approximately reach a constant rate of

---

**TABLE 10.1**

**Experimental Field Observations on Infiltration in Troupe Sand**

| Time from Start of Rain | Infiltration Rate | | Time from Start of Rain | Infiltration Rate | | Time from Start of Rain | Infiltration Rate | |
|---|---|---|---|---|---|---|---|---|
| (min) | (in./h) | (cm/h) | (min) | (in./h) | (cm/h) | (min) | (in./h) | (cm/h) |
| 5 | 6.25 (at 6 min) | 15.87 | 35 | 1.734 | 4.40 | 65 | 1.785 | 4.53 |
| 10 | 1.690 | 4.29 | 40 | 1.751 | 4.45 | 70 | 1.845 | 4.69 |
| 15 | 1.653 | 4.20 | 45 | 1.755 | 4.47 | 75 | 1.744 | 4.43 |
| 20 | 1.879 | 4.78 | 50 | 1.785 | 4.53 | 80 | 1.843 | 4.68 |
| 25 | 1.888 | 4.80 | 55 | 1.763 | 4.47 | 85 | 1.720 | 4.39 |
| 30 | 1.724 | 4.38 | 60 | 1.756 | 4.46 | 90 | 1.630 | 4.14 |

*Source:* After Rawls, W. et al., Calibration of selected infiltration equations for the Georgia plain, ARS-S-113, U.S. Department of Agriculture, Agricultural Research Service, New Orleans, LA, 1976.

infiltration = 110 min; $I_c$ is the constant (steady) rate of infiltration at the end of infiltration experiment = 4.40 cm/h; $I_0$ is the initial infiltration capacity rate given a few minutes later than the start of infiltration ($t = 0$) = 11.60 cm/h; $S = 3.12$ cm. Compute the infiltration rate using the Horton equation. Show the parameter values. Also, compute the Horton equation parameter values using the least square method and compute infiltration rate. Then, compare the computed infiltration rates with observed values.

**Solution**

The Horton equation has three parameters $I_c$, $I_0$, and $k$ as shown in Equation 10.29. For the entropy theory, parameters $I_c$ and $I_0$ are obtained from observations and the value of $S$ is also obtained from observations where it is the difference between the maximum soil moisture minus the initial soil moisture. Using these observed values of $I_c$, $I_0$, and $S$, parameter $k$ is computed using Equation 10.30. Thus, no calibration is done to obtain parameters $I_c$, $I_0$, and $k$. The three parameters are also obtained using the least square method in which the sum of squares of deviations between observed and computed infiltration rates is minimized. The Horton parameters obtained by calibration and entropy method are: $k = 0.43$ h, $I_0 = 11.6$ cm/h, $I_c = 4.4$ cm/h; and least square method: $k = 0.38$ h, $I_0 = 12.44$ cm/h, $I_c = 4.52$ cm/h. With parameter values so obtained, the Horton equation is applied to the observed data set, and the infiltration rates computed using these two methods and observed capacity rates are shown in Figure 10.2. The infiltration capacity rates obtained using the calibrated parameter values are in close agreement with observed values.

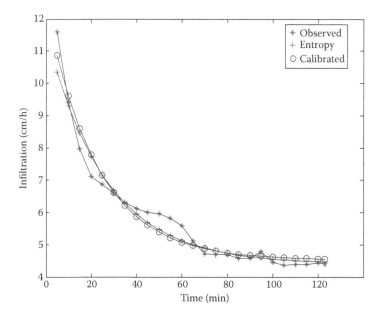

**FIGURE 10.2** Comparison of infiltration rates computed using the Horton equation with parameters determined using entropy theory and by least square method with observed infiltration rates for Trope sand.

## 10.5   KOSTIAKOV EQUATION

Let the constraints be defined as Equation 10.3 and

$$\int_{I_c}^{\infty} I^{-2(m-1)} f(I)dI = E[I^{-2(m-1)}] = \overline{I^{-2(m-1)}} \tag{10.33}$$

where $I_c$ is some small value equal to steady infiltration but tending to 0. Using POME and the method of Lagrange multipliers, the Lagrange function $L$ becomes

$$L = \frac{1}{m-1} \int_{I_c}^{\infty} f(I)\{1-[f(I)]^{m-1}\}dI + \lambda_0 \left[ \int_{I_c}^{\infty} f(I)dI - 1 \right] + \lambda_1 \left[ \int_{I_c}^{\infty} I^{-2(m-1)} f(I)dI - \overline{I^{-2(m-1)}} \right] \tag{10.34}$$

Differentiating Equation 10.34 with respect to $f$ and equating the derivative to 0, one obtains

$$\frac{\partial L}{\partial f} \Rightarrow 0 = \frac{1}{m-1} \left\{ \int_{I_c}^{\infty} [1-[f(I)]^{m-1} -[f(I)]^{m-1}] \right\} dI + \lambda_0 \left[ \int_{I_c}^{\infty} dI \right] + \lambda_1 \left[ \int_{I_c}^{\infty} I^{-2(m-1)} dI \right] \tag{10.35}$$

Solution of Equation 10.35 yields $f(I)$ as

$$f(I) = \left[ \frac{1}{m} + \frac{m-1}{m}(\lambda_0 + \lambda_1 I^{-2(m-1)}) \right]^{1/(m-1)} \tag{10.36}$$

Let $\lambda_* = \lambda_0 + (1/(m-1))$, $A = ((m-1)/m)\lambda_*$, and $B = ((m-1)/m)\lambda_1$. Introducing these quantities in Equation 10.36, one obtains

$$f(I) = [A + BI^{-2(m-1)}]^{1/(m-1)} \tag{10.37}$$

If it is assumed that $A = 0$ and $m = 2$, then

$$f(I) = \frac{B}{I^2} \tag{10.38}$$

Equation 10.36 will satisfy the total probability given by Equation 10.3 if $B = I_c$. This means that $\lambda_1 = mI_c/(m-1)$. If $m = 2$, then $\lambda_1 = 2I_c$.

Combining Equation 10.38 with Equation 10.14, the result with limits on $I$ from $I$ to $\infty$ and on $J$ from $J$ to 0 is

$$\frac{I_c S}{I} = J \tag{10.39}$$

Recalling that $I = dJ/dt$, Equation 10.39 can be expressed as

$$\frac{dJ}{dt} = \frac{I_c S}{J} \Rightarrow J = (2I_c S)^{0.5} t^{0.5} \tag{10.40}$$

Integration of Equation 10.38 yields

$$J = (2I_c S)^{0.5} t^{0.5} \tag{10.41}$$

Differentiating Equation 10.41, one obtains the rate of infiltration:

$$I = \frac{1}{2}(2I_c S)^{0.5} t^{-0.5} \tag{10.42}$$

Equation 10.41 can be recast as

$$J = at^{0.5} \tag{10.43}$$

and Equation 10.44 as

$$I(t) = 0.5at^{-0.5} \tag{10.44}$$

which is the Kostiakov equation with $a$ as parameter expressed as

$$a = 2I_c S \tag{10.45}$$

Thus, parameter $a$ has physical meaning that is twice the product of steady infiltration rate ($I_c$) and maximum soil moisture retention ($S$) both of which can be determined for a given soil. This means that parameter $a$ can be obtained from observations and does not need to be calibrated.

The PDF of infiltration rate given by the Kostiakov equation can be expressed as

$$f(I) = \frac{I_c}{I^2} \tag{10.46}$$

Substituting Equation 10.46 in Equation 10.2 and with $m = 2$, the entropy of the Kostiakov equation can be written as

$$H = 1 - \frac{1}{3I_c} \tag{10.47}$$

### Example 10.2

Using the data given in Table 10.1, compute the infiltration rate using the Kostiakov equation. Show the parameter values. Also, compute the Kostiakov equation parameter values using the least square method and compute the infiltration rate. Then,

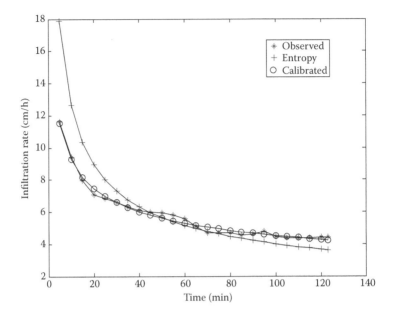

**FIGURE 10.3** Comparison of infiltration rates computed using the Kostiakov with parameters determined using entropy theory and by least square method with observed infiltration rates for Troupe sand.

compare the computed infiltration rates with observed values. Characteristics of infiltration observations are as in Example 10.1, but the value of $S$ in this case is 12.14 cm.

**Solution**

This equation has only one parameter $a$, which is obtained by the least square method as well as directly from observations using Equation 10.45 due to the entropy theory, as for entropy: $a = 10.34$ and $b = -0.5$; and for least square method: $a = 5.31$ and $b = -0.31$. Figure 10.3 compares observed infiltration rates and the rates computed using the entropy theory and calibration for data set IV. It may be noted that the value of parameter $a$ as estimated for the entropy theory may be less than accurate, for the value of $S$ as given in the data does not match the accumulated infiltration, that is the value of $S$ is significantly less than the accumulated infiltration at the time when the rate of infiltration became almost constant.

## 10.6   PHILIP TWO-TERM EQUATION

Let the infiltration rate be defined as $i = I - a$, where $a$ is some constant value. Let the constraints be defined by Equation 10.3 with limits as $a$ to $\infty$, and

$$\int_{a}^{\infty} i^{-2(m-1)} f(i)\,di = E[i^{-2(m-1)}] = \overline{i^{-2(m-1)}} \tag{10.48}$$

Using POME and the method of Lagrange multipliers, the Lagrangian function $L$ is

$$L = \frac{1}{m-1}\int_a^\infty f(i)\{1-[f(i)]^{m-1}\}di + \lambda_0\left[\int_a^\infty f(i)di - 1\right] + \lambda_1\left[\int_a^\infty i^{-2(m-1)}f(i)di - \overline{i^{-2(m-1)}}\right]$$

(10.49)

Differentiating Equation 10.49 with respect to $f$ and equating the derivative to 0, one gets

$$\frac{\partial L}{\partial f} \Rightarrow 0 = \frac{1}{m-1}\left\{\int_a^\infty [1-[f(i)]^{m-1} - (m-1)[f(i)]^{m-1}]di + \lambda_0\left[\int_a^\infty di\right] + \lambda_1\left[\int_a^\infty i^{-2(m-1)}di\right]\right.$$

(10.50)

Solution of Equation 10.50 yields $f(i)$ as

$$f(i) = [c + di^{-2(m-1)}]^{1/(m-1)}$$

(10.51)

where
$c = ((m-1)/m)\lambda_*$
$d = ((m-1)/m)\lambda_1$
$\lambda_* = (1/(m-1)) + \lambda_0$

Equation 10.51 is the PDF of infiltration rate given by the Philip equation. If $c$ is assumed zero, and $m = 2$, then

$$f(i) = di^{-2}$$

(10.52)

Substituting Equation 10.50 in Equation 10.3, one gets

$$\int_a^\infty di^{-2}di = 1 \Rightarrow d = a; \quad \lambda_1 = 2a$$

(10.53)

Combining Equation 10.52 with Equation 10.14, and integrating with limits on $i$ from $i$ equal to $\infty$ and on $J$ from $J$ equal to 0 the result is

$$\frac{aS}{i} = J$$

(10.54)

Recalling that $i = dJ/dt$, Equation 10.54 can be expressed as

$$\frac{dJ}{dt} = \frac{aS}{J} \Rightarrow J = (2aS)^{0.5}t^{0.5} \tag{10.55}$$

Differentiating Equation 10.55, one obtains the rate of infiltration:

$$i = \frac{1}{2}(2aS)^{0.5}t^{-0.5} \tag{10.56}$$

Equation 10.56 can be written in original terms as

$$I(t) = a + 0.5(2aS)^{0.5}t^{-0.5} = a + bt^{-0.5}, \quad b = 0.5(2aS)^{0.5} \tag{10.57}$$

which is the Philip two-term equation with $a$ and $b$ as parameters. Parameter $a$ is analogous to steady infiltration rate (or saturated hydraulic conductivity or a fraction thereof) and can be obtained without having any calibration. In general, $a$ is between 0.5 and 0.7 of $I_c$. Parameter $b$ can be expressed in terms of $a$ and maximum soil moisture retention $S$ and can be obtained from observations, as shown by Equation 10.57. Thus, parameters $a$ and $b$ have physical meaning and need no calibration.

Using Equation 10.53 in Equation 10.2, one obtains the entropy of infiltration rate by the Philip equation:

$$H = 1 - \frac{1}{3a} \tag{10.58}$$

## Example 10.3

Using the data given in Table 10.1, compute the infiltration rate using the Philip two-term equation. Show the parameter values. Also, compute the Philip equation parameter values using the least square method and compute infiltration rate. Then, compare the computed infiltration rate with observed values. Characteristics of infiltration observations are as in Example 10.1, but the value of $S$ in this case 12.14 cm.

### Solution

The Philip equation has two parameters $a$ and $I_c$ as shown in Equation 10.57. These parameters are estimated by the least square method and from observations using Equation 10.57 for the entropy theory and their values are as follows: for the entropy method: $a = 2.20$ and $b = 3.65$; and for the least square method: $a = 2.53$ and $b = 2.72$. Figure 10.4 compares observed infiltration capacity rate and the capacity rates computed using the entropy theory and least square method. The figure shows that the entropy theory overestimates infiltration rate for the entire duration of the experiment, and the calibration method underestimates up to about 62 min and overestimates for the remainder of the duration of the experiment. Considering that there is no calibration for the entropy theory, it compares reasonably well with the least square method.

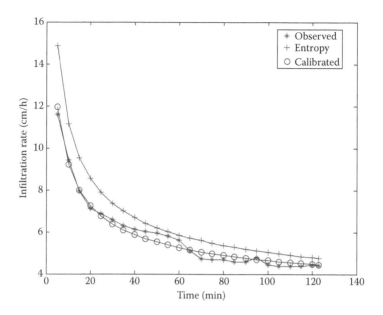

**FIGURE 10.4** Comparison of infiltration rates computed using the Philip two-term equation with parameters determined using entropy theory and by least square method with observed infiltration rates for Troupe sand.

## 10.7 GREEN–AMPT EQUATION

Let the constraints be defined by Equation 10.3 with limits as $b$ to $c$ where $b$ would tend to $\infty$, and $c$ to $I_c$, and

$$\int_c^\infty (I-I_c)^{-2(m-1)} f(I)dI = \overline{(I-I_c)^{-2(m-1)}} \tag{10.59}$$

Using POME and the method of Lagrange multipliers, the Lagrangian function $L$ is

$$L = \frac{1}{m-1}\int_c^\infty f(I)\{1-[f(I)]^{m-1}\}dI + \lambda_0 \left[\int_c^\infty f(I)dI - 1\right]$$

$$+ \lambda_1 \left[\int_c^\infty (I-I_c)^{-2(m-1)} f(I)dI - \overline{(I-I_c)^{-2(m-1)}}\right] \tag{10.60}$$

Differentiating Equation 10.60 with respect to $f$ and equating the derivative to 0, one gets

$$\frac{\partial L}{\partial f} \Rightarrow 0 = \frac{1}{m-1} \left\{ \int_c^\infty [1 - [f(I)]^{m-1} - (m-1)[f(I)]^{m-1}] \right\} dI$$

$$+ \lambda_0 \left[ \int_c^\infty dI \right] + \lambda_1 \left[ \int_c^\infty (I - I_c)^{-2(m-1)} dI \right] \qquad (10.61)$$

Equation 10.61 yields $f(I)$ as

$$f(I) = \left[ \frac{m-1}{m} (\lambda_* + \lambda_1 (I - I_c)^{-2(m-1)}) \right]^{1/(m-1)} \qquad (10.62)$$

Let $a = ((m-1)/m)\lambda_*, b = ((m-1)/m)\lambda_1, \lambda_* = (1/(m-1)) + \lambda_0$. Equation 10.62 becomes

$$f(I) = [a + b(I - I_c)^{-2(m-1)}]^{1/(m-1)} \qquad (10.63)$$

Taking $a = 0$ and $m = 2$, Equation 10.63 reduces to

$$f(I) = \frac{b}{(I - I_c)^2} \qquad (10.64)$$

In order for $f(I)$ to satisfy Equation 10.3, $b = I_c$ or $\lambda_1 = mI/(m-1)$. If $m = 2$, then

$$f(I) = \frac{I_c}{(I - I_c)^2} \qquad (10.65)$$

Equation 10.65 is the PDF of infiltration rate due to the Green–Ampt equation. It should, however, be noted that this density function is valid only for $2I_c \leq I < \infty$, not for the entire first quadrant.

Combining Equation 10.65 with Equation 10.14, the result is

$$\frac{I_c dI}{(I - I_c)^2} = -\frac{1}{S} \frac{dJ}{dI} \qquad (10.66)$$

Integrating with limits for $I$ from $I$ to $\infty$ and for $J$ from $J$ to 0,

$$\frac{I_c}{(I - I_c)} = -\frac{J}{S} \qquad (10.67)$$

Recalling that $I = dJ/dt$, Equation 10.67 can be expressed as

$$\frac{dJ}{dt} = \frac{SI_c}{J} + I_c \qquad (10.68)$$

Solution of Equation 10.68, with the condition that $t = 0$, $J = 0$, is

$$t = \frac{1}{I_c}\left[J - S\log\left(1 + \frac{J}{S}\right)\right]$$ (10.69)

Equation 10.69 can be expressed as

$$t = \frac{1}{I_c}\left[J - \frac{a}{I_c}\log\left(1 + \frac{J}{a/I_c}\right)\right]$$ (10.70)

where

$$a = SI_c$$ (10.71)

Equation 10.70 is the Green–Ampt equation in which parameter $I_c$ is the steady-state rate of infiltration and can be interpreted as equal to saturated hydraulic conductivity. Parameter $S$ equals the product of the capillary suction at the wetting front and the initial moisture deficit and can be interpreted as the maximum soil moisture retention and $S = a/I_c$. Since $I_c$ and $S$ can be obtained from observations, $a = SI_c$ can also be obtained from observations. In the hydrologic literature, $S$ is interpreted as equal to the product of the capillary suction at the wetting front and the initial moisture deficit (Singh, 1989). The entropy theory provides another interpretation of parameter $S$ and hence the G–A parameters can be estimated without calibration.

Entropy of infiltration rate given by the Green–Ampt equation can be written by substituting Equation 10.65 in Equation 10.2 as

$$H = 1 - \frac{1}{3I_c}$$ (10.72)

## Example 10.4

Using the data given in Table 10.1, compute the infiltration rate using the Green–Ampt equation. Show the parameter values. Also, compute the Green–Ampt equation parameter values using the least square method and compute infiltration rate. Then, compare the computed infiltration rate with observed values. Characteristics of infiltration observations are as in Example 10.1, but the value of $S$ in this case is 2.59 cm.

### Solution

The G–A equation has two parameters $a$ and $S$ as shown in Equation 10.70. These parameters are estimated by calibration and from observations using Equation 10.71 for the entropy theory, and their values are as follows: for the entropy method: $I_c = 4.4$ cm/h, $I_0 \times S = 11.4$ cm/h; and for the least square method:

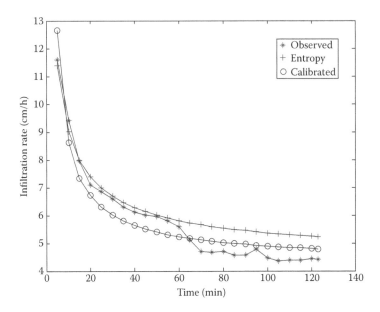

**FIGURE 10.5** Comparison of infiltration rates computed using the Green–Ampt equation with parameters determined using entropy theory and by least square method with observed infiltration rates for Troupe sand.

$I_c$ = 4.1 cm/h and $I_0 \times S$ = 4.1 cm/h. Figure 10.5 compares observed infiltration capacity rates and the capacity rates computed using the entropy theory and the least square method. The figure shows that the entropy theory consistently overestimates and the least square method underestimates infiltration up to about 62 min and then it overestimates. However, considering that there was no calibration of parameters, the performance is within error bounds that can be reduced.

## 10.8   OVERTON MODEL

Let the constraints be defined by Equation 10.3 and

$$\int_0^{i_0} i^{-0.5(m-1)} f(i)di = E[i^{-0.5(m-1)}] = \overline{i^{-0.5(m-1)}}$$                    (10.73)

where $i = I - I_c$. Using POME and the method of Lagrange multipliers, the Lagrangian function $L$ is

$$L = \frac{1}{m-1}\int_0^{i_0} f(i)\{1-[f(i)]^{m-1}\}di + \lambda_0\left[\int_0^{i_0} f(i)di - 1\right] + \lambda_1\left[\int_0^{i_0} i^{-0.5(m-1)} f(i)di - \overline{i^{-0.5(m-1)}}\right]$$

(10.74)

Differentiating Equation 10.74 with respect to $f$ and equating the derivative to 0, one gets

$$\frac{\partial L}{\partial f} \Rightarrow 0 = \frac{1}{m-1} \left\{ \int_0^{i_0} [1 - [f(i)]^{m-1} - (m-1)[f(i)]^{m-1}] \, di + \lambda_0 \left[ \int_0^{i_0} di \right] + \lambda_1 \left[ \int_c^{\infty} i^{-0.5(m-1)} di \right] \right\}$$

(10.75)

Solution of Equation 10.75 yields $f(i)$ as

$$f(i) = \left[ \frac{m-1}{m} (\lambda_* + \lambda_1 i^{-0.5(m-1)}) \right]^{1/(m-1)}$$

(10.76)

Let $\lambda_* = \lambda_0 + (1/(m-1))$, $A = ((m-1)/m)\lambda_*$, $B = ((m-1)/m)\lambda_1$. Equation 10.76 becomes

$$f(i) = [A + Bi^{-0.5(m-1)}]^{1/(m-1)}$$

(10.77)

Assuming $A = 0$ and $m = 2$, Equation 10.77 becomes

$$f(i) = Bi^{-0.5}$$

$$\int_{I_c}^{I_0} B(I - I_c)^{-0.5} dI = 1 \Rightarrow B = \frac{1}{2(I_0 - I_c)^{0.5}}$$

(10.78)

Inserting Equation 10.78 in Equation 10.76

$$f(i) = \frac{i^{-0.5}}{2i_0^{0.5}}$$

(10.79)

Equation 10.79 can be cast as

$$f(I) = \frac{(I - I_c)^{-0.5}}{2(I_0 - I_c)^{0.5}}$$

(10.80)

Equation 10.80 is the PDF of infiltration rate by the Overton model.
  Substituting Equation 10.80 in Equation 10.13, one obtains

$$-\frac{1}{S} dJ = \frac{0.5(I - I_c)^{-0.5}}{(I_0 - I_c)^{0.5}} dI$$

(10.81)

Integration of Equation 10.81 yields

$$I = \frac{(I_0 - I_c)}{S^2} J^2 + I_c \tag{10.82}$$

Recalling the continuity equation (Equation 10.10), Equation 10.82 with limits on $t$ from $t$ to $t_c$ and on $J$ from $J$ to $J_c$ (constant) gives

$$J = J_c - S \sqrt{\frac{I_c}{(I_0 - I_c)}} \tan \left[ \frac{\sqrt{I_c(I_0 - I_c)}}{S} (t_c - t) \right] \tag{10.83}$$

Differentiating Equation 10.83 leads to

$$I(t) = I_c \sec^2 \left[ \frac{\sqrt{I_c(I_0 - I_c)}}{S} (t_c - t) \right] \tag{10.84}$$

Let

$$(I_0 - I_c) = aS^2 \tag{10.85}$$

Equation 10.84 becomes

$$I(t) = I_c \sec^2 \left[ \sqrt{aI_c} (t_c - t) \right] \tag{10.86}$$

Equation 10.86 is the Overton model where $t_c$ is the time to steady-state infiltration rate $I_c$; this time may be much smaller than the duration of the infiltration experiment or observations and can be obtained from observations. Parameter $a$ is expressed as $(I_0 - I_c) = aS^2$ in which $I_0$ is the initial infiltration capacity. Thus, parameters of the Overton equation can be obtained from observations and calibration of these parameters may not be needed.

Using Equation 10.80 in Equation 10.2, one obtains the entropy of infiltration rate given by the Overton equation:

$$H = 1 - \frac{1}{3} \frac{(I_c)^2}{(I_0 - I_c)^3} \tag{10.87}$$

## Example 10.5

Using the data given in Table 10.1, compute the infiltration rate using the Overton equation. Show the parameter values. Also, compute the Overton equation parameter values using the least square method and compute infiltration rate.

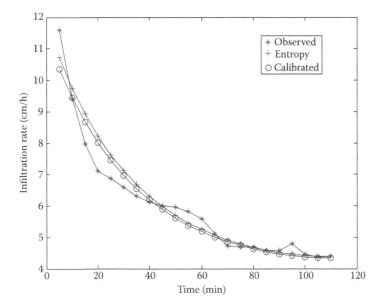

**FIGURE 10.6** Comparison of infiltration rates computed using the Overton equation with parameters determined using entropy theory and by the least square method with observed infiltration rates for Troupe sand.

Then, compare the computed infiltration rate with observed values. Characteristics of infiltration observations are as in Example 10.1, but the value of $S$ in this case 11.21 cm.

**Solution**

The Overton equation has actually three parameters $a$, $I_c$ and $t_c$, as shown in Equation 10.86. These parameters are estimated by calibration and from observations using Equation 10.85 for the entropy theory, and their parameter values are as follows: $I_c = 4.38$ cm/h and $a = 0.06$; for the least square method: $I_c = 4.34$ cm/h, $a = 0.06$. Figure 10.6 compares observed infiltration rates and the rates computed using the entropy theory and the least square method. The figure shows that the entropy theory consistently overestimates infiltration capacity rate and the least square method underestimates between $t = 20$ min and $t = 62$ min, and then overestimates. Considering that there was no calibration for the entropy theory, it compares reasonably well with the least square method. Reducing the value of $a$ through $S$ and $I_c$ would lead to improved infiltration estimates.

## 10.9  HOLTAN MODEL

Analogous to the Horton equation, let $i$ define the excess infiltration rate $(I - I_c)$ varying from 0 to $i_0$ where $i_0 = I_0 - I_c$. Then, the constraints can be defined by Equations 10.3 and 10.73 (with proper infiltration rate in mind). Using POME and the method of Lagrange multipliers, $f(i)$ is obtained as Equation 10.76 and eventually Equation 10.79:

$$f(i) = [A + Bi^{((1-n)/n)(m-1)}]^{1/(m-1)} \tag{10.88}$$

where

$$A = \frac{m-1}{m}\lambda_*, \quad B = \frac{m-1}{m}\lambda_1, \quad \lambda_* = \frac{1}{m-1} + \lambda_0 \tag{10.89}$$

Let $m = 2$. Equation 10.88 becomes

$$f(i) = [A + Bi^{((1-n)/n)}] \tag{10.90}$$

If $A = 0$, then Equation 10.90 can be recast as

$$f(i) = Bi^{((1-n)/n)} \tag{10.91}$$

Substituting Equation 10.91 in Equation 10.3 yields

$$B = \frac{1}{n(i_0)^{1/n}} \tag{10.92}$$

Equation 10.91, in concert with Equation 10.92, is the PDF of infiltration rate from the Holtan equation.

Substituting Equation 10.88 in Equation 10.14, one obtains

$$dJ = -Sbi^{((1-n)/n)}di \tag{10.93}$$

Integration of Equation 10.93 yields

$$S - J = Sbni^{1/n} \tag{10.94}$$

Equation 10.94 can be expressed as

$$(nbS)^n (S-J)^{-n} dJ = dt \tag{10.95}$$

Integrating Equation 10.95, one obtains

$$J = S - \left[ S^{1-n} - \frac{(1-n)}{(Sbn)^n} t \right]^{1/(1-n)} \tag{10.96}$$

Differentiation of Equation 10.96 with respect to $t$ and simplification yield

$$i = a[S^{1-n} - a(1-n)t]^{n/(1-n)} \tag{10.97}$$

where

$$a = \frac{i_0}{S^n} \tag{10.98}$$

Equation 10.97 can be written in original terms as

$$I = I_c + a[S^{1-n} - a(1-n)t]^{n/(1-n)}, \quad a = \frac{(I_0 - I_c)}{S^n} \tag{10.99}$$

Equation 10.99 is the Holtan equation with parameter $a$ given by Equation 10.98 and

$$\overline{I - I_c} = B\frac{n}{1+n}(I_0 - I_c)^{(1+n)/n} \tag{10.100}$$

with

$$B = \frac{1}{n(I - I_c)^{1/n}} \tag{10.101}$$

Parameters $a$ and $n$ can be obtained from observations as Equations 10.100 and 10.101 show and calibration may therefore not be needed. Thus, $n$ can also be expressed in terms of physically measurable quantities. Through simulation, Singh (2010) found parameter $n$ to be 1.5.

Substituting Equation 10.91 in Equation 10.2 yields

$$H = 1 + \frac{1}{(2-n)(I_0 - I_c)} \tag{10.102}$$

## Example 10.6

Using the data given in Table 10.1, compute the infiltration rate using the Holtan equation. Show the parameter values. Also, compute the Holtan equation parameter values using the least square method and compute infiltration rate. Then, compare the computed infiltration rate with observed values. Characteristics of infiltration observations are as in Example 10.1, but the value of $S$ in this case 3.12 cm.

### Solution

The Holtan equation has three parameters $a$, $I_c$ and $n$, as shown in Equation 10.99. These parameters are estimated by calibration and from observations using Equation 10.99 for the entropy theory, and their values are as follows: for

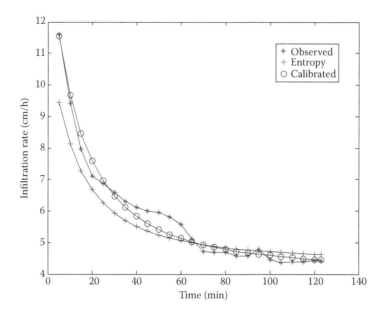

**FIGURE 10.7** Comparison of infiltration rates computed using the Holtan equation with parameters determined using the entropy theory with $n = 1.5$ and by the least square method with observed infiltration rates for Troupe sand.

the entropy method: $I_c = 4.40$ cm/h, $a = 1.30$, and $n = 1.5$; and for the least square method: $I_c = 4.26$ cm/h, $a = 1.21$, and $n = 1.5$. Figure 10.7 compares observed infiltration capacity rate and the capacity rates computed using the entropy theory and the least square method. The figure shows that both the entropy theory and the least square method are comparable up to $t = 62$ min, first underestimating and then overestimating infiltration a little bit, whereas the least square method first overestimates and then underestimates. In this case, the entropy theory does not yield not as good estimates as does the least square method. However, considering that there is no calibration of parameters, the theory performs remarkably well. Reducing the value of $a$ through $S$ and $I_c$ would lead to improved infiltration estimates.

## REFERENCES

Burgy, R.H. and Luthin, J.N. (1956). A test of the single- and double-ring types of infiltrometers Transactions, American Geophysical Union, 37(2), 189–192.

Crawford, N.H. and Linsley, R.K. (1966). Digital simulation in hydrology: Stanford watershed model IV. Technical Report No. 39, Stanford University, Palo Alto, CA.

Donigian, A.S. and Imhoff, J. (2006). History and evolution of watershed modeling derived from the Stanford watershed model (Chapter 2). In: *Watershed Models*, eds. V.P. Singh and D.K. Frevert. CRC Press, Boca Raton, FL, pp. 21–45.

Green, W.H. and Ampt, C.A. (1911). Studies on soil physics, I. Flow of air and water through soils. *Journal of Agricultural Sciences*, 4, 1–24.

Holtan, H.N. (1961). A concept of infiltration estimates in watershed engineering. ARS41-51, U.S. Department of Agriculture, Agricultural Research Service, Washington, DC.

Horton, R.I. (1938). The interpretation and application of runoff plot experiments with reference to soil erosion problems. *Soil Science Society of America Proceedings*, 3, 340–349.

Jaynes, E.T. (1957a). Information theory and statistical mechanics I. *Physical Review*, 106(4), 620–630.

Jaynes, E.T. (1957b). Information theory and statistical mechanics II. *Physical Review*, 108(2), 171–190.

Jaynes, E.T. (1982). On the rationale of maximum entropy methods. *Proceedings of IEEE*, 70, 939–952.

Kostiakov, A.N. (1932). On the dynamics of the coefficient of water percolations in soils. *Meeting of the Sixth Commission of the International Society of Soil Science, Part A*, pp. 15–21, Moscow, USSR.

Overton, D.E. (1964). Mathematical refinement of an infiltration equation for watershed engineering. ARS 41-99, U.S. Department of Agriculture, Agricultural Research Service, Washington, DC.

Philip, J.R. (1957). Theory of infiltration, Parts 1 and 4. *Soil Science*, 85(5), 345–357.

Rawls, W., Yates, P., and Asmussen, L. (1976). Calibration of selected infiltration equations for the Georgia plain. ARS-S-113, U.S. Department of Agriculture, Agricultural Research Service, New Orleans, LA.

Singh, V.P. (1989). *Hydrologic Systems*: Vol. II. *Watershed Modeling*. Prentice Hall, Englewood Cliffs, NJ.

Singh, V.P. (ed.). (1995). *Computer Models of Watershed Hydrology*. Water Resources Publications, Littleton, CO.

Singh, V.P. (1998). *Entropy-Based Parameter Estimation in Hydrology*. Kluwer Academic Publishers, Dordrecht, the Netherlands.

Singh, V.P. (2010). Tsallis entropy theory for derivation of infiltration equations. *Transactions of the ASABE*, 53(2), 447–463.

Singh, V.P. (2013). *Entropy Theory in Environmental and Water Engineering*. Wiley-Blackwell, Sussex, U.K.

Singh, V.P. and Frevert, D.K. (eds.). (2002a). *Mathematical Models of Large Watershed Hydrology*. Water Resources Publications, Highlands Ranch, CO.

Singh, V.P. and Frevert, D.K. (eds.). (2002b). *Mathematical Models of Small Watershed Hydrology and Applications*. Water Resources Publications, Highlands Ranch, CO.

Singh, V.P. and Frevert, D.K. (eds.). (2006). *Watershed Models*. CRC Press, Boca Raton, FL.

Singh, V.P. and Woolhiser, D.A. (2002). Mathematical modeling of watershed hydrology. *Journal of Hydrologic Engineering*, 7(4), 270–292.

Singh, V.P. and Yu, F.X. (1990). Derivation of infiltration equation using systems approach. *Journal of Irrigation and Drainage Engineering*, 116(6), 837–858.

Tsallis, C. (1988). Possible generalization of Boltzmann–Gibbs statistics. *Journal of Statistical Physics*, 52(1/2), 479–480.

Tsallis, C. (2002). Entropic nonextensivity: A possible measure of complexity. *Chaos, Solitons & Fractals*, 13, 371–391.

Tsallis, C. (2004). Nonextensive statistical mechanics: Construction and physical interpretation. In: *Nonextensive Entropy: Interdisciplinary Applications*, eds. M. Gell-Mann and C. Tsallis. Oxford University Press, New York, pp. 1–52.

# 11 Movement of Soil Moisture

Soil moisture occupies a central position in the hydrological cycle, interfacing between land surface hydrological processes and atmospheric processes, on one hand and between land surface processes and lithosphere (groundwater zone), on the other hand. The zone of soil moisture (also called vadose zone) is often called the gate-keeper in hydrology. Soil moisture is fundamental to analysis and evaluation of droughts; estimation of soil erosion and sediment yield; determination of infiltration, evapotranspiration, and generation of runoff; irrigation scheduling and management; maintaining salt balance and reducing water logging; tactical military encampment and mobility; sustaining ecological health; and spread of bacterial and viral activities. Because of its ubiquitous use, recent years have witnessed a considerable emphasis on measurement of soil moisture using, as for example, neutron probes, TDR probes, and satellite and other remote sensing techniques. In the case of remote sensing, soil moisture estimates are obtained within a depth of no more than 5 cm (Ulaby et al., 1996) and modeling methods are needed to estimate the entire soil moisture profile. The objective of this chapter is to discuss the construction of soil moisture using the Tsallis entropy.

## 11.1 SOIL MOISTURE ZONE

The porous medium below the land surface can be divided into two zones: one between the water table and the land surface, and the other below the water table. The water table is defined as the surface on which the fluid pressure in the pores of the medium is exactly atmospheric. This means that the hydraulic head at any point on the water table must equal the elevation of the water table at that point. The porous medium below the water table is saturated, that is the pores are filled with water, and can be referred to as ground water or saturated geologic zone. As shown in Figure 11.1, the porous medium above the water table is often divided into three zones: (1) capillary fringe, (2) intermediate zone, and (3) soil moisture zone (also called rootzone).

There exists a narrow zone immediately above the water table, called capillary zone or fringe, where the porous medium is tension-saturated but the pressure head is negative. This zone is also called tension-saturated zone. This pressure is the air entry pressure or bubbling pressure. The medium above the capillary fringe is called unsaturated zone or vadose zone or zone of aeration. In this zone, pores are partially filled with water, and partially filled with air. This means that water in the soil pores is under surface tension forces and thus the pressure will be negative. In this zone, both the moisture content ($\theta$) and the hydraulic conductivity ($K$) are functions of

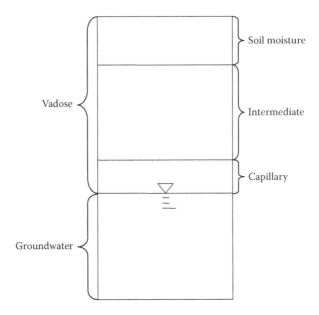

**FIGURE 11.1** Subsurface zones below the land surface: vadose zone (soil moisture, intermediate, and capillary) and groundwater zone.

the pressure head ($\psi$). Furthermore, the $\theta$–$\psi$ relationship is hysteretic and the same is true of the $K$–$\psi$ relationship. This means that these relationships during wetting are somewhat different from those during drying. From an agricultural standpoint, the vadose zone can be further divided into two zones. The zone below the land surface is the zone in which agricultural crops grow and it may thus be called root zone. This is also referred to as soil moisture zone. Below this zone is intermediate zone or percolation zone or transmission zone.

## 11.2 SOIL MOISTURE PHASES

When water is applied to the land surface either artificially by irrigation or naturally by rainfall, the movement of moisture upon entry at the surface depends on the duration for which water is applied at the surface and the moisture existing beforehand. The soil surface first gets saturated at the surface and the depth of saturation moves downward until it reaches the water table. This is called the wetting phase. Above the saturation front the soil is saturated and below the front the soil is unsaturated. In this phase, the distribution of soil moisture monotonically decreases from the surface to the water table or up to a point of concern.

When the supply of water is cut off, the downward movement of moisture continues and the soil starts draining resulting in drying. This is called the drying phase. Above the drying front from the soil is unsaturated and below it the soil is saturated. In this case, the distribution of moisture monotonically increases downward. Between the wetting and drying phases there exists a situation where the distribution of moisture monotonically increases downward up to a point (zone one) but then

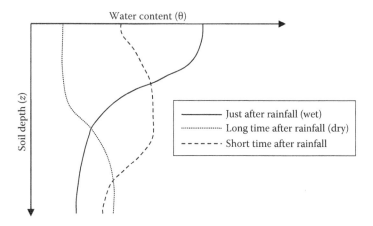

**FIGURE 11.2**   Moisture distributions in three phases.

decreases downward (zone two). In this case, one can divide the unsaturated zone into these two zones. These three cases or phases are shown in Figure 11.2. This chapter presents the derivation of the one-dimensional distribution of soil moisture under three phases.

## 11.3   ESTIMATION OF SOIL MOISTURE PROFILE

The soil moisture profile using near-surface soil moisture observations has been estimated using a range of approaches (Schmugge et al., 1980) that can be classified into four groups. The first group includes theoretical (Russo, 1988) approaches that include solution of equations governing flow of water in soils. The solution, of course, requires the knowledge of soil hydraulic characteristics, including a hydraulic conductivity function and a water retention function that needs to be determined. Thus, in this sense this becomes an inverse problem.

Approaches in the second group are probabilistic wherein the soil structure is hypothesized to evolve from a random fragmentation process. Assouline et al. (1998) have presented a conceptual model using a probabilistic approach in which the fragmentation process leads to the determination of the soil particle size distribution. Particle volumes are converted into pore volumes using a power function. Then, a capillarity equation is employed to obtain an expression for the water retention curve. Or et al. (2000) developed a stochastic model coupling the probabilistic nature of pore-space distributions with physically based soil deformations employing the Fokker–Planck equation. They addressed three features of pore space evolution: reduction of total porosity, reduction of mean pore radius, and changes in the variance of pore size distribution. This model permits computation of temporal variation of near-surface soil hydraulic properties. Pachepsky et al. (2006) and Al-Hamdan and Cruise (2010) used the Shannon entropy, whereas Singh (2010) used the Tsallis entropy for describing the movement of soil moisture. The entropy theory permits a probabilistic description of soil moisture. Pachepsky et al. (2006) also compared and evaluated different soil water models using information measures.

Approaches in the third group are based on the water balance equation (Singh, 1989), that incorporates soil moisture as output in the water balance (De Troch et al., 1996). This approach entails modeling infiltration, including redistribution and rein-filtration (Melone et al., 2006). In recent years, soil moisture observations have been assimilated into hydrological models (Das and Mohanty, 2006) and integrating soil moisture observations with hydrological models seems a more promising approach (Kostov and Jackson, 1993).

The fourth group includes approaches that are based on regression (Arya et al., 1983). These approaches are curve fitting, relating near-surface soil moisture observations to wetting and drying separately at specific locations. For shallow depths, Arya et al. (1983), Bruckler et al. (1988), Srivastava et al. (1997), among others, found regression techniques to yield satisfactory estimates, but the development of regression relations needs sufficient observations at each location and that these relations cannot be transferred to other locations.

In the inverse approaches (Kostov and Jackson, 1993), remotely sensed brightness temperature is employed for estimating soil moisture (Kostov and Jackson, 1993; Jackson, 1994). Intelligence techniques are based on artificial neural networks (Koekkoek and Booltink, 1999; Jain et al., 2004), genetic algorithms, fuzzy logic, artificial intelligence, and the like. Using a priori information on the hydrological properties of soils, soil moisture content is determined at different depths. Methods of determination include correlations between surface soil moisture and that at lower layers (Kondratyev et al., 1977), or energy-based methods with radiative properties of soil at different soil moisture states (Reutov and Shutko, 1986), or models using hydrostatic principles (Jackson et al., 1987).

## 11.4 SOIL MOISTURE AS A RANDOM VARIABLE

In the stochastic approach, the soil moisture content is considered as a random variable. Let there be a soil column of length $L$. The moisture in this soil column can vary from a very low value $\Theta_0$ to soil porosity $n$. Let the effective saturation $\theta$ be defined as

$$\theta = \frac{\Theta - \Theta_0}{n - \Theta_0} \tag{11.1}$$

where
$\Theta$ is the moisture content
$\Theta_0$ is the initial moisture content or the moisture content that cannot be extracted by plants
$n$ is the porosity

From now onward, soil moisture content will be denoted by $\theta$. The effective saturation at any point in space varies in time. It is assumed that at any value of $z$ between 0 and $L$, and all values of $\theta$ are equally likely. Thus, the effective saturation is considered as a random variable with a probability density function (PDF) as $f(\theta)$.

## 11.5 METHODOLOGY FOR DERIVING SOIL MOISTURE DISTRIBUTION USING TSALLIS ENTROPY

The procedure for deriving soil moisture profiles using entropy comprises four parts (Singh, 2010): (1) Tsallis entropy, (2) principle of maximum entropy (POME), (3) specification of constraints for the maximization of the Tsallis entropy in accord with POME, (4) maximization of entropy, and (5) soil moisture profiles for different phases. Each part is now discussed.

### 11.5.1 TSALLIS ENTROPY

The Tsallis entropy (Tsallis, 1988), $H$, can be expressed as

$$H(\theta) = \frac{1}{m-1}\left[1 - \int_a^b (f(\theta))^m \, d\theta\right] = \frac{1}{m-1}\int_a^b f(\theta)\{1 - [f(\theta)]^{m-1}\} \, d\theta \qquad (11.2)$$

where
$f(\theta)$ is the PDF of $\theta$
$m$ is a real number
$a$ and $b$ are limits of $\theta$

$H$ describes the uncertainty associated with $f(\theta)$. Quantity $\{1 - [f(\theta)]^{m-1}\}/(m - 1)$ is a measure of uncertainty of $f(\theta)$ or $\theta$ and $H(\theta)$ expresses the mean uncertainty of $\theta$.

### 11.5.2 PRINCIPLE OF MAXIMUM ENTROPY

The principle of maximum entropy formulated by Jaynes (1958) says that the least-biased probability distribution of $\theta$, $f(\theta)$, will be the one that maximizes Equation 11.2, subject to the given information on $\theta$ expressed as constraints.

### 11.5.3 SPECIFICATION OF CONSTRAINTS

Information on $\theta(z)$ can be expressed as constraints that can be defined as

$$\int_a^b f(\theta) d\theta = 1 \qquad (11.3)$$

$$\int_a^b g_r(\theta) f(\theta) d\theta = \overline{g_r(\theta)}, \quad r = 1, 2, ..., n \qquad (11.4)$$

where $g_r(\theta)$, $r = 1, 2, ..., n$, represent some functions of $\theta$. For example, $r = 1$, Equation 11.4 would correspond to the mean effective saturation; likewise, $r = 2$ would denote the variance of $\theta$. For most moisture profiles, more than two constraints may not be needed.

### 11.5.4  MAXIMIZATION OF TSALLIS ENTROPY

The entropy given by Equation 11.2, subject to Equations 11.3 and 11.4, can be maximized using the method of Lagrange multipliers, which would lead to the probability distribution of $\theta$ in terms of the given constraints:

$$f(\theta) = \left\{ \frac{m-1}{m} \left[ \frac{1}{m-1} + \lambda_0 - \sum_{r=1}^{n} \lambda_r g_r(\theta) \right] \right\}^{1/(m-1)} \tag{11.5}$$

where $\lambda_r$s are the Lagrange multipliers that can be determined with the use of Equations 11.3 and 11.4. Now the entropy theory is applied to the derivation of soil moisture profiles for three phases: wetting, drying, and mixed.

### 11.5.5  SOIL MOISTURE PROFILE FOR WETTING PHASE

This phase occurs during and immediately after rainfall and is designated as wet case. The moisture content is highest near the surface and decreases downward. It is hypothesized that

$$F(\theta) = 1 - \frac{z}{L}, \quad f(\theta) = -\frac{1}{L}\frac{dz}{d\theta} \tag{11.6}$$

where $F(\theta)$ is the cumulative probability distribution (CDF). In order to derive the moisture content profile using the entropy theory, the following constraints are defined:

$$\int_{\theta_L}^{\theta_u} f(\theta)d\theta = 1 \tag{11.7}$$

$$\int_{\theta_L}^{\theta_u} \theta f(\theta)d\theta = \bar{\theta} \tag{11.8}$$

where $\theta_L$ and $\theta_u$ are the values of effective saturation at $z = L$ and $z = 0$, respectively.

Applying POME and the method of Lagrange multipliers, one gets the Lagrangian function $L_a$:

$$L_a = \int_{\theta_L}^{\theta_u} \frac{f(\theta)}{m-1}\{1-[f(\theta)]^{m-1}\}d\theta + \lambda_0\left[\int_{\theta_L}^{\theta_u} f(\theta)d\theta - 1\right] + \lambda_1\left[\int_{\theta_L}^{\theta_u} \theta f(\theta)d\theta - \bar{\theta}\right] \tag{11.9}$$

Differentiating Equation 11.9 with respect to $f(\theta)$ and equating the derivative to 0, one obtains

$$f(\theta) = \left\{ \frac{m-1}{m} \left[ \frac{1}{m-1} + \lambda_0 + \lambda_1 \theta \right] \right\}^{1/(m-1)} \tag{11.10}$$

The maximum entropy becomes

$$H(\theta) = \frac{1}{m-1} - \frac{1}{m} \left( \frac{1}{m-1} + \lambda_0 + \lambda_1 \overline{\theta} \right) \tag{11.11}$$

Let $\lambda_* = \lambda_0 + 1/m - 1$. Then, Equation 11.10 becomes

$$f(\theta) = \left\{ \frac{m-1}{m} [\lambda_* + \lambda_1 \theta] \right\}^{1/(m-1)} \tag{11.12}$$

where $\lambda_0$ and $\lambda_1$ are the Lagrange multipliers that can be determined using Equations 11.7 and 11.8. Equation 11.12 is the PDF of soil moisture content. Substitution of Equation 11.12 in Equation 11.7 yields

$$\int_{\theta_L}^{\theta_u} \left\{ \frac{m-1}{m} [\lambda_* + \lambda_1 \theta] \right\}^{1/(m-1)} d\theta = 1 \tag{11.13}$$

This results in

$$(\lambda_* + \lambda_1 \theta_u)^{m/(m-1)} = \lambda_1 \left( \frac{m}{m-1} \right)^{m/(m-1)} + (\lambda_* + \lambda_1 \theta_L)^{m/(m-1)} \tag{11.14}$$

Substituting Equation 11.12 in Equation 11.8, one gets

$$\int_{\theta_L}^{\theta_u} \theta \left\{ \frac{m-1}{m} [\lambda_* + \lambda_1 \theta] \right\}^{1/(m-1)} d\theta = \overline{\theta} \tag{11.15}$$

Equation 11.15 leads to

$$\theta_u (\lambda_* + \lambda_1 \theta_u)^{m/(m-1)} + \frac{m-1}{2m-1} \frac{1}{\lambda_1} (\lambda_* + \lambda_1 \theta_L)^{(2m-1)/(m-1)} - \theta_L (\lambda_* + \lambda_1 \theta_L)^{m/(m-1)}$$

$$- \frac{1}{\lambda_1} \frac{m-1}{2m-1} (\lambda_* + \lambda_1 \theta_u)^{(2m-1)/(m-1)} = \overline{\theta} \left[ \lambda_1 \left( \frac{m}{m-1} \right)^{m/(m-1)} \right] \tag{11.16}$$

Equations 11.14 and 11.16 contain two unknowns $\lambda_*$ and $\lambda_1$ and can be solved numerically.

Coupling Equations 11.12 and 11.6

$$\left[\frac{m-1}{m}(\lambda_* + \lambda_1\theta)\right]^{1/(m-1)} d\theta = -\frac{1}{L}dz \tag{11.17}$$

Integrating Equation 11.17 from $\theta = \theta_u$ to $\theta$, and $z = 0$ to $z$, one gets

$$\theta = -\frac{\lambda_*}{\lambda_1} + \frac{1}{\lambda_1}\left[\frac{z}{L} + \left(\frac{m-1}{m}\right)^{m/(m-1)}\frac{1}{\lambda_1}(\lambda_* + \lambda_1\theta_u)^{m/(m-1)}\right]^{m/(m-1)} \tag{11.18}$$

or

$$\theta = \frac{1}{\lambda_1}\left[(\lambda_* + \lambda_1\theta_u)^{m/(m-1)} - \frac{z}{L\lambda_1}\left(\frac{m}{m-1}\right)^{(m-1)/m}\right]^{(m-1)/m} - \frac{\lambda_*}{\lambda_1} \tag{11.19}$$

Equation 11.19 describes the moisture profile as a function of $z$ for the wetting phase, where the maximum soil moisture occurs at the surface and the moisture decreases downward.

If $\theta_L = 0$, Equation 11.14 simplifies to

$$(\lambda_* + \lambda_1\theta_u)^{m/(m-1)} = \lambda_1\left(\frac{m}{m-1}\right)^{m/(m-1)} + \lambda_*^{m/(m-1)} \tag{11.20}$$

Likewise, Equation 11.16 simplifies to

$$\theta_u(\lambda_* + \lambda_1\theta_u)^{m/(m-1)} + \frac{m-1}{2m-1}\frac{1}{\lambda_1}(\lambda_*)^{(2m-1)/(m-1)}$$

$$-\frac{1}{\lambda_1}(\lambda_* + \lambda_1\theta_u)^{(2m-1)/(m-1)} = \bar{\theta}\left[\lambda_1\left(\frac{m}{m-1}\right)^{m/(m-1)}\right] \tag{11.21}$$

## Example 11.1

Compute and plot the soil moisture profile as a function of $z$ for two wet cases: 1 ($\theta_u = 0.86$) and case 2 ($\theta_u = 0.46$) for different values of parameter $m$ for given $\theta_u$, $\lambda_1$, $\lambda_*$. For computing and plotting, consider the following scenarios: (1) $\theta_u = 0.86$, $\lambda_1 = 2$, $\lambda_* = 3$, $m = (3/4)–2$; (2) $\theta_u = 0.86$, $\lambda_1 = 1–8$, $\lambda_* = 1.48$, $m = 3/4$; (3) $\theta_u = 0.46$, $\lambda_1 = -1$ to $-6$, $\lambda_* = 3.19$, $m = 3/4$; (4) $\theta_u = 0.86$, $\lambda_1 = 2.10$, $\lambda_* = 1–10$, $m = 3/4$; (5) $\theta_u = 0.46$, $\lambda_1 = 2.61$, $\lambda_* = 1–3.2$, $m = 3/4$; and (6) $\theta_u = 0.46$, $\theta_m = 0.30$, $\lambda_1$, $\lambda_*$, $m = (3/4)–2$.

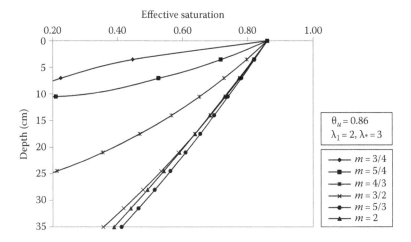

**FIGURE 11.3** Soil moisture profile as a function of $z$ for wet case 1 for different values of the $m$ parameter.

### Solution

Soil moisture profiles as a function of $z$ for the wet case are computed using Equation 11.19. For different values of the $m$ parameter, $\lambda_1 = 2.0$, and $\lambda_* = 3$, the soil moisture profiles are shown in Figure 11.3. This figure shows that the moisture profile is quite sensitive to $m$. It seems that $m = 3/4$ would be more realistic.

For different values of, $\lambda_1$, and $\lambda_* = 1.48$, and $m = 3/4$, the soil moisture profiles are shown in Figure 11.4. This figure shows that the moisture profile is quite

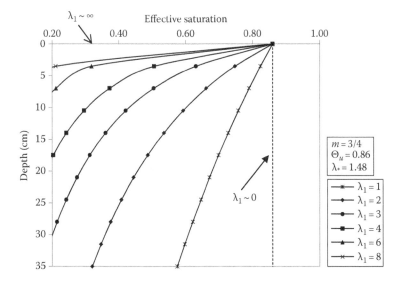

**FIGURE 11.4** Soil moisture profiles for different values of $\lambda_1$ (wet case 1).

sensitive to $\lambda_1$. Figure 11.5 shows that soil moisture profile moves slowly for larger values of $\lambda_1$ and more rapidly for smaller values of $\lambda_1$. It is seen from Figure 11.6 that the soil moisture profile is quite sensitive to the Lagrange multiplier $\lambda_*$ and the sensitivity increases with decreasing value of $\lambda_*$. Figure 11.7 also shows similar sensitivity to $\lambda_*$. The depth profile does not seem sensitive to the value of $m$, as shown in Figures 11.8 and 11.9.

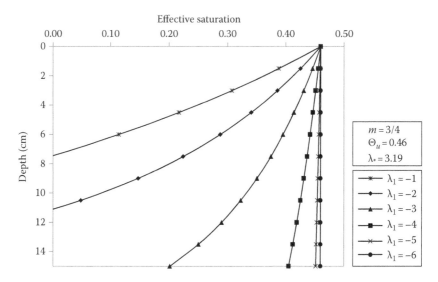

**FIGURE 11.5** Soil moisture profiles for different values of $\lambda_1$ (wet case 2).

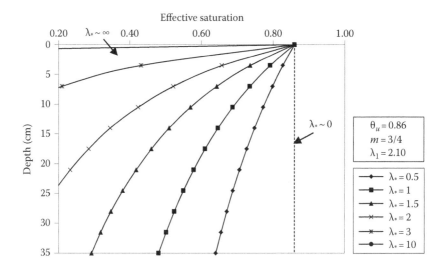

**FIGURE 11.6** Soil moisture profiles for different values of $\lambda_*$ (wet case 1).

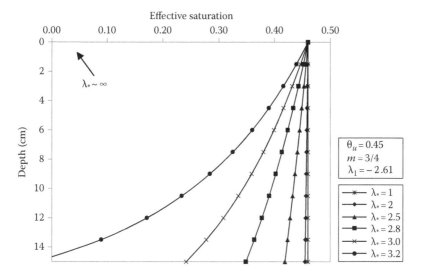

**FIGURE 11.7** Soil moisture profiles for different values of $\lambda_*$ (wet case 2).

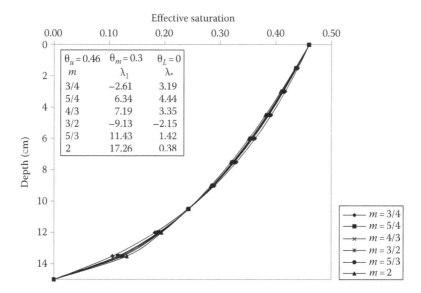

**FIGURE 11.8** Soil moisture profiles for different values of $m$ (wet case 1).

### 11.5.6 MOISTURE PROFILE FOR DRYING PHASE

In this phase, the lowest moisture occurs at $z = 0$ and highest at $z = L$. Therefore, it is hypothesized that

$$F(\theta) = \frac{z}{L}, \quad f(\theta) = \frac{1}{L}\frac{dz}{d\theta} \tag{11.22}$$

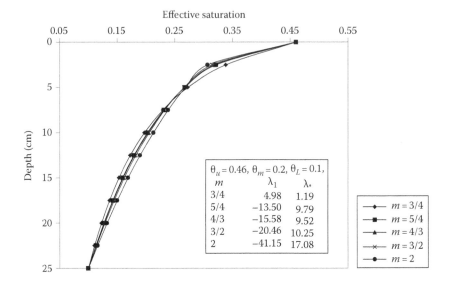

**FIGURE 11.9**   Soil moisture profiles for different values of $m$ (wet case 2).

The constraints for this case can be expressed as

$$\int_{\theta_u}^{\theta_L} f(\theta)d\theta = 1 \tag{11.23}$$

$$\int_{\theta_u}^{\theta_L} \theta f(\theta)d\theta = \bar{\theta} \tag{11.24}$$

The entropy-based probability distribution $f(\theta)$ becomes the same as Equation 11.12. Substituting Equation 11.23 in Equation 11.12, one obtains

$$(\lambda_* + \lambda_1\theta_L)^{m/(m-1)} - (\lambda_* + \lambda_1\theta_u)^{m/(m-1)} = \lambda_1\left(\frac{m}{m-1}\right)^{m/(m-1)} \tag{11.25}$$

Substituting Equation 11.12 in Equation 11.24, one obtains

$$\frac{1}{\lambda_1}\left(\frac{m-1}{m}\right)^{m/(m-1)}\left[\theta_L(\lambda_* + \lambda_1\theta_L)^{m/(m-1)} - \theta_L(\lambda_* + \lambda_1\theta_u)^{m/(m-1)}\right]$$

$$-\frac{1}{\lambda_1^2}\left(\frac{m-1}{2m-1}\right)\left(\frac{m-1}{m}\right)^{m/(m-1)}\left[(\lambda_* + \lambda_1\theta_L)^{(2m-1)/(m-1)} - (\lambda_* + \lambda_1\theta_u)^{(2m-1)/(m-1)}\right] = \bar{\theta}$$

$$\tag{11.26}$$

Equations 11.25 and 11.26 can be employed to determine $\lambda_0$ and $\lambda_1$.

If $\theta_u = 0$, Equations 11.25 and 11.26 simplify to

$$(\lambda_* + \lambda_1 \theta_L)^{m/(m-1)} = (\lambda_*)^{m/(m-1)} + \lambda_1 \left(\frac{m}{m-1}\right)^{m/(m-1)} \tag{11.27}$$

$$\frac{1}{\lambda_1}\left(\frac{m-1}{m}\right)^{m/(m-1)}\left[\theta_L(\lambda_* + \lambda_1\theta_L)^{m/(m-1)}\right] - \frac{1}{\lambda_1^2}\left(\frac{m-1}{2m-1}\right)\left(\frac{m-1}{m}\right)^{m/(m-1)}$$

$$\left[(\lambda_* + \lambda_1\theta_L)^{(2m-1)/(m-1)} - (\lambda_*)^{(2m-1)/(m-1)}\right] = \bar{0} \tag{11.28}$$

Substitution of Equation 11.12 in Equation 11.22 yields

$$\left[\frac{m-1}{m}(\lambda_* + \lambda_1\theta)\right]^{1/(m-1)} = \frac{1}{L}dz \tag{11.29}$$

Solution of Equation 11.29 can be expressed as

$$\theta = \frac{1}{\lambda_1}\left[\frac{z}{L}\lambda_1\left(\frac{m}{m-1}\right)^{(m-1)/m} + \lambda_*^{m/(m-1)}\right]^{m/(m-1)} - \frac{\lambda_*}{\lambda_1} \tag{11.30}$$

Equation 11.30 yields the soil moisture profile as a function of $z$.

## Example 11.2

Compute and plot the soil moisture profile as a function of $z$ for two drying cases: 1 ($\theta_u = 0$) and 2 ($\theta_u = 0.45$) for different values of parameter $m$ for given $\theta_u$, $\lambda_1$, $\lambda_*$. For computing and plotting consider the following cases: (1) $\theta_u = 0.0$, $\lambda_1 = 3$, $\lambda_* = 2$, $m = 3/4$–2; (2) $\theta_u = 0.45$, $\lambda_1 = 0.5$–1.7, $\lambda_* = 1.69$, $m = 3/4$; (3) $\theta_u = 0.0$, $\lambda_1 = -1$ to $-30$, $\lambda_* = 2.96$, $m = 3/4$; (4) $\theta_u = 0.45$, $\lambda_1 = 1.14$, $\lambda_* = 0.5$–1.8, $m = 3/4$; (5) $\theta_u = 0.0$, $\lambda_1 = 1.55$, $\lambda_* = 1$–4, $m = 3/4$; (6) $\theta_u = 0.45$, $\theta_m = 0.64$, $\theta_L = 0.89$, $\lambda_1$, $\lambda_*$, $m = 3/4$–2; and (7) $\theta_u = 0.0$, $\theta_m = 0.21$, $\theta_L = 0.25$, $\lambda_1$, $\lambda_*$, $m = 3/4$–2.

## Solution

Soil moisture profiles as a function of $z$ for the dry case are computed using Equation 11.30. For different values of the $m$ parameter and $\lambda_1 = 3$ and $\lambda_* = 2$, the profiles are shown in Figure 11.10. This figure shows the variation of the effective saturation for various values of $m$ for given $\theta_u$, $\lambda_1$, $\lambda_*$. The moisture profile is quite sensitive to $m$. It seems that $m = 3/4$ would be more realistic. Figure 11.11 shows that soil moisture profile is quite sensitive to the Lagrange multiplier $\lambda_1$ and the sensitivity increases with decreasing $\lambda_1$. Figure 11.12 also shows a similar behavior of soil moisture movement. Figure 11.13 shows that soil moisture profile is highly sensitive to $\lambda_*$ with increasing sensitivity to decreasing $\lambda_*$. A similar behavior of soil moisture profile is observed in Figure 11.14. Both Figures 11.15 and 11.16 exhibit that the soil moisture profile is relatively insensitive to $m$.

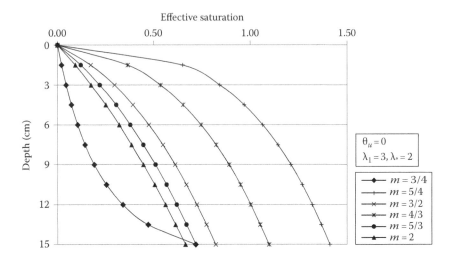

**FIGURE 11.10**　Soil moisture profile as a function of $z$ for case 1 for different values of the $m$ parameter for the drying phase.

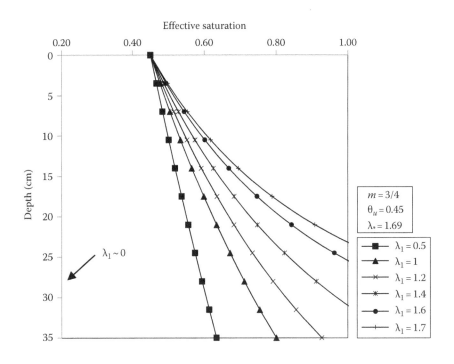

**FIGURE 11.11**　Soil moisture profiles with different $\lambda_1$ (drying case 1). [Note $\lambda 1 = \lambda_1$]

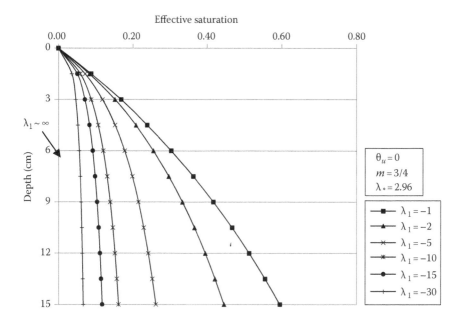

**FIGURE 11.12**  Soil moisture profiles with different $\lambda_1$ (drying case 2). [Note $\lambda 1 = \lambda_1$]

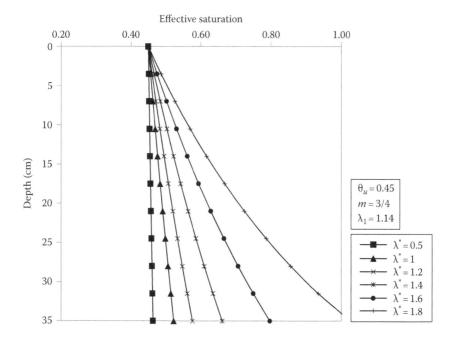

**FIGURE 11.13**  Soil moisture profiles with different $\lambda_*$ (drying case 1). [Note $\lambda^* = \lambda_*$]

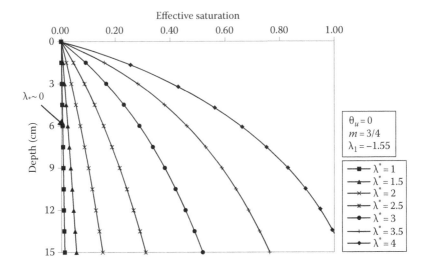

**FIGURE 11.14** Soil moisture profiles with different $\lambda_*$ (drying case 2). [Note $\lambda^* = \lambda_*$]

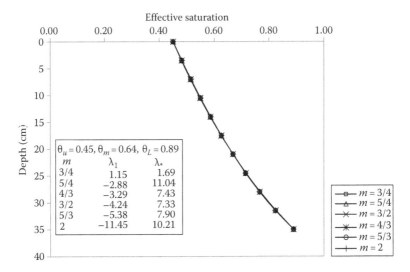

**FIGURE 11.15** Soil moisture profiles with different $m$ values (drying case 1).

### 11.5.7 Moisture Profiles for Mixed Phase

This case can be considered to consist of two parts. For part I, $0 \leq z \leq d$, the PDF of $\theta$, $f(\theta)$, is the same as in case 2 with the proviso that $\theta_u \leq \theta \leq \theta_d$, and $\theta = \theta_d$ at $z = d$. It may therefore be noted that

$$\overline{\theta_{up}} = \frac{1}{d}\int_0^d \theta(z)dz \tag{11.31}$$

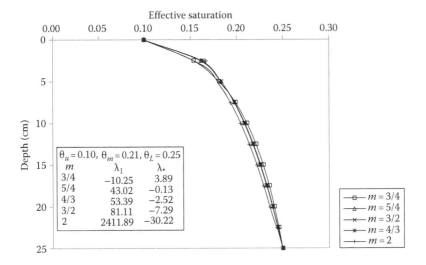

**FIGURE 11.16** Soil moisture profiles with different $m$ values (drying case 2).

$$\overline{\theta_{Low}} = \frac{1}{L-d}\int_{d}^{L}\theta(z)dz \tag{11.32}$$

For part I, the moisture profile becomes

$$\left(\frac{m-1}{m}\right)^{m/(m-1)}\frac{1}{\lambda_1}\left[(\lambda_* + \lambda_1\theta)^{m/(m-1)} - (\lambda_* + \lambda_1\theta_u)^{m/(m-1)}\right] = \frac{z}{L} \tag{11.33}$$

The Lagrange multipliers can be determined using the constraints given by Equations 11.23 and 11.24. Substituting Equation 11.12 in Equation 11.23, one gets

$$\int_{\theta_u}^{\theta_d}\left[\frac{m-1}{m}(\lambda_* + \lambda_1\theta)\right]^{1/(m-1)}d\theta = 1 \tag{11.34}$$

Equation 11.34 yields

$$(\lambda_* + \lambda_1\theta_d)^{m/(m-1)} - (\lambda_* + \lambda_1\theta_u)^{m/(m-1)} = \lambda_1\left(\frac{m}{m-1}\right)^{m/(m-1)} \tag{11.35}$$

Substitution of Equation 11.12 in Equation 11.24 yields

$$\int_{\theta_u}^{\theta_d}\theta\left[\left(\frac{m-1}{m}\right)(\lambda_* + \lambda_1\theta)\right]^{1/(m-1)} = \overline{\theta_{up}} \tag{11.36}$$

This can be solved as before.

For part II, $d \leq z \leq L$. The moisture profile can be expressed as

$$\theta = \frac{1}{\lambda_1} \left[ \frac{d-z}{L} + \left( \frac{m-1}{m} \right)^{m/(m-1)} \frac{1}{\lambda_1} (\lambda_* + \lambda_1 \theta_d)^{m/(m-1)} \right]^{m/(m-1)} - \frac{\lambda_*}{\lambda_1} \quad (11.37)$$

$$\int_{\theta_d}^{\theta_L} \left[ \frac{m-1}{m} (\lambda_* + \lambda_1 \theta) \right]^{1/(m-1)} d\theta = 1 \quad (11.38)$$

This leads to

$$(\lambda_* + \lambda_1 \theta_d)^{m/(m-1)} = (\lambda_* + \lambda_1 \theta_L)^{m/(m-1)} = \lambda_1 \left( \frac{m}{m-1} \right)^{m/(m-1)} \quad (11.39)$$

$$\theta_d (\lambda_* + \lambda_1 \theta_d)^{m/(m-1)} + \frac{m-1}{2m-1} \frac{1}{\lambda_1} (\lambda_* + \lambda_1 \theta_L)^{(2m-1)/(m-1)} - \theta_L [(\lambda_* + \lambda_1 \theta_L)^{m/(m-1)}$$

$$- \frac{1}{\lambda_1} \frac{m-1}{2m-1} (\lambda_* + \lambda_1 \theta_d)^{(2m-1)/(m-1)} = \overline{\theta_{Low}} \left[ \lambda_1 \left( \frac{m}{m-1} \right)^{m/(m-1)} \right]$$

$$(11.40)$$

### Example 11.3

Compute and plot the soil moisture profile as a function of $z$ for two mixed cases: 1 and 2 for different values of parameter $m$ for given $\theta_u$, $\theta_d$, $\theta_L$, $\lambda_1$, $\lambda_*$: (1) $\theta_u = 0.20$, $\theta_d = 0.46$, $\theta_L = 0.0$, $m = 3/4$–2 and corresponding values of $\lambda_1$ and $\lambda_*$; and (2) $\theta_u = 0.13$, $\theta_d = 0.78$, $\theta_L = 0.10$, $m = 5/4$–2 and corresponding values of $\lambda_1$ and $\lambda_*$.

#### Solution

Figures 11.17 and 11.18 show that soil moisture profiles are not sensitive to the $m$ value. This means that any value of $m$ between 0.75 and 2.00 would be adequate.

## 11.6 SOIL MOISTURE PROFILE IN TIME

In order to determine the soil moisture profile as a function of time, the boundary condition needs to be specified. The same applies to the constraints. The mean value of $\theta$ can be calculated as follows (Al-Hamdan and Cruise, 2010):

$$\overline{\theta} = \frac{w/(nL)}{n - \theta_0} \quad (11.41)$$

where $\theta_0$ is the initial soil moisture, and $w$ is the applied water to the soil surface that can be computed using the water balance as

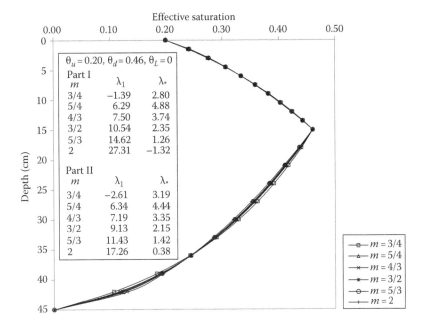

**FIGURE 11.17**   Soil moisture profiles for different *m* values (mixed case 1).

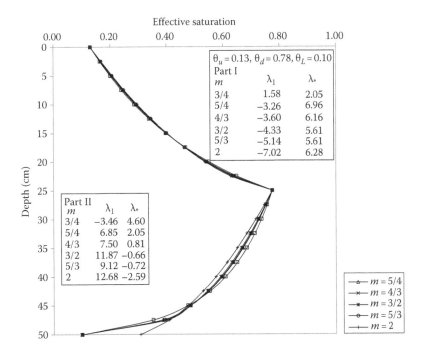

**FIGURE 11.18**   Soil moisture profiles for different values of *m* (mixed case 2).

$$w = w_{5i-1} - w_{5i} + P_i + \Delta R - ET \tag{11.42}$$

where all quantities are measured in units of depth, subscript $i$ denotes the $i$th time, $w$ is the water depth applied to the soil within a time step, $w_5$ is the water content (in units of depth) for the 5 cm deep surface that can be measured by remote sensing, $P$ is the amount of precipitation (in units of depth), $ET$ is the amount of evapotranspiration (in units of depth), and $\Delta R$ is the difference between the amount of runoff leaving a particular grid cell and the amount entering that grid cell (in units of depth).

Parameters $\theta_u$, $\theta_L$, and $\bar{\theta}$ need to be characterized. Also need to be characterized are $\theta_0$ and $n$. The depth $d$ also needs to be determined, which can be done using an infiltration model or a kinematic wave model requiring only the value of soil moisture at the soil surface ($z = 0$). Following Singh (1997) and Singh and Joseph (1994), the wetting front depth $z_f$ can be given as

$$z_f = \frac{K}{\theta_u} t \tag{11.43}$$

where
  $K$ is the hydraulic conductivity (treated as parameter)
  $t$ is time

For each time step, $d = z_f$.

## Example 11.4

Using a soil tray 152 cm long, 122 cm wide, and 78 cm deep, Melone et al. (2006) conducted experiments on loamy soil and sandy clay loam. Beneath the bottom of the soil column, a 7 cm deep gravel layer was created to allow for the outflow of percolated water. Experiments were conducted on a uniform soil moisture content and under uniform rainfall. For sandy clay loam soil, rainfall = 2.4 cm/h for 8 h, $K = 2.1$ cm/h, $n = 0.485$, $w$ = accumulated rainfall, $\theta_0 = 0.043$, $ET = 0$, and no lateral runoff. The time step was as follows: $\Delta t = 1$ h. Therefore, $w = 2.4$, 4.8 (= 2.4 + 2.4), 7.2, 9.6, 12, 14.4, 16.8, and 19.2 cm at $t = 1$, 2, 3, 4, 5, 6, 7, and 8 h, respectively. The value of $L = 55$ cm = the effective soil column depth during the rainfall event. After the rainfall event, $w = 19.2$ cm during the time of redistribution of soil moisture, assuming no deep percolation. Compute the values of the Lagrange multipliers for the experimental data and then compute the soil moisture content at different depths and compare with observed values.

### Solution

First, the Lagrange multipliers are computed using Equations 11.14 and 11.16. This is done numerically, and parameter values are as shown in Table 11.2. Computed soil moisture values for all four cases are shown in Table 11.1. Also shown are the relative errors in computed values. Computed values of soil moisture are plotted in Figure 11.19. Parameters were obtained by fitting soil moisture profile equations to experimental data sets discussed earlier. The parameter values are given in Table 11.2.

**TABLE 11.1**

**Experimental Data for Wet Case**

| | Soil Moisture Experiment 1 | | |
|---|---|---|---|
| Depth (cm) | $\theta$ (Observed) | $\theta$ (Computed) | Relative Error |
| 15 | 0.860 | 0.860 | 0.000 |
| 25 | 0.803 | 0.700 | 0.128 |
| 35 | 0.509 | 0.460 | 0.096 |
| 45 | 0.000 | 0.000 | 0.000 |

*Source:* Melone, F. et al., *Hydrol. Process.*, 20, 439, 2006.

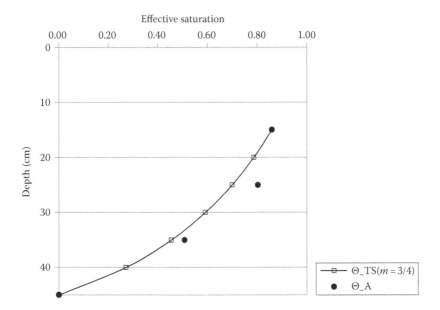

**FIGURE 11.19** Soil moisture profile for soil moisture experiment 1.

**TABLE 11.2**

**Parameter Estimation for Four Sets of Data**

| $\Theta_u$ | $\Theta_m$ | $\Theta_l$ | $\lambda_1$ | $\lambda_*$ | $m$ | Data Source |
|---|---|---|---|---|---|---|
| 0.860 | 0.543 | 0.000 | −1.388 | 3.585 | 3/4 | Experiment 1 |

*Source:* After Melone, F. et al., *Hydrol. Process.*, 20, 439, 2006.

The soil moisture profile computed using Equation 11.19 with $m = 0.75$ compared well with the observed profile for experimental data set 1 as shown in Figure 11.19. In the middle portion, the computed moisture is higher than the observed value.

## Example 11.5

For a dry case soil moisture, data are given in Table 11.3. Compute the values of the Lagrange multipliers for the experimental data and then compute the soil moisture content at different depths and compare with observed values.

### Solution

First, the Lagrange multipliers are computed using Equations 11.25 and 11.26. This is done numerically, and parameter values are as shown in Table 11.4. Computed soil moisture values are shown in Table 11.3. Also shown are the relative errors in computed values. Computed values of soil moisture are plotted in Figure 11.20. Parameters are obtained by fitting soil moisture profile equations to experimental data sets discussed earlier. The parameter values are given in Table 11.4.

## Example 11.6

For a mixed case, soil moisture data are given in Table 11.5. Compute the values of the Lagrange multipliers for the experimental data and then compute the soil moisture content at different depths and compare with observed values.

### Solution

First, the Lagrange multipliers are computed using Equations 11.35 and 11.36 for the first phase and Equations 11.37 and 11.39 for the second phase. This is done numerically, and parameter values are as shown in Table 11.6. Computed soil

### TABLE 11.3
### Experimental Soil Moisture Data for a Dry Case

| Soil Moisture Experiment 1 for Dry Case | | | |
|---|---|---|---|
| Depth (cm) | θ (Observed) | θ (Computed) | Relative Error (%) |
| 10 | 0.130 | 0.130 | 0.000 |
| 15 | 0.175 | 0.180 | −0.029 |
| 20 | 0.220 | 0.240 | −0.091 |
| 25 | 0.293 | 0.300 | −0.024 |
| 35 | 0.650 | 0.480 | 0.262 |
| 45 | 0.775 | 0.770 | 0.006 |

### TABLE 11.4
### Parameter Estimation for Experimental Data for Dry Case

| $\theta_L$ | $\theta_M$ | $\Theta_U$ | $\lambda_1$ | $\lambda_*$ | $M$ | $K$ | Data Source |
|---|---|---|---|---|---|---|---|
| 0.775 | 0.374 | 0.130 | 1.584 | 2.050 | 3/4 | −3 | Soil experiment 1 |

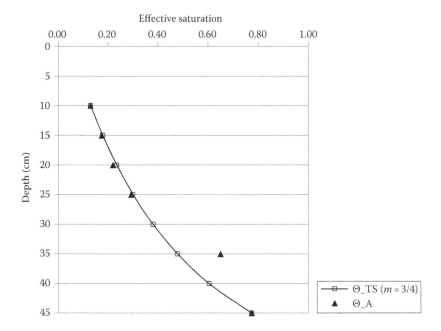

**FIGURE 11.20** Soil moisture profile for soil experiment 1 for dry case.

**TABLE 11.5**
**Soil Moisture Experimental Data**

| Soil Moisture Experiment for a Mixed Case | | | |
|---|---|---|---|
| Depth (cm) | θ (Observed) | θ (Computed) | Relative Error |
| 10 | 0.351 | 0.351 | 0.00 |
| 20 | 0.417 | 0.417 | 0.00 |
| 30 | 0.341 | 0.346 | −0.01 |
| 45 | 0.338 | 0.338 | 0.00 |
| 60 | 0.337 | 0.337 | 0.00 |

**TABLE 11.6**
**Parameter Estimation for a Mixed Case**

| $\Theta_U$ | $\Theta_{M\_I}$ | $\Theta_d$ | $\lambda_1$ | $\lambda_*$ | $m$ | $K$ | Data Source |
|---|---|---|---|---|---|---|---|
| 0.351 | 0.364 | 0.417 | −28.375 | 8.989 | 3/4 | −3 | Soil experiment 1 |
| $\Theta_d$ | $\Theta_{M\_II}$ | $\Theta_L$ | $\lambda_1$ | $\lambda_*$ | $m$ | $K$ | |
| 0.417 | 0.340 | 0.337 | 104.621 | 34.621 | 3/4 | −3 | |

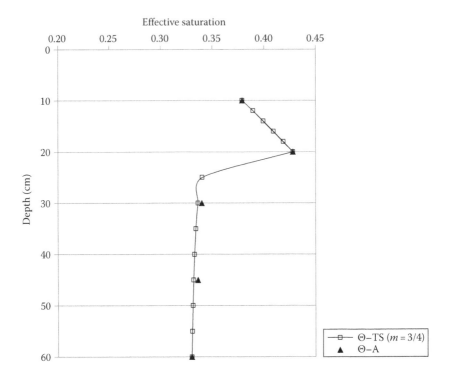

**FIGURE 11.21**　Soil moisture profile for a mixed case.

moisture values are shown in Table 11.5. Also shown are the relative errors in computed values. Computed values of soil moisture are plotted in Figure 11.21. Parameters were obtained by fitting soil moisture profile equations to experimental data sets discussed earlier. The parameter values are given in Table 11.6.

## REFERENCES

Al-Hamdan, O.Z. and Cruise, J.F. (2010). Soil moisture profile development from surface observations by principle of maximum entropy. *Journal of Hydrologic Engineering*, 15(5), 327–337.

Arya, L.M., Richter, J.C., and Paris, J.F. (1983). Estimating profile water storage from surface zone soil moisture measurements under bare field conditions. *Water Resources Research*, 19(2), 403–412.

Assouline, S., Tessier, D., and Bruand, A. (1998). A conceptual model of the soil water retention curve. *Water Resources Research*, 34(2), 223–231.

Das, N.N. and Mohanty, B.P. (2006). Rootzone soil moisture assessment using remote sensing and vadose zone modeling. *Vadose Zone Journal*, 5, 296–307.

De Troch, F.P., Troch, P.A., Su, Z., and Lin, D.S. (1996). Application of remote sensing for hydrological modeling. In: *Distributed Hydrological Modeling*, eds. M.B. Abbott and J. Refsgaard. Kluwer Academic Publishers, Dordrecht, the Netherlands, pp. 165–191.

Heathman, G.C. (1992). *Data Report: Profile Soil Moisture*, Chapter XIV. U.S. Department of Agriculture, Washington, DC.

Jackson, T.J. (1994). *Ground Measurements of Surface Soil Moisture*, Chapter VII. U.S. Department of Agriculture, Washington, DC.

Jackson, T.J., Hawley, M.E., and O'Neill, P.E. (1987). Preplanting soil moisture using passive microwave sensors. *Water Resources Bulletin*, 23(1), 11–19.

Jaynes, E.T. (1958). Probability Theory in Science and Engineering. *Colloquium Lectures in Pure and Applied Science*, Vol. 4, Field Research Laboratory, Socony Mobil Oil Company, Inc., U.S.

Jain, S.K., Singh, V.P., and van Genuchten, M.T. (2004). Analysis of soil water retention data using artificial neural networks. *Journal of Hydrologic Engineering*, 9(5), 415–420.

Koekkoek, E.J.W. and Booltink, H. (1999). Neural network models to predict soil water retention. *European Journal of Soil Science*, 50, 489–495.

Kondratyev, K.Y., Melentyev, V.V., Rabinovich, Y.I., and Shulgina, E.M. (1977). Passive microwave remote sensing of soil moisture. *Proceedings of the, 11th Symposium on Remote Sensing of the Environment*, Environmental Research Institute of Michigan, Ann Arbor, MI, pp. 1641–1661.

Kostov, K.G. and Jackson, T.J. (1993). Estimating profile soil moisture from surface layer measurements—A review. *Proceedings of the International Society for Optical Engineering*, 1941, 125–136.

Melone, F., Corradini, C., Morbidelli, R., and Saltalippi, C. (2006). Laboratory experimental check of a conceptual model for infiltration under complex patters. *Hydrological Processes,* 20, 439–452.

Or, D., Leij, F.J., Snyder, V., and Ghezzehei, A. (2000). Stochastic model for posttillage soil pore space evolution. *Water Resources Research*, 36(7), 1641–1652.

Pachepsky, Y., Guber, A., Jacques, D., Simunek, J., Genuchten, M.T.V., Nicholson, T., and Cady, R. (2006). Information content and complexity of simulated soil water fluxes. *Geoderma*, 134(3–4), 253–266.

Reutov, E.A. and Shutko, A.M. (1986). Prior-knowledge based soil moisture determination of microwave radiometry. *Soviet Journal of Remote Sensing*, 5(1), 100–125.

Russo, D. (1988). Determining soil hydraulic properties by parameter estimation: On the selection of a model for the hydraulic properties. *Water Resources Research*, 24(3), 453–459.

Schmugge, T.J., Jackson, T.J., and McKim, H.L. (1980). Survey of methods for soil moisture determination. *Water Resources Research*, 16(6), 961–979.

Singh, V.P. (1989). *Hydrologic Systems, Vol. 2: Watershed Modeling*. Prentice Hall, Englewood Cliffs, NJ.

Singh, V.P. (1997). *Kinematic Wave Modeling in Water Resources: Environmental Hydrology*. John Wiley & Sons, New York.

Singh, V.P. (2010). Entropy theory for movement of moisture in soils. *Water Resources Research*, 46(W03516), 1–12. doi:10.1029/2009WR008288.

Singh, V.P. and Joseph, E. (1994). Kinematic wave model for soil moisture movement with plant root extraction. *Irrigation Science*, 14, 188–191.

Srivastava, S.K., Yograjan, N., Jayaraman, V., Nageswara, P.P., and Chandrsekhar, M.G. (1997). On the relationship between ERS-1 SAR/backscatter and surface/subsurface soil moisture variations in vertisols. *Acta Astronautica*, 40(10), 693–699.

Tsallis, C. (1988). Possible generalization of Boltzmann–Gibbs statistics. *Journal of Statistical Physics*, 52(1/2), 479–487.

Ulaby, F.T., Dubois, P.C., and Zyl, J.V. (1996). Radar mapping of surface soil moisture. *Journal of Hydrology*, 184, 57–84.

# 12 Flow Duration Curve

The flow duration curve (FDC) is employed for the prediction of distribution of future flows, forecasting of future recurrence frequencies, comparison of watersheds, construction of load duration curves, and determination of low flow thresholds. Usually, the FDC is constructed empirically for a given set of flow data and the FDC so constructed is found to vary from one year to the other and from one gaging station to another within the same watershed. The objective of this chapter is to present the derivation of FDC using the Tsallis entropy. The entropy-based derivation permits a probabilistic characterization of the FDC and hence a quantitative assessment of its uncertainty.

## 12.1 DEFINITION OF FLOW DURATION CURVE

For a gaging station, an FDC is a plot of streamflow values from high to low against the percent of time these values are either equaled or exceeded. The plot considers the full range of flows and is constructed over a specified period of time scaled between 0% and 100%. The time interval for constructing an FDC depends on the need, but daily average discharge values are usually used; sometimes weekly, monthly, or seasonally averaged values can also be used. However, averaging over longer time intervals obscures details of the variations in flow and the effect of varying time interval is not the same for all streams. The difference between an FDC based on daily discharge values and that based on monthly values can be as high as 35%, as noted by Foster (1934). For large streams where flow from day to day is almost uniform, weekly FDC may be almost the same as daily FDC, whereas for flashy streams with sudden floods lasting for a few hours in a day, daily and weekly FDCs will be greatly different.

## 12.2 USE OF FLOW DURATION CURVE

FDCs are constructed using the entire range of flow conditions for any given stream. If the FDC of a stream is based upon long-term flow, then it can be employed for predicting the distribution of future flows for water supply (Mitchell, 1957), hydropower (Hickox and Wassenauer, 1933), sediment load (Miller, 1951), and pollutant load (Searcy, 1959). It can also be utilized to compare watersheds and hence their clustering, construct load duration curves for total maximum daily loads (TMDLs) (U.S. EPA, 2007), forecast future recurrence frequencies, determine the low-flow threshold for defining droughts, and construct power duration curves.

## 12.3 CONSTRUCTION OF FLOW DURATION CURVE

FDCs are generally constructed empirically. A typical semilog FDC exhibits a sigmoidal shape, curving upward near the flow duration of 0 and downward at a frequency near 100%, with nearly a constant slope in between, as shown in Figure 12.1. The overall slope of an FDC is an indication of streamflow variability at the gage, reflecting, in turn, the integrated effect of watershed characteristics. For practical applications, U.S. EPA (2007) classified the flow region into five different classes: 0%–10% interval for high flows, 10%–40% for moist conditions, 40%–60% for mid-range conditions, 60%–90% for dry conditions, and 90%–100% for low flows, as shown in Figure 12.2.

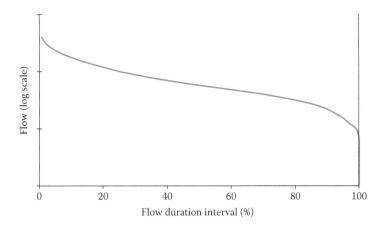

**FIGURE 12.1**   General form of FDC.

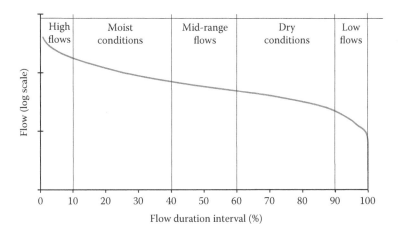

**FIGURE 12.2**   Flow regime classified into five classes.

## 12.4   DERIVATION OF FLOW DURATION CURVE

It is assumed that temporally averaged discharge $Q$ is a random variable, varying from a minimum value $Q_{min}$ to a maximum value $Q_{max}$, with a probability density function (PDF) denoted as $f(Q)$. The time interval for which the discharge is averaged depends on the purpose of constructing an FDC but frequently it is taken as 1 day. The procedure for deriving the FDC entails essentially the following main steps (Singh et al., 2014): (1) defining the Tsallis entropy, (2) specification of constraints, (3) optimization of the Tsallis entropy using the method of Lagrange multipliers, (4) derivation of the probability distribution of discharge, (5) and determination of the Lagrange multipliers, (6) hypothesizing the cumulative distribution of discharge in terms of time, and (7) derivation of FDC. Each of these steps is now discussed.

### 12.4.1   DEFINING TSALLIS ENTROPY

If discharge values are available as a discrete series, the Tsallis entropy (Tsallis, 1988), denoted as $H$, takes on the form

$$H(Q) = H[P] = \frac{1}{m-1} \sum_{i=1}^{N} p(Q_i)\left\{1 - [p(Q_i)]^{m-1}\right\} \qquad (12.1)$$

where
  $p(Q_i) = p_i$ is the probability that $Q = Q_i$, $P = \{p_i, i = 1, 2, ..., N\}$ is the probability distribution of $Q$
  $N$ is the number of values that $Q$ takes on between its maximum $(Q_{max})$ and minimum $(Q_{min})$

Equation 12.1 expresses a measure of uncertainty about $p(Q_i)$ measured by $\{[1 - [p(Q_i)]^{m-1}]/(m-1)\}$ or the average information content of sampled $Q$.

Since $Q$ is often represented as a continuous series, the Tsallis entropy for discharge $Q \in (Q_{min}, Q_{max})$ can be defined in continuous form as

$$H(Q) = H[f(Q)] = \frac{1}{m-1} \int_{Q_{min}}^{Q_{max}} f(Q)\left\{1 - [f(Q)]^{m-1}\right\} dQ \qquad (12.2)$$

where
  $m$ is the entropy index
  $f(Q)$ is the PDF of discharge

Equation 12.2 expresses a measure of uncertainty about $f(Q)$ measured by $\{[1 - [f(Q)]^{m-1}]/(m-1)\}$ or the average information content of sampled $Q$. The continuous form of the Tsallis entropy will be utilized in the discussion that follows. First, $f(Q)$ must be derived, which can be accomplished by maximizing $H(Q)$, subject to specified constraints.

### 12.4.2 Specification of Constraints

In order to determine the $f(Q)$ that is least biased toward what is not known and most biased toward what is known (with regard to discharge), the principle of maximum entropy (POME), developed by Jaynes (1957, 1982), is invoked. POME requires the specification of certain information on discharge, expressed in terms of what is called constraints and leads to the most appropriate probability distribution that has the maximum entropy or uncertainty.

Since $f(Q)$ is a PDF, it must satisfy

$$C_1 = \int_{Q_{min}}^{Q_{max}} f(Q)dQ = 1 \tag{12.3}$$

For purposes of simplicity, it is assumed that all that is known about discharge is the mean value that can be expressed as

$$C_1 = \int_{Q_{min}}^{Q_{max}} Qf(Q)dQ = \bar{Q} = Q_m = Q_{mean} \tag{12.4}$$

Equation 12.3 is easy to use from a practical standpoint. The mean discharge is a relatively stable quantity and its value can be obtained directly from measurements.

### 12.4.3 Maximization of Tsallis Entropy and Probability Distribution

For maximizing the Tsallis entropy defined by Equation 12.2, the method of Lagrange multipliers can be employed. To that end, the Lagrangian function can be constructed using Equations 12.2 through 12.4 as

$$L = \frac{1}{m-1} \int_{Q_{min}}^{Q_{max}} f(Q)\left\{1-[f(Q)]^{m-1}\right\}dQ - \lambda_0\left[\int_{Q_{min}}^{Q_{max}} f(Q)dQ - 1\right] - \lambda_1\left[\int_{Q_{min}}^{Q_{max}} Qf(Q)dQ - \bar{Q}\right] \tag{12.5}$$

Differentiating Equation 12.5 with respect to $f(Q)$ and equating the derivative to 0, one obtains

$$\frac{\partial L}{\partial f(Q)} = 0 = \frac{1}{m-1}\left\{-(m-1)[f(Q)]^{m-2} + 1 - [f(Q)]^{m-1}\right\} - \lambda_0 - \lambda_1 Q \tag{12.6}$$

### 12.4.4 DERIVATION OF DISCHARGE PDF

Equation 12.6 yields the entropy-based PDF of $Q$ as

$$f(Q) = \left[ \frac{m-1}{m} \left( \frac{1}{m-1} - \lambda_0 - \lambda_1 Q \right) \right]^{1/(m-1)}$$ (12.7)

It is interesting to note that at $Q = 0$, $f(Q)$ becomes

$$f(Q)\big|_{Q=0} = \left\{ \frac{m-1}{m} \left[ \left( \frac{1}{m-1} \right) - \lambda_0 \right] \right\}^{1/(m-1)}$$ (12.8)

For purposes of simplification, let

$$\lambda_* = \frac{1}{m-1} - \lambda_0$$ (12.9)

With the use of Equation 12.8, the PDF given by Equation 12.6 can be expressed as

$$f(Q) = \left[ \frac{m-1}{m} (\lambda_* - \lambda_1 Q) \right]^{1/(m-1)}$$ (12.10)

The cumulative probability distribution function (CDF) of $Q$ can be obtained by integrating Equation 12.7 from $Q_{min}$ to $Q$ as

$$F(Q) = \left( \frac{m-1}{m} \right)^{m/(m-1)} \frac{1}{\lambda_1} \left[ (\lambda_* - \lambda_1 Q_{min})^{m/(m-1)} - (\lambda_* - \lambda_1 Q)^{m/(m-1)} \right]$$ (12.11)

If $Q_{min} = 0$, Equation 12.10 reduces to

$$F(Q) = \left( \frac{m-1}{m} \right)^{m/(m-1)} \frac{1}{\lambda_1} \left[ \lambda_*^{m/(m-1)} - (\lambda_* - \lambda_1 Q)^{m/(m-1)} \right]$$ (12.12)

### 12.4.5 MAXIMUM ENTROPY

The maximum Tsallis entropy or uncertainty of discharge can be obtained by substituting Equation 12.10 in Equation 12.1:

$$H(Q) = \frac{1}{m-1} \left\{ (Q_{max} - Q_{min}) + \left( \frac{m-1}{m} \right)^{m/(m-1)} \right.$$

$$\left. \times \frac{1}{(2m-1)\lambda_1} \left[ (\lambda_* - \lambda_1 Q_{max})^{(2m-1)/(m-1)} - (\lambda_* - \lambda_1 Q_{min})^{(2m-1)/(m-1)} \right] \right\}$$ (12.13)

### 12.4.6 Determination of the Lagrange Multipliers

The PDF of $Q$ has two unknown Lagrange multipliers $\lambda_0$ (or $\lambda_*$) and $\lambda_1$ that can be determined using Equations 12.3 and 12.4. Substituting Equation 12.10 in Equation 12.3, one obtains

$$\int_{Q_{min}}^{Q_{max}} \left[ \frac{m-1}{m} \left( \frac{1}{m-1} - \lambda_0 - \lambda_1 Q \right) \right]^{1/(m-1)} dQ = 1 \qquad (12.14)$$

Solution of Equation 12.14 can be written as

$$\left( \frac{1}{m-1} - \lambda_0 - \lambda_1 Q_{max} \right)^{m/(m-1)} - \left( \frac{1}{m-1} - \lambda_0 - \lambda_1 Q_{min} \right)^{m/(m-1)} = - \left( \frac{m}{m-1} \right)^{m/(m-1)} \lambda_1$$

$$(12.15)$$

Likewise, substitution of Equation 12.7 in Equation 12.4 yields

$$\int_{Q_{min}}^{Q_{max}} Q \left[ \frac{m-1}{m} \left( \frac{1}{m-1} - \lambda_0 - \lambda_1 Q \right) \right]^{1/(m-1)} dQ = \bar{Q} \qquad (12.16)$$

Integration of Equation 12.16 results in

$$-\lambda_1 \bar{Q} \left( \frac{m}{m-1} \right)^{m/(m-1)}$$

$$= Q_{max} \left( \frac{1}{m-1} - \lambda_0 - \lambda_1 Q_{max} \right)^{m/(m-1)} - Q_{min} \left( \frac{1}{m-1} - \lambda_0 - \lambda_1 Q_{min} \right)^{m/(m-1)}$$

$$+ \frac{m-1}{2m-1} \frac{1}{\lambda_1} \left[ \left( \frac{1}{m-1} - \lambda_0 - \lambda_1 Q_{max} \right)^{(2m-1)/(m-1)} - \left( \frac{1}{m-1} - \lambda_0 - \lambda_1 Q_{min} \right)^{(2m-1)/(m-1)} \right]$$

$$(12.17)$$

Equations 12.15 and 12.17 can be cast, respectively, as

$$(\lambda_* - \lambda_1 Q_{max})^{m/(m-1)} - (\lambda_* - \lambda_1 Q_{min})^{m/(m-1)} = - \left( \frac{m}{m-1} \right)^{m/(m-1)} \lambda_1 \qquad (12.18)$$

$$-\lambda_1 \bar{Q} \left( \frac{m}{m-1} \right)^{m/(m-1)} = Q_{max} (\lambda_* - \lambda_1 Q_{max})^{m/(m-1)} - Q_{min} (\lambda_* - \lambda_1 Q_{min})^{m/(m-1)}$$

$$+ \frac{m-1}{2m-1} \frac{1}{\lambda_1} \left[ (\lambda_* - \lambda_1 Q_{max})^{(2m-1)/(m-1)} - (\lambda_* - \lambda_1 Q_{min})^{(2m-1)/(m-1)} \right]$$

$$(12.19)$$

Equations 12.18 and 12.19 are implicit in the Lagrange multipliers $\lambda_0$ (or $\lambda_*$) and $\lambda_1$ and do not have an explicit closed form solution but can be solved numerically without any difficulty.

### Example 12.1

A set of values of mean discharge and maximum discharge for a specified site on the Pee Dee River are given in Table 12.1. Determine the Lagrange multipliers. Plot $\lambda_*$ versus mean discharge, and plot $\lambda_1$ versus maximum discharge. Also, plot $\lambda_*$ versus $\lambda_1$ for values of maximum discharge as well as for mean discharge.

### Solution

The Lagrange multipliers are computed using Equations 12.18 and 12.19. The value of $m$ is taken as 3. The Lagrange multiplier $\lambda_*$ is plotted against mean discharge in Figure 12.3 that shows the variation of $\lambda_*$ with mean discharge for different values of $\lambda_1$ and Figure 12.4 shows the variation of $\lambda_1$ with maximum discharge. It can be seen from the figures that the Lagrange multipliers retain the same sign. For positive values, $\lambda_*$ decreases with increasing $Q_{mean}$ and $\lambda_1$ decreases with increasing $Q_{max}$, while for negative values their behavior is opposite. Comparing the two figures it is seen from that $\lambda_*$ has a wider distribution than $\lambda_1$; further, $\lambda_1$ drops quickly under the value of 0.1. The relations between the two Lagrange multipliers for different values of maximum discharge and mean discharge are shown in Figures 12.5 and 12.6. In both figures, $\lambda_*$ increases with $\lambda_1$ but with different slopes. The slope is milder for $Q_{mean} = 10$ m³/s and 100 m³/s but much faster for 500 m³/s and 1000 m³/s. The Lagrange multiplier $\lambda_*$ also increases with $Q_{mean}$ or $Q_{max}$. For constant $\lambda_1$, $\lambda_*$ is larger for higher $Q_{mean}$ or $Q_{max}$.

The Lagrange multipliers are computed by solving Equations 12.18 and 12.19 that involve $Q_{mean}$ and $Q_{max}$. Singh et al. (2014) computed for a number of river basins the Lagrange multipliers, whose histograms are plotted in Figures 12.7 and 12.8. It can be seen from Figure 12.8 that the value of $\lambda_1$ is highly concentrated within the values between 0 and 0.025, whereas $\lambda_*$ is distributed widely. The mean values for all basins obtained for $\lambda_1$ and $\lambda_*$ are, respectively, 0.012 and 0.175, with standard deviations of 0.051 and 0.167.

### 12.4.7  HYPOTHESIS ON CUMULATIVE DISTRIBUTION FUNCTION OF DISCHARGE

In order to derive the FDC, it is assumed that all temporally averaged values of discharge $Q$ measured at the gaging station under consideration between $Q_{min}$ and $Q_{max}$ are equally likely. In reality, this is not highly unlikely because at different times different values of discharge do occur and each value is equally likely. Then, the cumulative probability distribution of discharge can be expressed as one minus the percent time (or the ratio of time to the period of time under consideration, say 365 days for daily discharge). The probability of discharge being equal to or less than a given value of $Q$, or the cumulative probability distribution function of discharge (CDF), $F(Q) = P$ (discharge $\leq a$ given value of $Q$), $P =$ probability, can be expressed as

**TABLE 12.1**

**Maximum and Mean Discharges for Station 02131000 on the Pee Dee River**

| $Q_{max}$ (m³/s) | $Q_{mean}$ (m³/s) | $Q_{max}$ (m³/s) | $Q_{mean}$ (m³/s) |
|---|---|---|---|
| 1854 | 310 | 1024 | 350 |
| 979 | 159 | 1970 | 410 |
| 484 | 173 | 2827 | 402 |
| 954 | 214 | 1203 | 440 |
| 974 | 301 | 532 | 149 |
| 1435 | 310 | 957 | 327 |
| 1254 | 338 | 1296 | 432 |
| 852 | 233 | 1797 | 489 |
| 1930 | 401 | 835 | 193 |
| 1353 | 383 | 869 | 217 |
| 724 | 238 | 2694 | 385 |
| 597 | 178 | 569 | 205 |
| 1740 | 290 | 1316 | 325 |
| 1124 | 281 | 1412 | 466 |
| 1684 | 239 | 1449 | 498 |
| 894 | 207 | 1010 | 247 |
| 608 | 177 | 1322 | 518 |
| 849 | 230 | 906 | 303 |
| 1463 | 443 | 1047 | 289 |
| 815 | 244 | 824 | 345 |
| 1961 | 569 | 928 | 329 |
| 1288 | 300 | 1220 | 435 |
| 1092 | 335 | 634 | 174 |
| 1511 | 319 | 611 | 214 |
| 1330 | 316 | 504 | 114 |
| 1740 | 457 | 441 | 91 |
| 1593 | 268 | 2759 | 471 |
| 733 | 158 | 1058 | 201 |
| 1039 | 251 | 889 | 282 |
| 911 | 284 | 543 | 190 |
| 886 | 222 | 1039 | 285 |
| 1339 | 354 | 543 | 132 |
| 1302 | 412 | 705 | 227 |
| 2043 | 505 | 1474 | 346 |
| 993 | 334 | 487 | 160 |
| 2363 | 472 | 594 | 184 |
| 645 | 255 | | |

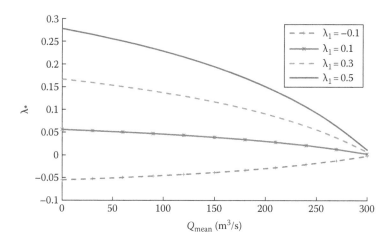

**FIGURE 12.3** Relation between Lagrange multiplier $\lambda_*$ and $Q_{\text{mean}}$ for different values of $\lambda_1$.

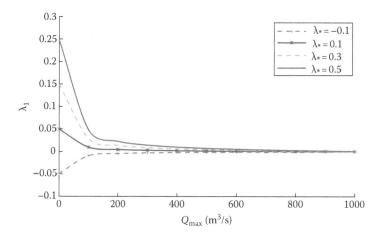

**FIGURE 12.4** Relation between Lagrange multiplier $\lambda_1$ and $Q_{\text{max}}$ for different values of $\lambda_*$.

$$F(Q) = 1 - \left(\frac{t}{T}\right) = 1 - \tau, \quad \tau = t/T \tag{12.20}$$

where
  $t$ time (say in days)
  $\tau$ is dimensionless time
  $T$ is the duration under consideration (say, 365 days)

It should be noted that on the left side the argument of function $F$ in Equation 12.20 is variable $Q$, whereas on the right side the variable is $\tau$. The CDF of $Q$ is not linear

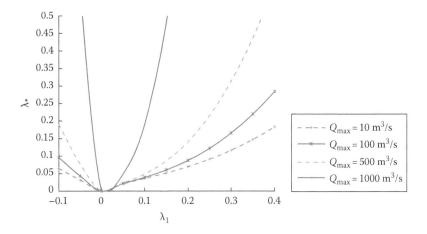

**FIGURE 12.5**   Relation between Lagrange multipliers $\lambda_*$ and $\lambda_1$ for different values of $Q_{max}$.

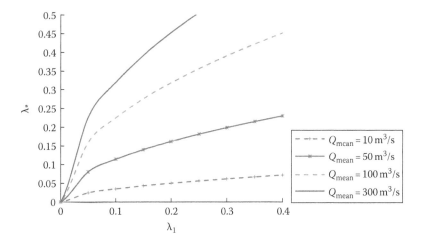

**FIGURE 12.6**   Relation between Lagrange multipliers $\lambda_*$ and $\lambda_1$ for different values of $Q_{mean}$.

in terms of $Q$, unless $Q$ and $\tau$ are linearly related. It may also be noted that a similar hypothesis has been employed in the hydrologic literature.

### 12.4.8 FLOW DURATION CURVE

Equating Equations 12.20 through 12.11, $Q$ can be explicitly written as

$$Q = \frac{\lambda_*}{\lambda_1} - \frac{1}{\lambda_1}\frac{m}{m-1}\left\{ -\lambda_1 \frac{m-1}{m}F(Q) + \left[\frac{m-1}{m}(\lambda_* - \lambda_1 Q_{min})\right]^{m/(m-1)}\right\}^{(m-1)/m} \tag{12.21}$$

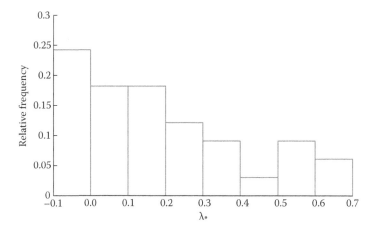

**FIGURE 12.7** Relative frequency histogram of Lagrange multiplier $\lambda_*$.

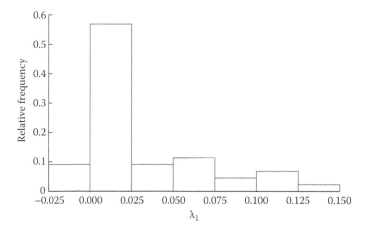

**FIGURE 12.8** Relative frequency histogram of Lagrange multiplier $\lambda_1$.

Likewise, equating Equations 12.20 through 12.12, $Q$ can be written as

$$Q = \frac{\lambda_*}{\lambda_1} - \frac{1}{\lambda_1}\frac{m}{m-1}\left\{-\lambda_1\frac{m-1}{m}F(Q)+\left(\frac{m-1}{m}\lambda_*\right)^{m/(m-1)}\right\}^{(m-1)/m} \quad (12.22)$$

Equations 12.21 and 12.22 are discharge quantile–probability relationships. The value of entropy index $m > 0$, but the question arises as to what the value of $m$ is or should be. This is illustrated using an example.

**TABLE 12.2**
**FDC for Year 2006 Observed at Station 02131000**

| Flow Duration Interval (%) | Flow | Flow Duration Interval (%) | Flow |
|---|---|---|---|
| 0 | 19,500 | 55 | 10,900 |
| 5 | 17,400 | 60 | 10,500 |
| 11 | 16,200 | 66 | 10,200 |
| 16 | 14,400 | 71 | 9,730 |
| 22 | 13,400 | 77 | 9,310 |
| 27 | 12,600 | 82 | 9,120 |
| 33 | 12,300 | 88 | 8,770 |
| 38 | 11,900 | 93 | 8,470 |
| 44 | 11,500 | 99 | 8,040 |
| 49 | 11,200 | 100 | 7,760 |

**Example 12.2**

Considering the flow data given in Table 12.2 for the Pee Dee River for year 2006, construct FDCs for $m = 4/3$, 2, 5/2, 3, 13/4, and 4. Determine which value of $m$ best corresponds to the observed FDC?

**Solution**

The Lagrange multipliers are computed using Equations 12.18 and 12.19. Then, the FDC is determined for $m = 4/3$, 2, 5/2, 3, 13/4, and 4, as shown in Figure 12.9. It is seen from the figure that the high discharge part of the FDC is closer to the observations for $m = 5/2$, 3, and 13/4, while the low discharge part of the FDC is closer to the observations for $m = 4$. The estimated sum of squared errors for

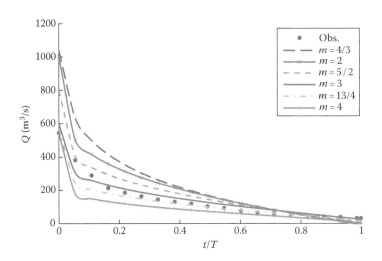

**FIGURE 12.9** FDC of year 2006 for different $m$ values.

the FDC corresponding to $m = 4/3$, 2, 5/2, 3, 13/4, and 4, respectively, is 65.57, 48.39, 31.71, 12.52, 17.33, and 21.24 m³/s. Thus, $m = 3$ can be selected for this data set and it is plausible that this value of $m$ may be satisfactory for other data sets as well.

## 12.5   REPARAMETERIZATION

It is possible to simplify the derived FDC using a dimensionless parameter $M$ defined as

$$M = \frac{\lambda_1 Q_{max}}{\lambda_1 Q_{max} - \lambda_*} \quad \text{or} \quad Q_{max} = \frac{\lambda_* M}{\lambda_1 (M-1)} \tag{12.23}$$

Considering $Q_{min} = 0$, the ratio of $f(0)$ to $f(Q_{max})$ using Equation 12.10 can be expressed in terms of $M$ given by Equation 12.23 as

$$\frac{f(0)}{f(Q_{max})} = \frac{\left[ m - 1/m(\lambda_*) \right]^{1/(m-1)}}{\left[ m - 1/m(\lambda_* - \lambda_1 Q_{max}) \right]^{1/(m-1)}} = (1-M)^{1/(m-1)} \quad \text{or} \quad M = 1 - \left[ \frac{f(0)}{f(Q_{max})} \right]^{m-1} \tag{12.24}$$

Equation 12.24 defines $M$ and shows that if $M = 0$, $f(0) = f(Q_{max})$ and the distribution of discharge would tend to be uniform. On the contrary, if $M = 1$, $f(0) = 0$, and $f(Q_{max})$ would tend to infinity, which means that the probability distribution of discharge would be highly nonuniform. Thus, $M$ can be used as an index of the uniformity of the probability distribution of discharge.

When discharge tends to reach $Q_{max}$, $F(Q_{max}) = 1$ and Equation 12.21 with $Q_{min}$ considered as 0 yields

$$Q_{max} = \frac{\lambda_*}{\lambda_1} - \frac{1}{\lambda_1} \frac{m}{m-1} \left[ -\lambda_1 \frac{m-1}{m} + \left( \frac{m-1}{m} \lambda_* \right)^{m/(m-1)} \right]^{(m-1)/m} \tag{12.25}$$

Dividing Equation 12.20 by $Q_{max}$, one obtains

$$\frac{Q}{Q_{max}} = \frac{\lambda_*}{\lambda_1 Q_{max}} - \frac{1}{\lambda_1 Q_{max}} \frac{m}{m-1} \left\{ -\lambda_1 \frac{m-1}{m} F(Q) + \left[ \frac{m-1}{m} \lambda_* \right]^{m/(m-1)} \right\}^{(m-1)/m}$$

$$= \frac{\lambda_*}{\lambda_1 Q_{max}} \left( 1 - \frac{1}{\lambda_*} \left\{ -\lambda_1 \left( \frac{m-1}{m} \right)^{1/(m-1)} F(Q) + \lambda_*^{m/(m-1)} \right\}^{(m-1)/m} \right) \tag{12.26}$$

Note from Equation 12.23 that

$$\frac{\lambda_*}{\lambda_1 Q_{max}} = 1 - \frac{1}{M} \tag{12.27}$$

When Equation 12.27 is substituted in Equation 12.26, the result is given as follows:

$$\frac{Q}{Q_{max}} = 1 - \frac{1}{M} - \left(1 - \frac{1}{M}\right) \left\{ -\left(\frac{1}{\lambda_*}\right)^{m/(m-1)} \lambda_1 \left(\frac{m}{m-1}\right)^{1/(m-1)} F(Q) + 1 \right\}^{(m-1)/m}$$

$$= 1 - \frac{1}{M} \left( 1 - \left\{ \left(\frac{m}{m-1}\right)^{m/(m-1)} \frac{M}{Q_{max}} [F(Q)-1]+1 \right\}^{(m-1)/m} \right) \tag{12.28}$$

Substituting $Q = 0$, at $F(Q) = 0$ Equation 12.28 reduces to

$$0 = 1 - \frac{1}{M} \left\{ 1 - \left[ 1 - \left(\frac{m}{m-1}\right)^{m/(m-1)} \frac{M}{Q_{max}} \right]^{(m-1)/m} \right\} \tag{12.29}$$

Rearranging Equation 12.29, one obtains

$$\left(\frac{m}{m-1}\right)^{m/(m-1)} \frac{M}{Q_{max}} = 1 - (1-M)^{m/(m-1)} \tag{12.30}$$

Inserting Equation 12.30, Equation 12.28 can be simplified as

$$\frac{Q}{Q_{max}} = 1 - \frac{1}{M} (1 - \{(1-M)^{m/(m-1)} + [1-(1-M)^{m/(m-1)}]F(Q)\}^{(m-1)/m}) \tag{12.31}$$

In Equation 12.31, the Lagrange multipliers are replaced with $M$ and hence the FDC can be determined with only one parameter, $M$.

## Example 12.3

Compute the FDC for data in Table 12.1. Also, compute the 95% confidence intervals.

### Solution

First, with $m = 3$, Lagrange multipliers are solved for from Equations 12.18 and 12.19, which give $\lambda_1 = -0.50$, and $\lambda_* = 0.585$. Thus, $M = 0.99$ is obtained using Equation 12.23. Then, flow discharge is calculated using Equation 12.31,

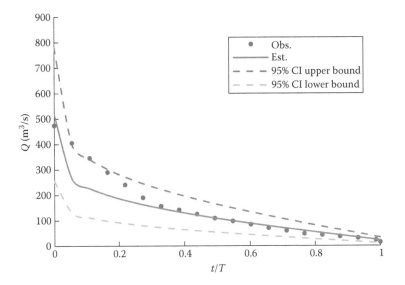

**FIGURE 12.10**  FDCs computed for station 02131000.

as shown in Figure 12.10. Referring to the probability of a parameter being in an interval, the 95% confidence intervals of $Q$ from $[Q_{low}, Q_{up}]$ are computed from repeated simulations. Then the confidence interval is obtained from the cumulative distribution of parameters conditioned on the observed data, so that $P(Q_{low} \leq Q \leq Q_{up}) = 0.95$. The Tsallis entropy–based flow duration fits the observations closely but has about 9% error in estimation as compared with maximum observed values.

## 12.6  MEAN FLOW AND RATIO OF MEAN TO MAXIMUM FLOW

The mean flow can be determined by taking the first moment of Equation 12.10 as

$$
\begin{aligned}
\overline{Q} &= \int_{Q_{min}}^{Q_{max}} Q f(Q) dQ = \int_{Q_{min}}^{Q_{max}} \left[ \frac{m-1}{m} (\lambda_* - \lambda_1 Q) \right]^{1/(m-1)} Q \, dQ \\
&= -\frac{1}{\lambda_1} \left( \frac{m-1}{m} \right)^{m/(m-1)} \left\{ Q_{max} (\lambda_* - \lambda_1 Q_{max})^{m/(m-1)} \right. \\
&\quad \left. + \frac{m-1}{2m-1} \frac{1}{\lambda_1} \left[ (\lambda_* - \lambda_1 Q_{max})^{(2m-1)/(m-1)} - \lambda_*^{(2m-1)/(m-1)} \right] \right\}
\end{aligned}
\tag{12.32}
$$

Now the ratio between the mean flow to the maximum flow can be expressed as a function of parameter $M$ as

$$\frac{\overline{Q}}{Q_{max}} = -\left(1-\frac{1}{M}\right)\left(\frac{m-1}{m}\right)^{m/(m-1)}\left\{\lambda_*^{1/(m-1)}\frac{M\lambda_*}{\lambda_1(M-1)}(1-M)^{m/(m-1)}\right.$$

$$\left.+\frac{m-1}{2m-1}\frac{\lambda_*^{m/(m-1)}}{\lambda_1}\left[(1-M)^{(2m-1)/(m-1)}-\lambda_*^{(2m-1)/(m-1)}\right]\right\} \qquad (12.33)$$

Equation 12.33 can be cast as

$$\frac{\overline{Q}}{Q_{max}} = \Psi(M) \qquad (12.34)$$

In Equation 12.34, the right-hand side depends only on $M$. In order to establish this relationship, the $M$ value can be computed from Equation 12.23 by solving for the Lagrange multipliers with the use of Equations 12.18 and 12.19. Singh et al. (2014) computed the $M$ values from annual mean and maximum discharge for recent 5 years collected from 13 gaging stations of Pee Dee River and plotted Equation 12.34, as shown in Figure 12.11. The figure shows that $M$ is linearly related to the ratio between the mean flow and maximum flows, which using regression can be written as

$$M = 2.246 - 4.891\frac{\overline{Q}}{Q_{max}} \qquad (12.35)$$

which has a coefficient of determination of 0.9972.

To determine the FDC for a given year, the first step is to compute the $M$ value from the given values of mean and maximum discharges. It is noted that Equation 12.30 is derived by assuming $Q_{min} = 0$, however, $Q_{min}$ may not be small enough to be neglected. Thus, the modified discharge using $Q' = Q - Q_{min}$ is preferred to compute $M$.

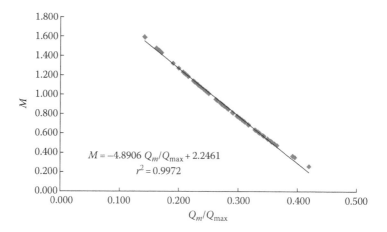

FIGURE 12.11    Relationship between $M$ and the ratio of mean to maximum discharge.

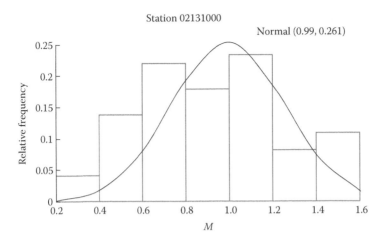

**FIGURE 12.12**   Histogram of *M* values for a gaging station.

Then, the *M* value is computed using Equation 12.30 for each year of record and then the mean, standard deviation, and coefficient of variance of the computed values are calculated. Singh et al. (2014) computed the *M* values from Equation 12.30 and found it to have different ranges for each station and a histogram of *M* is plotted for each station, as shown in Figure 12.12 for a sample station; the histogram varies from one station to another but seems to fit the normal distribution in all cases. In general, *M* varies from 0.2 to 1.6 and its standard deviation is around 0.2–0.4. Combining the values of five stations, a histogram of the *M* values is plotted, as shown in Figure 12.13. Again, parameter *M* seems to follow a bell-shaped distribution, with a mean value of 0.798 and a standard deviation of 0.493. The average values of *M* are also plotted against the drainage area, but the relationship is weak.

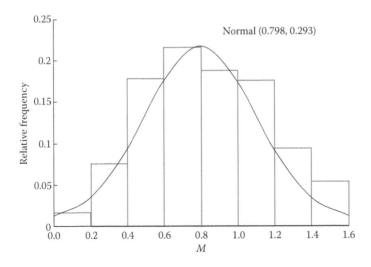

**FIGURE 12.13**   Histogram of *M* values for combined stations.

## Example 12.4

Consider the mean and maximum values of discharge for the gaging station for different years as given in Table 12.3. Using Equation 12.35, compute the $M$ value. Then, for given mean value compute the maximum discharge and compare it with the observed discharge.

### Solution

With given mean and maximum values, $M$ is computed from Equation 12.35 and is given in Table 12.3. Then $Q'_{max}$ is computed from $Q_{mean}$ with the $M$ value and tabulated in Table 12.3. The relative error of estimated maximum value from the observed values is less than 0.009.

## Example 12.5

Considering an average value of $M$ as 0.8, compute the FDC for data in Table 12.1 and compare the curve with the observed curve. How well the two FDCs match?

### Solution

With $M$ estimated as 0.8, the FDC is computed from Equation 12.31 and plotted in Figure 12.14. It is seen from the figure that using $M = 0.8$, the estimated flow is

### TABLE 12.3
### Mean and Maximum Values for Different Years

| Year | $Q_{max}$ (m³/s) | $Q_{min}$ (m³/s) | $Q_{mean}$ (m³/s) | $M$ | $Q'_{max}$ (m³/s) |
|------|------|------|------|------|------|
| 1991 | 35,700 | 1460 | 8,733 | 1.036 | 35,315 |
| 1992 | 46,700 | 1860 | 18,297 | 0.366 | 47,596 |
| 1993 | 32,000 | 1680 | 10,692 | 0.611 | 31,983 |
| 1994 | 37,000 | 1300 | 10,220 | 0.880 | 36,593 |
| 1995 | 29,100 | 1380 | 12,198 | 0.257 | 30,002 |
| 1996 | 32,800 | 1660 | 11,642 | 0.520 | 32,987 |
| 1997 | 43,100 | 1180 | 15,363 | 0.513 | 43,367 |
| 1998 | 22,400 | 743 | 6,155 | 0.887 | 22,151 |
| 1999 | 21,600 | 1100 | 7,548 | 0.544 | 21,685 |
| 2000 | 17,800 | 767 | 4,033 | 1.129 | 17,659 |
| 2001 | 15,600 | 664 | 3,232 | 1.231 | 15,572 |
| 2002 | 97,500 | 1200 | 16,646 | 1.429 | 99,685 |
| 2003 | 37,400 | 1260 | 7,120 | 1.321 | 37,655 |
| 2004 | 31,400 | 854 | 9,975 | 0.685 | 31,255 |
| 2005 | 19,200 | 977 | 6,721 | 0.541 | 19,279 |
| 2006 | 36,700 | 653 | 10,087 | 0.887 | 36,293 |
| 2007 | 19,200 | 1030 | 4,678 | 1.042 | 18,995 |
| 2008 | 24,900 | 1390 | 8,007 | 0.667 | 24,807 |
| 2009 | 52,100 | 1350 | 12,212 | 1.089 | 51,607 |
| 2010 | 17,200 | 1160 | 5,658 | 0.634 | 17,167 |
| 2011 | 21,000 | 1630 | 6,513 | 0.719 | 20,869 |

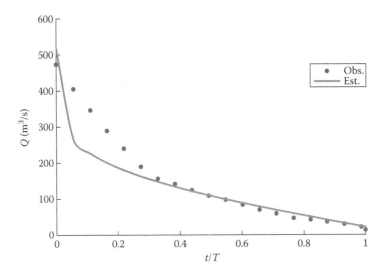

**FIGURE 12.14**   Flow duration estimated using $M = 0.8$.

60–114 m³/s lower than the observed value for low duration, where $t/T$ is less than 0.3, while for higher duration the estimated values fit the observed values well. Thus, the overall $r^2$ turns out to be 0.877.

## 12.7   PREDICTION OF FLOW DURATION CURVE FOR UNGAGED SITES

It may be interesting to investigate the behavior of $Q_{min}$, $Q_{mean}$, and $Q_{max}$ in relation to drainage area. The values of mean, minimum, and maximum flows are obtained for each year of record. The discharge values for 13 stations show significant differences; for example, the mean value varies from 6 cfs [0.17 m³/s] to greater than $10^4$ cfs [283 m³/s]. For a sample station 02131000, histograms of $Q_{min}$, $Q_{mean}$, and $Q_{max}$ are plotted in Figures 12.15 through 12.17. Singh et al. (2014) plotted the average values of minimum, mean, and peak flows against drainage area, as shown in Figures 12.18 through 12.20 that show on the log–log plot a power relationship of $Q_{min}$, $Q_{mean}$, and $Q_{max}$ values with drainage area, and the power law fitted well with a coefficient of determination around 0.9. Ogden and Dawdy (2003) and Gupta et al. (2010) showed a power law relating peak discharge to drainage area. These power relationships are expressed as

$$Q_{min} = 0.004A^{0.9136}, \quad r^2 = 0.8878 \tag{12.36}$$

$$Q_{mean} = 0.0218A^{0.8785}, \quad r^2 = 0.9408 \tag{12.37}$$

$$Q_{max} = 0.134A^{0.8187}, \quad r^2 = 0.885 \tag{12.38}$$

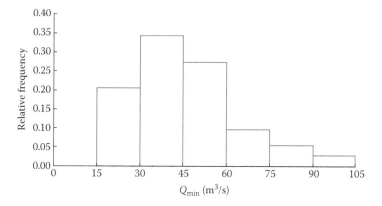

**FIGURE 12.15**   Frequency histogram of the $Q_{min}$ values of station 02131000.

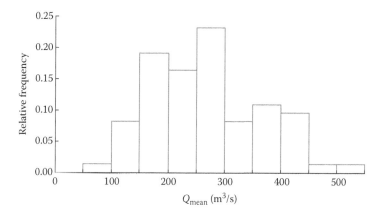

**FIGURE 12.16**   Frequency histogram of the $Q_{mean}$ values of station 02131000.

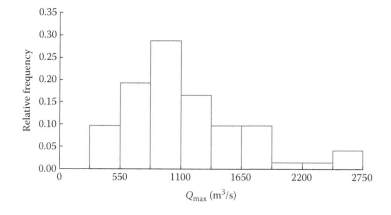

**FIGURE 12.17**   Frequency histogram of the $Q_{max}$ values of station 02131000.

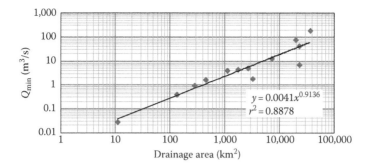

**FIGURE 12.18** Relation between $Q_{min}$ and drainage area ($y = Q_{min}$ and $x =$ drainage area).

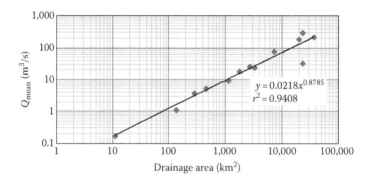

**FIGURE 12.19** Relation $Q_{mean}$ and drainage area ($y = Q_{mean}$ and $x =$ drainage area).

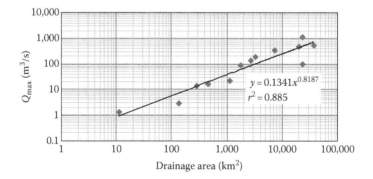

**FIGURE 12.20** Relation between $Q_{max}$ and drainage area ($y = Q_{max}$ and $x =$ drainage area).

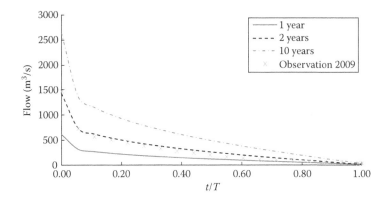

**FIGURE 12.21**   Prediction of FDC for gaging station 02131000 for year 2009.

where area $A$ is km$^2$ and discharge is in m$^3$/s. These relationships can be employed for constructing FDCs for ungagged sites, that is, sites without any information on historical discharge.

### Example 12.6

Compute the FDC for gaging station 02131000 with a drainage area of 2790 miles$^2$ (7726 km$^2$). Assume there are no discharge records collected at this station.

#### Solution

First, $Q_{min}$, $Q_{mean}$, and $Q_{max}$ are computed from regression Equations 12.36 through 12.38. Then, $M$ is computed using Equation 12.29, which equals 0.243. Thus, the FDC is predicted and is compared with observations as shown in Figure 12.21. It can be seen from the figure that the predicted FDC is in close agreement with the observed FDC and the RMS value is only 3.39 m$^3$/s.

## 12.8   FORECASTING OF FLOW DURATION CURVE

The FDC can be forecasted ahead of time for a given station, once the entropy parameter has been determined. To forecast the FDC, $Q_{min}$, $Q_{mean}$, and $Q_{max}$ need to be forecasted. Since the future values of peak, minimum, and mean discharge are subject to uncertainty, they can only be predicted for given probability values. For any gaging station, the observed data can be used as past information, from which the distributions of $Q_{min}$, $Q_{mean}$, and $Q_{max}$ as well as $M$ are obtained. The data from 2007 to 2011 are used for forecasting. To that end, $Q_{min}$, $Q_{mean}$, and $Q_{max}$ of 1-, 2-, 10-, and 50-year recurrence intervals are computed from the given information.

### Example 12.7

Compute the values of $Q_{min}$, $Q_{mean}$, and $Q_{max}$ for USGS station 02131000 for 1-, 2-, 10-, and 50-year recurrence intervals. Then, forecast the FDCs.

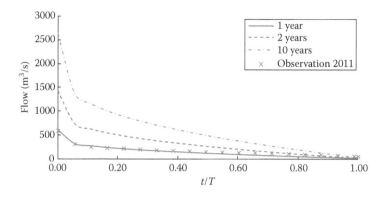

**FIGURE 12.22** Prediction of FDC for gaging station 02131000 for year 2011.

**Solution**

The $Q_{max}$ values of 1-, 2-, 10-, and 50-year recurrence intervals are 441.7, 1039.2, 1915.9, and 2732.2 m³/s, respectively. Then, the $M$ values are computed and FDCs are constructed for different recurrence intervals. The predicted FDCs of 1-, 2-, and 10-year recurrence intervals are shown in Figure 12.22 with observations of 2011. It can be seen from the figure that the observed FDC for year 2011 is in close agreement with the predicted 1-year recurrence interval FDC.

## 12.9 VARIATION OF ENTROPY WITH TIME SCALE

Singh et al. (2014) computed entropy using Equation 12.2 considering the entire flow series for 13 gaging stations. The variation of entropy with the computation interval is shown for a sample station in Figure 12.23. It can be seen from the figure that the entropy increases with increasing time interval. However, the rate of increase is high during the first phase for about 15 days, but as interval increases the rate significantly

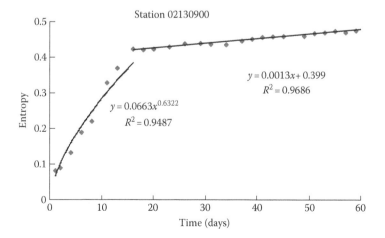

**FIGURE 12.23** Variation of entropy with time scale.

decreases. After about 15 days, the rate of entropy increase declines significantly and then entropy almost reaches a constant value. This suggests that at higher intervals, the flow regime becomes more complex, reflecting the reduced influence of anthropogenic changes on the flow regime. The opposite is true for smaller intervals. The entropy of the first phase can be fitted by a power law. The exponent for the first part is 0.63 with a sample exponent average of 0.24. The entropy of second phase is fitted by a linear equation, where the slope is less than 0.0005. However, the change point between two phases may be different for different years, but Singh et al. (2014) found it to be occurring around the 15th day. Entropy of flow varies with the time interval of flow. In general, entropy increases with increasing time interval, eventually more or less a constant value.

## REFERENCES

Foster, H.A. (1934). Duration curves. *Transaction of ASCE*, 99, 1213–1267.

Gupta, V.K., Mantilla, R., Troutman, B.M., Dawdy, D., and Krajewski, W.F. (2010). Generalizing a nonlinear geophysical flood theory to medium-sized river networks. *Geophysical Research Letter*, 37(11), 1–6.

Hickox, G.H. and Wessenauer, G.O. (1933). Application of duration curves to hydroelectric studies. *Transactions of ASCE*, 98, 1276–1308.

Jaynes, E.T. (1957). Information theory and statistical mechanics, I. *Physical Review*, 106, 620–630.

Jaynes, E.T. (1982). On the rationale of maximum entropy methods. *Proceedings of the IEEE*, 70, 939–952.

Miller, C.R. (1951). *Analysis of Flow Duration, Sediment-Rating Curve Method of Computing Sediment Yield*. U.S. Department of Interior, Bureau of Reclamation, Sedimentation Section, Hydrology Branch, Denver, CO.

Mitchell, W.D. (1957). *Flow-Duration of Illinois Streams*. Illinois Department of Public Works and Buildings, Division of Waterways, Springfield, Illinois, pp. 189.

Ogden, F.L. and Dawdy, D.R. (2003). Peak discharge scaling in small Hortonian watersheds. *Journal of Hydrologic Engineering*, 8(2), 64–73.

Searcy, J.K. (1959). Flow-duration curves. USGS Water-Supply Paper 1542-A, U.S. Government Printing Office, Washington, DC, pp. 33.

Singh, V.P., Cui, H., and Byrd, A.R. (2014). Tsallis entropy-based flow duration curve. *Transactions of the American Society of Agricultural and Biological Engineers (ASABE)*, 57(3), 837–849.

Tsallis, C. (1988). Possible generalizations of Boltzmann-Gibbs statistics. *Journal of Statistical Physics*, 52(½), 479–487.

U.S. Environmental Protection Agency. (2007). An approach for using load duration curves in the development of TMDL. EPA-841-B-07-006, Watershed Branch, Office of Wetlands, Oceans and Watersheds, U.S.EPA, Washington, DC.

# Section IV

*Water Resources Engineering*

# 13 Eco-Index

Freshwater resources are becoming increasingly limited as the population multiplies. Over half of the world's accessible runoff presently is appropriate for human use, and that fraction is projected to grow to 70% by 2025. A recent projection of water demand through 2025 indicated that ensuring sustainable water supply would become increasingly challenging for large areas of the globe (Vorosmarty et al., 2000). Four hundred and fifty million people in thirty-one countries already face serious shortages of water. These shortages occur almost exclusively in developing countries, some of which are ill-equipped to address water shortages. By the year 2025, one-third of the world's population is expected to face severe to chronic water shortages. Allocating water for diverse and often competing traditional uses, such as industry, agriculture, urban, energy, waste disposal, and recreation, is now even more complex due to society's expectation that for their health and integrity ecosystems receive adequate attention and accommodation. Freshwater and freshwater-dependent ecosystems provide a range of services for humans, including fish, flood protection, wildlife, etc. To maintain these services, water needs to be allocated to ecosystems, as it is allocated to other users. In the face of limited freshwater supply, there are multiple competing and conflicting demands.

With increasing concern for eco-needs, the scientific field of "eco-flows" has prospered in recent years with the result that there are more than 200 methods for their computation. These methods can be grouped into four categories: hydrological rules, hydraulic rating methods, habitat simulation methods, and holistic methodologies (Naiman et al., 2002; Dyson et al., 2003; Postel and Richter, 2003; Tharme, 2003). Past studies include those based on the percentages of natural mean or median annual flow, percentages of total divertible annual flow allocated to wet and dry seasons, and eco-flow prescriptions based on a percentage of total annual base flow plus a high-flow component derived as a percentage of mean annual runoff (Smakhtin et al., 2004). However, such guides have no documented empirical basis and the temptation to adopt them may represent a risk to the future integrity and biodiversity of riverine ecosystems (Arthington, 1998). It is now recognized in the literature that the structure and function of a riverine ecosystem and many adaptations of its biota are dictated by the patterns of temporal variation in river flow or the "natural flow-regime paradigm reflected by Indicators of Hydrologic Alteration" (Richter et al., 1996; Poff et al., 1997; Stromberg, 1997; Lytle and Poff, 2004).

The objective of this chapter is to outline a Tsallis entropy-based hydrological alteration assessment of biologically relevant flow regimes using gauged flow data. The maximum entropy ordered weighted averaging (OWA) method is used to aggregate noncommensurable biologically relevant flow regimes to fit an eco-index such that the harnessed level of the ecosystem is reflected. The methodology can serve as a guide for eco-managers when allocating water resources among potential users and where to concentrate their attention, while mitigating the man-induced effects on natural flow regimes to have a sustainable development (Kim and Singh, 2014).

## 13.1   INDICATORS OF HYDROLOGICAL ALTERATION

Indicators of hydrological Alteration (IHA) aim to protect a range of flows in a river. Richter et al. (1996) proposed 32 biologically relevant parameters, as shown in Table 13.1, which jointly reflect different aspects of flow variability (magnitude, frequency, duration, and timing of flows). These parameters are estimated from a natural daily flow time series at a site of interest (often times at a gaging site). The parameters consider intra- and inter-annual variation of hydrological regime, which is necessary to sustain the ecosystem. In other words, a range of flow regime is considered to define the state of the ecosystem such that hydrological requirements for all aquatic species are met. The ecosystem alteration is then assessed by comparing with the natural system that is relatively unharnessed (Richter et al., 1996).

**TABLE 13.1**
**Hydrological Parameters Used in IHA**

| Group | Regime Characteristics | 32 Parameters | Number of Parameters |
|---|---|---|---|
| Group 1: Magnitude of monthly water conditions | Magnitude | Mean value for each calendar month | 12 |
| | Timing | | |
| Group 2: Magnitude and duration of annual extreme water conditions | Magnitude | Annual minimum/maximum of 1-day means | 10 |
| | Duration | Annual minimum/maximum of 3-day means | |
| | | Annual minimum/maximum of 7-day means | |
| | | Annual minimum/maximum of 30-day means | |
| | | Annual minimum/maximum of 90-day means | |
| Group 3: Timing of annual extreme water conditions | Timing | Julian date of each annual 1-day minimum and maximum | 2 |
| Group 4: Frequency and duration of high and low pulses | Frequency | Number of high and low pulses each year | 2 + 2 |
| | Duration | Mean duration of high and low pulses | |
| Group 5: Rate/frequency of consecutive water-condition changes | Rate of change | Means of all positive differences between daily values | 1 + 1 + 1 + 1 |
| | | Means of all negative differences between daily values | |
| | | Number of rises | |
| | | Number of falls | |

*Source:* Richter, B.D. et al., *Conserv. Biol.*, 10, 1163, 1996.

Richter et al. (1996), Poff et al. (1997), Lytle and Poff (2004), among others, have emphasized why these 32 biologically relevant parameters are required to represent an ecosystem. However, there is a need for a tool that allows the transmission of technical information in a summarized format. Such package may preserve the original meaning of data by using only the variables that best reflect the desired objective. Information on the 32 biologically relevant parameters and their values may not show the dependence amongst parameters and the importance of each of these parameters. Somehow it appears difficult to visualize the 32 parameters spatially. In addition, the increasing tendency of diversifying stakeholders (public participation) in water-related issues requires the results of technical analyses to be presented in a way that can be understood and shared by all stakeholders, including those with little technical background. Therefore, narrowing the result to a single value which characterizes the ecosystem may be more desirable.

## 13.2 PROBABILITY DISTRIBUTIONS OF IHA PARAMETERS

Each of the 32 biologically relevant hydrological parameters, proposed by Richter et al. (1996), can be considered as a random variable. Then, for each variable the least-biased probability distribution can be obtained by maximizing entropy (Singh, 1998). Tsallis (1988) proposed a formula for entropy computation, now called as the Tsallis entropy $S_m$, which can be expressed as

$$S_m = \frac{1}{m-1} \sum_{i=1}^{n} p_i \left(1 - p_i^{m-1}\right) \tag{13.1}$$

where
  $m$ is the Tsallis entropy index or parameter
  $S_m$ is the Tsallis entropy
  $p_1, p_2,..., p_n$ are the values of probabilities corresponding to the specific values $x_i$,
      $i = 1, 2,..., n$ of the hydrological parameter $X$
  $n$ is the number of values

The Tsallis entropy has been widely applied in water resources studies (Tsallis and Brigatti, 2004; Papalexiou and Koutsoyiannis, 2012). In this chapter we use a discrete probability distribution $P = \{p_1, p_2,..., p_n\}$ for a parameter $X:\{x_i, i = 1, 2,..., n\}$ and a value of $m = 2$.

## 13.3 COMPUTATION OF NONSATISFACTION ECO-LEVEL

The nonsatisfaction level (NSL) for a "$j$"th parameter can be defined in absolute difference terms as

$$NSL_j = \left|(S_{pre} - S_{post})\right|_j, \quad j = 1, 2,..., 32 \tag{13.2}$$

The *NSL* can also be defined in relative difference terms as

$$NSL_j = \frac{S_{pre} - S_{post}}{S_{pre}}, \quad j = 1, 2, \ldots, 32 \tag{13.3}$$

where $S_{pre}$ and $S_{post}$ are the Tsallis entropies for parameter "*j*" for pre- and postchange conditions, respectively. The change may be represented by a dam or levee or even a land use change such as urbanization. Equation 13.2 or 13.3 relates the lack of information about the ecosystem to the level of nonsatisfaction. In the examples discussed in this chapter, we will be using the absolute terms of *NSL* (Equation 13.2). The satisfaction level can be seen as how much the system is unharnessed. The *NSL* values are computed for all IHA parameters separately.

## 13.4 COMPUTATION OF ECO-INDEX

Eco-index can be computed using the steps shown in Figure 13.1. The values of the nonsatisfaction level of biological parameters are aggregated based on Yager's (1999) finding such that the final aggregation maximizes the information associated with

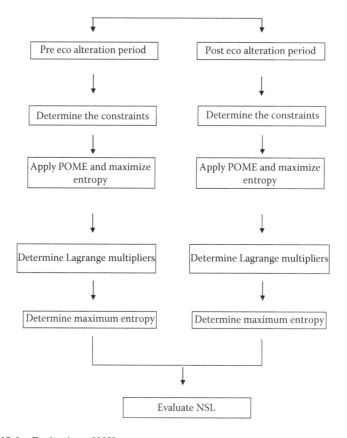

**FIGURE 13.1** Evaluation of NSL.

each *NSL*. The OWA operator introduced by Yager (1999) is a general type of operator that provides flexibility in the aggregation process such that the aggregated value
is bounded between minimum and maximum values of input parameters. The OWA
operator is defined as

$$F(a_1, a_2, ..., a_n) = \sum_{j=1}^{n} w_j b_j \qquad (13.4)$$

where
   the computed value of *NSL* for each of the 32 parameters is the argument $(a_i)$
   $b_j$ is the *j*th largest of $a_i$
   $w_j$ are a collection of weights such that $w_j \in [0, 1]$ $\sum_{j=1}^{n} w_j = 1$

Equation 13.4 can also be written as

$$\text{Eco-index} = F(a_1, ..., a_{32}) = F(NSL_1, NSL_2, ..., NSL_{32}) = \sum_{j=1}^{32} w_j b_j \qquad (13.5)$$

The methodology used for obtaining the OWA weighting vector is based on Lamata
(2004). This approach, which only requires the specification of just the Orness
value (1-Andness), generates a class of OWA weights that are called Maximum
Entropy Operator Weighted Averaging (ME-OWA) weights. The determination of
these weights, $w_1, ..., w_{32}$, from a degree of optimism Orness given by the decision
maker requires the solution of an optimization problem formulated in the following. The objective function used for optimization is one of trying to maximize the
dispersion or entropy of weights, which calculates the weights to be the ones that
use as much information as possible about the values of *NSL* in the aggregation.
It is assumed here that weights are considered as values of probability having the
probability distribution $W = \{w_i, i = 1, 2, ..., n\}$. Then, the Tsallis entropy can be
maximized as

   maximize:

$$S(W) = \frac{1}{m-1} \sum_{i=1}^{n} w_i \left(1 - w_i^{m-1}\right) \qquad (13.6)$$

subject to the constraint defined as

$$Orness(W) = \frac{1}{n-1} \sum_{i=1}^{n} (n-i) w_i \qquad (13.7)$$

and

$$\sum_{i=1}^{n} w_i = 1 \tag{13.8}$$

Here

$n = 32$

$w_i \in [0, 1]$

O'Hagan (1988) suggested this approach to obtain the OWA operators with maximal entropy of the OWA weights for a given level of Orness, which he expressed as Equations 13.6 and 13.7 rewritten as

$$\alpha = \frac{1}{n-1} \sum_{i=1}^{n} (n-i)w_i, \quad 0 \le \alpha \le 1 \tag{13.9}$$

The objective is to determine the values of $w_i$ by maximizing the Tsallis entropy. Applying the method of Lagrange multipliers, the Lagrangian function can be expressed as

$$L = \frac{1}{m-1} \sum_{i=1}^{n} w_i \left(1 - w_i^{m-1}\right) + \lambda_0 \left[\sum_{i=1}^{n} w_i - 1\right] + \lambda_1 \left[\frac{1}{n-1} \sum_{i=1}^{n} (n-i)w_i - \alpha\right] \tag{13.10}$$

where $\lambda_0$ and $\lambda_1$ are the Lagrange multipliers. Differentiating Equation 13.10 with respect to $w_i$ and equating the derivative to zero yields

$$\frac{dL}{dw_i} \Rightarrow 0 = \frac{1}{m-1} - \frac{m}{m-1} w_i^{m-1} + \lambda_0 + \lambda_1 \frac{(n-i)}{n-1} \tag{13.11}$$

Equation 13.11 yields the distribution of weights:

$$w_i = \left\{ \frac{m-1}{m} \left[ \lambda_* + \lambda_1 \frac{(n-i)}{n-1} \right] \right\}^{1/(m-1)}, \quad \lambda_* = \lambda_0 + \frac{1}{m-1} \tag{13.12}$$

The Lagrange multipliers can be determined by substituting Equation 13.12 in Equations 13.8 and 13.9, respectively, as

$$\sum_{i=1}^{n} \left\{ \frac{m-1}{m} \left[ \lambda_* + \lambda_1 \frac{(n-i)}{n-1} \right] \right\}^{1/(m-1)} = 1 \tag{13.13}$$

and

$$\frac{1}{n-1}\sum_{i=1}^{n}(n-i)\left\{\frac{m-1}{m}\left[\lambda_* + \lambda_1 \frac{(n-i)}{n-1}\right]\right\}^{1/(m-1)} = \alpha \tag{13.14}$$

Equations 13.13 and 13.14 cannot be solved in closed form for an arbitrary value of $m$. However, for $m = 2$, they can be solved explicitly. Then, Equation 13.13 becomes

$$\sum_{i=1}^{n}\left[\lambda_* + \lambda_1 \frac{(n-i)}{n-1}\right] = 2 \Rightarrow \lambda_* n + \frac{n\lambda_1}{(n-1)}\left[n - \frac{(n+1)}{2}\right] \tag{13.15}$$

and Equation 13.14 becomes

$$\sum_{i=1}^{n}(n-i)\left[\lambda_* + \lambda_1 \frac{(n-i)}{n-1}\right] = 2\alpha(n-1) \Rightarrow \lambda_*\sum_{i=1}^{n}(n-i) + \frac{\lambda_1}{n-1}\sum_{i=1}^{n}(n-i)^2 \tag{13.16}$$

Equation 13.16 can be simplified as

$$\lambda_*\left[n^2 - \frac{n(n+1)}{2}\right] + \frac{\lambda_1}{n-1}\left[n^3 + \frac{n(n+1)(2n+1)}{6} - n^2(n+1)\right] = 2\alpha(n-1) \tag{13.17}$$

Equations 13.16 and 13.17 can be solved for $\lambda_0$ and $\lambda_1$. To simplify algebra, Equation 13.15 is recast as

$$a_0\lambda_* + a_1\lambda_1 = 2, \quad a_0 = n, \quad a_1 = \frac{n}{n-1}\left[n - \frac{(n+1)}{2}\right] \tag{13.18}$$

and Equation 13.17 as

$$b_0\lambda_* + b_1\lambda_1 = 2\alpha(n-1), \quad b_0 = n^2 - \frac{n(n+1)}{2},$$

$$b_1 = \frac{1}{n-1}\left[n^3 + \frac{n(n+1)(2n+1)}{6} - n^2(n+1)\right] \tag{13.19}$$

Equations 13.18 and 13.19 yield

$$\lambda_* = \frac{1}{a_0}(2 - a_1\lambda_1) = \frac{1}{n}\left\{2 - \frac{n\lambda_1}{n-1}\left[n - \frac{n+1}{2}\right]\right\} \tag{13.20}$$

and

$$\lambda_1 = \frac{2b_0 - 2\alpha(n-1)a_0}{b_0 a_1 - a_0 b_1} \tag{13.21}$$

or

$$\lambda_1 = \frac{2n^2 - n(n+1) - 2\alpha(n-1)n}{[n^2 - (n(n+1)/2)](n/(n-1))[n - ((n+1)/2)]} \\ - (n/(n-1))[n^3 + (n(n+1)(2n+1)/6) - (n^2(n+1))]$$ (13.22)

The value of $\alpha$ is to be specified, as expressed by Equation 13.9. To calculate the OWA operator using the method of Fullér and Majlender (2001), we set the optimum $\alpha = 0.60$ for illustrative purposes and this value is considered to calculate the OWA operator. Then, we get the Lagrange multipliers as

$$\lambda_* = \frac{217}{7688}$$ (13.23)

$$\lambda_1 = \frac{279}{3844}$$ (13.24)

The implication of using the Orness value of 0.60 for analysis is the assumption that the impact of NSL of all the IHA parameters is considered in the index development and to avoid assigning equal weights since some of the parameters may have more influence on defining the underlying ecosystem.

## Example 13.1

Obtain data for a river and then summarize hydrological parameters used in IHA for the river.

### Solution

We choose a region covering three subcatchments in the Texas Gulf watershed (Figure 13.2): the Trinity River basin (USGS Hydrologic Unit Code HUC 1203), the Neches river basin (HUC 1202) and the Sabine River basin (HUC 1201). For the past two decades, water has become a critical resource in Texas. Along with ground water, surface water is the main source of water in the state. We find it relevant to address surface water alteration in two of the three data bases, Sabine and Trinity basin, which have two major reservoirs. Although the Sabine River basin is relatively narrow, it sustains the Toledo Bend reservoir which is the United states' fifth largest in surface area, with water normally covering an area of about 200,000 acres and having a controlled storage capacity of 4,477,000 acre ft. The Trinity basin sustains the Livingstone reservoir. With 55 ft average depth, the Livingston reservoir has a normal capacity of 1,741,867 acre ft and covers 83,277 acres at its normal pool elevation of 131 ft above mean sea level. The lake drains an area of 16,616 square miles and is reported as the largest lake constructed for water supply purposes located entirely within the Texas territory. The targeted features in the watershed are represented in Figure 13.2. The daily stream flow series at each of the two gauges are employed in the eco-index analysis as shown in Table 13.2.

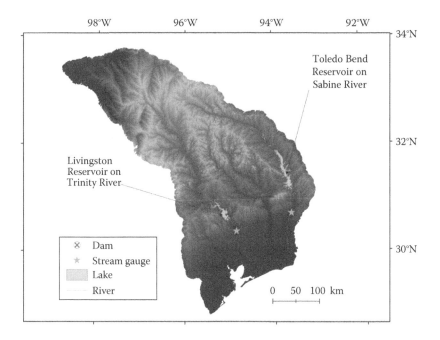

**FIGURE 13.2** Study area: the combined Trinity, Neches, and Sabine River basins. The locations of water bodies, stream gauges and dams are represented. The flow series at the stream gauges are employed in the eco-Index analysis.

## TABLE 13.2
## Stream Gauges Characteristics in the Trinity and Sabin River Basins

| Name | Primary Outflow | Administrative Region | Predam Period | Postdam Period | Stream Gauge ID | Latitude | Longitude |
|------|-----------------|----------------------|---------------|----------------|-----------------|----------|-----------|
| Toledo Bend | Sabine River | Texas/ Louisiana | 1930–1969 (40 years) | 1970–2012 (43 years) | USGS 0806500 | 30.43 | −94.85 |
| Livingston | Trinity River | Texas | 1930–1969 (40 years) | 1970–2012 (43 years) | USGS 08028500 | 30.75 | −93.61 |

The 33 biologically relevant parameters, proposed by Richter et al. (1996), are shown in Table 13.3. These data can be obtained using the IHA software. In the IHA software, parameters can be calculated using parametric (mean/standard deviation) or nonparametric (percentile) statistics. For most situations, nonparametric statistics constitute a better choice because of the skewed nature of many hydrological datasets (TNC, 2009). Only 12 parameters of monthly inflow are obtained from parametric statistics because the mean is a better representation of the hydrological characteristics than is the median. Annual minimum 1-day mean values at the Sabine and Trinity gauge stations are shown in Table 13.4.

**TABLE 13.3**

**Summary of IHA Parameters**

| IHA Parameter Group | Hydrological Parameters |
|---|---|
| 1: Magnitude of monthly water conditions | Mean or median value for each calendar month |
| | Subtotal: 12 parameters (No. 1–12) |
| 2: Magnitude and duration of annual extreme water conditions | Annual minima, 1-day mean |
| | Annual minima, 3-day mean |
| | Annual minima, 7-day mean |
| | Annual minima, 30-day mean |
| | Annual minima, 90-day mean |
| | Annual maxima, 1-day mean |
| | Annual maxima, 3-day mean |
| | Annual maxima, 7-day mean |
| | Annual maxima, 30-day mean |
| | Annual maxima, 90-day mean |
| | Number of zero-flow days |
| | Base flow index: 7-day minimum flow/mean flow for year |
| | Subtotal: 12 parameters (No. 13–24) |
| 3: Timing of annual extreme water conditions | Julian date of each annual 1-day minimum |
| | Julian date of each annual 1-day maximum |
| | Subtotal: 2 parameters (No. 25–26) |
| 4: Frequency and duration of high and low pulses | Number of low pulses within each water year |
| | Mean or median duration of low pulses (days) |
| | Number of high pulses within each water year |
| | Mean or median duration of high pulses (days) |
| | Subtotal: 4 parameters (No. 27–30) |
| 5: Rate and frequency of water-condition changes | Rise rates: Mean or median of all positive differences between consecutive daily values |
| | Fall rates: Mean or median of all negative differences between consecutive daily values |
| | Number of hydrologic reversals |
| | Subtotal: 3 parameters (No. 31–33) |

*Source:* TNC (The Nature Conservancy), User's manual for the Indicators of Hydrologic Alteration (IHA) Software, Ver 7.1, 2009.

## Example 13.2

Using the Tsallis entropy, derive and compute the PDFs of four IHA parameters (13–16) using the data in Example 13.1 and plot them.

### Solution

The selected parameters 13–16 are annual minimum 1-, 3-, 7-, and 30-day means. As shown in Tables 13.4 through 13.7, the inflow is improved by the dams. For analyzing the data obtained using the IHA software, histograms are constructed to determine the probability density function (PDF) which best represents the hydrological characteristics using a goodness fit test: the Kolmogorov–Sminrov (K–S) test and results are shown in Tables 13.8 through 13.11 as well as the fitting graphs that are presented in Figure 13.3a through c.

**TABLE 13.4**

**Annual Minimum 1-Day Means (X) at the Toledo and Livingston Gauge Stations**

**Sabine/Toledo Dam**

| Predam | | Predam | | Postdam | | Postdam | |
|---|---|---|---|---|---|---|---|
| Year | X (m³/s) | Year | X (m³/s) | Year | X (m³/s) | Year | X (m³/s) |
| 1930 | 9 | 1950 | 24 | 1970 | 11 | 1991 | 34 |
| 1931 | 7 | 1951 | 7 | 1971 | 9 | 1992 | 16 |
| 1932 | 8 | 1952 | 7 | 1972 | 11 | 1993 | 15 |
| 1933 | 20 | 1953 | 12 | 1973 | 22 | 1994 | 21 |
| 1934 | 8 | 1954 | 6 | 1974 | 15 | 1995 | 21 |
| 1935 | 17 | 1955 | 11 | 1975 | 15 | 1996 | 17 |
| 1936 | 8 | 1956 | 5 | 1976 | 12 | 1997 | 19 |
| 1937 | 12 | 1957 | 14 | 1977 | 10 | 1998 | 19 |
| 1938 | 10 | 1958 | 20 | 1978 | 9 | 1999 | 16 |
| 1939 | 6 | 1959 | 13 | 1979 | 12 | 2000 | 11 |
| 1940 | 15 | 1960 | 11 | 1980 | 13 | 2001 | 14 |
| 1941 | 29 | 1961 | 25 | 1981 | 14 | 2002 | 22 |
| 1942 | 22 | 1962 | 12 | 1982 | 15 | 2003 | 17 |
| 1943 | 10 | 1963 | 6 | 1983 | 18 | 2004 | 19 |
| 1944 | 13 | 1964 | 6 | 1984 | 16 | 2005 | 14 |
| 1945 | 23 | 1965 | 6 | 1985 | 17 | 2006 | 14 |
| 1946 | 28 | 1966 | 4 | 1986 | 22 | 2007 | 11 |
| 1947 | 11 | 1967 | 5 | 1987 | 23 | 2008 | 13 |
| 1948 | 9 | 1968 | 15 | 1988 | 17 | 2009 | 14 |
| 1949 | 21 | 1969 | 6 | 1989 | 19 | 2010 | 13 |
| | | | | 1990 | 22 | 2011 | 15 |
| | | | | | | 2012 | 13 |

**Trinity/Livingston Dam**

| Predam | | Predam | | Postdam | | Postdam | |
|---|---|---|---|---|---|---|---|
| Year | X (m³/s) | Year | X (m³/s) | Year | X (m³/s) | Year | X (m³/s) |
| 1930 | 7 | 1950 | 21 | 1970 | 10 | 1991 | 34 |
| 1931 | 4 | 1951 | 11 | 1971 | 9 | 1992 | 20 |
| 1932 | 15 | 1952 | 5 | 1972 | 8 | 1993 | 18 |
| 1933 | 11 | 1953 | 7 | 1973 | 16 | 1994 | 38 |
| 1934 | 5 | 1954 | 5 | 1974 | 33 | 1995 | 26 |
| 1935 | 21 | 1955 | 7 | 1975 | 22 | 1996 | 19 |
| 1936 | 8 | 1956 | 3 | 1976 | 28 | 1997 | 20 |
| 1937 | 7 | 1957 | 9 | 1977 | 16 | 1998 | 30 |
| 1938 | 10 | 1958 | 17 | 1978 | 13 | 1999 | 21 |
| 1939 | 6 | 1959 | 14 | 1979 | 23 | 2000 | 29 |
| 1940 | 11 | 1960 | 14 | 1980 | 15 | 2001 | 34 |
| 1941 | 31 | 1961 | 19 | 1981 | 16 | 2002 | 25 |
| 1942 | 21 | 1962 | 18 | 1982 | 17 | 2003 | 26 |
| 1943 | 11 | 1963 | 10 | 1983 | 24 | 2004 | 31 |
| 1944 | 13 | 1964 | 10 | 1984 | 13 | 2005 | 18 |
| 1945 | 22 | 1965 | 17 | 1985 | 21 | 2006 | 20 |
| 1946 | 24 | 1966 | 20 | 1986 | 37 | 2007 | 28 |
| 1947 | 20 | 1967 | 11 | 1987 | 12 | 2008 | 21 |
| 1948 | 13 | 1968 | 14 | 1988 | 16 | 2009 | 18 |
| 1949 | 11 | 1969 | 32 | 1989 | 24 | 2010 | 23 |
| | | | | 1990 | 28 | 2011 | 27 |
| | | | | | | 2012 | 27 |

**TABLE 13.5**
**Annual Minimum 3-Day Means (X) at the Toledo and Livingston Gauge Stations**

**Sabine/Toledo Dam**

| Predam | | | | Postdam | | | |
|---|---|---|---|---|---|---|---|
| Year | X (m³/s) | Year | X (m³/s) | Year | X (m³/s) | Year | X (m³/s) |
| 1930 | 9 | 1950 | 25 | 1970 | 12 | 1991 | 58 |
| 1931 | 7 | 1951 | 8 | 1971 | 9 | 1992 | 22 |
| 1932 | 8 | 1952 | 7 | 1972 | 13 | 1993 | 19 |
| 1933 | 21 | 1953 | 12 | 1973 | 22 | 1994 | 24 |
| 1934 | 8 | 1954 | 6 | 1974 | 15 | 1995 | 23 |
| 1935 | 18 | 1955 | 11 | 1975 | 15 | 1996 | 20 |
| 1936 | 8 | 1956 | 5 | 1976 | 13 | 1997 | 20 |
| 1937 | 13 | 1957 | 15 | 1977 | 10 | 1998 | 21 |
| 1938 | 10 | 1958 | 21 | 1978 | 9 | 1999 | 21 |
| 1939 | 6 | 1959 | 13 | 1979 | 13 | 2000 | 15 |
| 1940 | 15 | 1960 | 11 | 1980 | 15 | 2001 | 20 |
| 1941 | 33 | 1961 | 25 | 1981 | 15 | 2002 | 25 |
| 1942 | 22 | 1962 | 12 | 1982 | 20 | 2003 | 21 |
| 1943 | 11 | 1963 | 6 | 1983 | 18 | 2004 | 23 |
| 1944 | 13 | 1964 | 7 | 1984 | 19 | 2005 | 18 |
| 1945 | 24 | 1965 | 7 | 1985 | 19 | 2006 | 19 |
| 1946 | 28 | 1966 | 4 | 1986 | 23 | 2007 | 11 |
| 1947 | 12 | 1967 | 5 | 1987 | 25 | 2008 | 13 |
| 1948 | 9 | 1968 | 15 | 1988 | 24 | 2009 | 16 |
| 1949 | 22 | 1969 | 6 | 1989 | 23 | 2010 | 15 |
| | | | | 1990 | 24 | 2011 | 19 |
| | | | | | | 2012 | 14 |

**Trinity/Livingston Dam**

| Predam | | | | Postdam | | | |
|---|---|---|---|---|---|---|---|
| Year | X (m³/s) | Year | X (m³/s) | Year | X (m³/s) | Year | X (m³/s) |
| 1930 | 7 | 1950 | 21 | 1970 | 10 | 1991 | 35 |
| 1931 | 4 | 1951 | 11 | 1971 | 9 | 1992 | 21 |
| 1932 | 15 | 1952 | 5 | 1972 | 8 | 1993 | 18 |
| 1933 | 11 | 1953 | 7 | 1973 | 19 | 1994 | 38 |
| 1934 | 5 | 1954 | 5 | 1974 | 35 | 1995 | 27 |
| 1935 | 24 | 1955 | 7 | 1975 | 22 | 1996 | 20 |
| 1936 | 8 | 1956 | 3 | 1976 | 30 | 1997 | 20 |
| 1937 | 7 | 1957 | 10 | 1977 | 16 | 1998 | 30 |
| 1938 | 11 | 1958 | 18 | 1978 | 15 | 1999 | 21 |
| 1939 | 6 | 1959 | 14 | 1979 | 24 | 2000 | 29 |
| 1940 | 11 | 1960 | 14 | 1980 | 15 | 2001 | 38 |
| 1941 | 33 | 1961 | 21 | 1981 | 17 | 2002 | 26 |
| 1942 | 23 | 1962 | 19 | 1982 | 17 | 2003 | 26 |
| 1943 | 11 | 1963 | 10 | 1983 | 25 | 2004 | 32 |
| 1944 | 13 | 1964 | 10 | 1984 | 14 | 2005 | 21 |
| 1945 | 22 | 1965 | 17 | 1985 | 21 | 2006 | 21 |
| 1946 | 25 | 1966 | 20 | 1986 | 38 | 2007 | 28 |
| 1947 | 21 | 1967 | 12 | 1987 | 12 | 2008 | 21 |
| 1948 | 13 | 1968 | 21 | 1988 | 17 | 2009 | 20 |
| 1949 | 12 | 1969 | 33 | 1989 | 24 | 2010 | 23 |
| | | | | 1990 | 32 | 2011 | 28 |
| | | | | | | 2012 | 27 |

**TABLE 13.6**

**Annual Minimum 7-Day Mean at the Toledo and Livingston Gauge Stations**

| Sabine/Toledo Dam | | | | | | | | Trinity/Livingston Dam | | | | | | | |
| --- | --- | --- | --- | --- | --- | --- | --- | --- | --- | --- | --- | --- | --- | --- | --- |
| Predam | | | | Postdam | | | | Predam | | | | Postdam | | | |
| Year | X (m³/s) | Year | X (m³/s) | Year | X (m³/s) | Year | X (m³/s) | Year | X (m³/s) | Year | X (m³/s) | Year | X (m³/s) | Year | X (m³/s) |
| 1930 | 10 | 1950 | 27 | 1970 | 14 | 1991 | 80 | 1930 | 8 | 1950 | 22 | 1970 | 10 | 1991 | 36 |
| 1931 | 8 | 1951 | 8 | 1971 | 9 | 1992 | 33 | 1931 | 5 | 1951 | 11 | 1971 | 9 | 1992 | 21 |
| 1932 | 8 | 1952 | 7 | 1972 | 16 | 1993 | 26 | 1932 | 15 | 1952 | 6 | 1972 | 8 | 1993 | 19 |
| 1933 | 23 | 1953 | 13 | 1973 | 27 | 1994 | 34 | 1933 | 13 | 1953 | 7 | 1973 | 28 | 1994 | 38 |
| 1934 | 8 | 1954 | 6 | 1974 | 15 | 1995 | 27 | 1934 | 6 | 1954 | 5 | 1974 | 36 | 1995 | 27 |
| 1935 | 20 | 1955 | 12 | 1975 | 15 | 1996 | 21 | 1935 | 29 | 1955 | 7 | 1975 | 23 | 1996 | 20 |
| 1936 | 8 | 1956 | 5 | 1976 | 13 | 1997 | 25 | 1936 | 9 | 1956 | 3 | 1976 | 30 | 1997 | 21 |
| 1937 | 14 | 1957 | 16 | 1977 | 12 | 1998 | 23 | 1937 | 8 | 1957 | 10 | 1977 | 17 | 1998 | 30 |
| 1938 | 10 | 1958 | 22 | 1978 | 9 | 1999 | 24 | 1938 | 11 | 1958 | 19 | 1978 | 15 | 1999 | 22 |
| 1939 | 6 | 1959 | 15 | 1979 | 13 | 2000 | 18 | 1939 | 6 | 1959 | 14 | 1979 | 28 | 2000 | 29 |
| 1940 | 15 | 1960 | 13 | 1980 | 18 | 2001 | 21 | 1940 | 11 | 1960 | 15 | 1980 | 15 | 2001 | 42 |
| 1941 | 36 | 1961 | 26 | 1981 | 18 | 2002 | 28 | 1941 | 36 | 1961 | 22 | 1981 | 20 | 2002 | 27 |
| 1942 | 22 | 1962 | 13 | 1982 | 21 | 2003 | 23 | 1942 | 26 | 1962 | 23 | 1982 | 19 | 2003 | 27 |
| 1943 | 12 | 1963 | 6 | 1983 | 20 | 2004 | 26 | 1943 | 11 | 1963 | 10 | 1983 | 26 | 2004 | 39 |
| 1944 | 13 | 1964 | 7 | 1984 | 21 | 2005 | 19 | 1944 | 14 | 1964 | 10 | 1984 | 16 | 2005 | 22 |
| 1945 | 25 | 1965 | 7 | 1985 | 26 | 2006 | 21 | 1945 | 23 | 1965 | 18 | 1985 | 22 | 2006 | 22 |
| 1946 | 29 | 1966 | 4 | 1986 | 26 | 2007 | 12 | 1946 | 26 | 1966 | 21 | 1986 | 42 | 2007 | 29 |
| 1947 | 12 | 1967 | 5 | 1987 | 30 | 2008 | 16 | 1947 | 22 | 1967 | 12 | 1987 | 12 | 2008 | 24 |
| 1948 | 9 | 1968 | 16 | 1988 | 28 | 2009 | 18 | 1948 | 13 | 1968 | 26 | 1988 | 20 | 2009 | 21 |
| 1949 | 24 | 1969 | 7 | 1989 | 30 | 2010 | 17 | 1949 | 12 | 1969 | 36 | 1989 | 26 | 2010 | 23 |
| | | | | 1990 | 41 | 2011 | 20 | | | | | 1990 | 33 | 2011 | 28 |
| | | | | | | 2012 | 16 | | | | | | | 2012 | 28 |

**TABLE 13.7**

**Annual Minimum 30-Day Means (X) at the Toledo and Livingston Gauge Stations**

| | Sabine/Toledo Dam | | | | | | | | Trinity/Livingston Dam | | | | | | | |
|---|---|---|---|---|---|---|---|---|---|---|---|---|---|---|---|---|
| | Predam | | | | Postdam | | | | Predam | | | | Post-Dam | | | |
| Year | X (m³/s) | Year | X (m³/s) | Year | X (m³/s) | Year | X (m³/s) | Year | X (m³/s) | Year | X (m³/s) | Year | X (m³/s) | Year | X (m³/s) |
| 1930 | 14 | 1950 | 35 | 1970 | 19 | 1991 | 113 | 1930 | 10 | 1950 | 27 | 1970 | 10 | 1991 | 62 |
| 1931 | 9 | 1951 | 15 | 1971 | 14 | 1992 | 44 | 1931 | 9 | 1951 | 14 | 1971 | 10 | 1992 | 24 |
| 1932 | 9 | 1952 | 8 | 1972 | 32 | 1993 | 36 | 1932 | 16 | 1952 | 7 | 1972 | 9 | 1993 | 39 |
| 1933 | 27 | 1953 | 16 | 1973 | 100 | 1994 | 112 | 1933 | 17 | 1953 | 8 | 1973 | 81 | 1994 | 65 |
| 1934 | 10 | 1954 | 7 | 1974 | 28 | 1995 | 38 | 1934 | 7 | 1954 | 6 | 1974 | 38 | 1995 | 29 |
| 1935 | 34 | 1955 | 13 | 1975 | 23 | 1996 | 26 | 1935 | 39 | 1955 | 8 | 1975 | 28 | 1996 | 27 |
| 1936 | 12 | 1956 | 5 | 1976 | 25 | 1997 | 66 | 1936 | 10 | 1956 | 4 | 1976 | 34 | 1997 | 33 |
| 1937 | 17 | 1957 | 20 | 1977 | 31 | 1998 | 41 | 1937 | 11 | 1957 | 13 | 1977 | 18 | 1998 | 57 |
| 1938 | 12 | 1958 | 56 | 1978 | 16 | 1999 | 25 | 1938 | 13 | 1958 | 30 | 1978 | 15 | 1999 | 22 |
| 1939 | 8 | 1959 | 21 | 1979 | 41 | 2000 | 28 | 1939 | 7 | 1959 | 22 | 1979 | 37 | 2000 | 30 |
| 1940 | 20 | 1960 | 29 | 1980 | 28 | 2001 | 24 | 1940 | 12 | 1960 | 37 | 1980 | 18 | 2001 | 45 |
| 1941 | 68 | 1961 | 44 | 1981 | 26 | 2002 | 56 | 1941 | 72 | 1961 | 34 | 1981 | 22 | 2002 | 39 |
| 1942 | 33 | 1962 | 23 | 1982 | 28 | 2003 | 34 | 1942 | 68 | 1962 | 61 | 1982 | 20 | 2003 | 39 |
| 1943 | 14 | 1963 | 7 | 1983 | 33 | 2004 | 34 | 1943 | 24 | 1963 | 11 | 1983 | 28 | 2004 | 94 |
| 1944 | 19 | 1964 | 9 | 1984 | 37 | 2005 | 23 | 1944 | 21 | 1964 | 16 | 1984 | 24 | 2005 | 25 |
| 1945 | 35 | 1965 | 12 | 1985 | 45 | 2006 | 26 | 1945 | 74 | 1965 | 19 | 1985 | 25 | 2006 | 30 |
| 1946 | 45 | 1966 | 13 | 1986 | 58 | 2007 | 18 | 1946 | 27 | 1966 | 22 | 1986 | 54 | 2007 | 46 |
| 1947 | 17 | 1967 | 5 | 1987 | 37 | 2008 | 22 | 1947 | 26 | 1967 | 15 | 1987 | 18 | 2008 | 29 |
| 1948 | 10 | 1968 | 22 | 1988 | 31 | 2009 | 27 | 1948 | 16 | 1968 | 37 | 1988 | 24 | 2009 | 31 |
| 1949 | 32 | 1969 | 19 | 1989 | 36 | 2010 | 21 | 1949 | 24 | 1969 | 40 | 1989 | 27 | 2010 | 26 |
| | | | | 1990 | 44 | 2011 | 21 | | | | | 1990 | 40 | 2011 | 29 |
| | | | | | | 2012 | 19 | | | | | | | 2012 | 29 |

**TABLE 13.8**
**Parameter Estimates for the Tsallis Entropy-Based PDF of 1-Day Minimum Discharge (m³/s)**

| Parameters | Sabine River | | Trinity River | |
|---|---|---|---|---|
| | Predam | Postdam | Predam | Postdam |
| $a_0$ | 5.386 | 47.938 | 5.8500 | 7.69680 |
| $a_1$ | 0.008 | 0.129 | 0.0047 | 0.00002 |
| $c_1$ | 2.034 | 2.107 | 2.1821 | 3.43940 |
| $c_2$ | 2.093 | 8.858 | 2.1533 | 2.23800 |
| $m$ | 0.9 | 0.9 | 0.9 | 0.9 |
| $M_1$ | 12.525 | 16.225 | 13.375 | 21.925 |
| $M_2$ | 203.275 | 286.175 | 227.125 | 538.275 |
| $M_3$ | 3986.925 | 5516.125 | 4590 | 14,438 |

**TABLE 13.9**
**Parameter Estimates for the Tsallis Entropy-Based PDF of 3-Day Minimum Discharge (m³/s)**

| Parameters | Sabine River | | Trinity River | |
|---|---|---|---|---|
| | Predam | Postdam | Predam | Postdam |
| $a_0$ | 5.5331 | 111.8711 | 5.4297 | 12.9209 |
| $a_1$ | 0.0129 | 1.0784 | 0.0021 | 0.00002 |
| $c_1$ | 1.89792 | 1.94202 | 2.33654 | 3.45426 |
| $c_2$ | 2.1572 | 13.3082 | 1.9134 | 3.0406 |
| $M$ | 0.9 | 0.9 | 0.9 | 0.9 |
| $M_1$ | 12.95 | 19.125 | 14 | 22.8 |
| $M_2$ | 219.35 | 425.625 | 251.85 | 582.15 |
| $M_3$ | 4539.05 | 11,798.93 | 5393 | 16,239 |

The entropy theory allows obtaining the least-biased probability distribution by maximizing entropy subject to appropriate constraints. Using the Tsallis entropy, Koutsoyiannis (2005a,b) proposed a four-parameter distribution that fits a wide range of distributions (power or exponential). Assuming $x$ as a random value of discharge (m³/s), this distribution can be expressed as

$$f(x) = [1 + k(a_0 + a_1 x^{c_1})]^{-1-1/k} x^{c_2-1} \qquad (13.25)$$

where
   $a_1$ is the scale parameter
   $c_1$, $c_2$, and $k$ are the shape parameters

## TABLE 13.10
### Parameter Estimates for the Tsallis Entropy-Based PDF of 7-Day Minimum Discharge

| | Sabine River | | Trinity River | |
|---|---|---|---|---|
| Parameters | Predam | Postdam | Predam | Postdam |
| $a_0$ | 5.40890 | 64.67070 | 5.38750 | 11.27600 |
| $a_1$ | 0.00960 | 1.43960 | 0.00270 | 0.00001 |
| $c_1$ | 1.92876 | 1.60067 | 2.20374 | 3.61610 |
| $c_2$ | 2.02960 | 10.96670 | 1.85140 | 2.71100 |
| $m$ | 0.9 | 0.9 | 0.9 | 0.9 |
| $M_1$ | 13.68 | 22.93 | 15.03 | 24.28 |
| $M_2$ | 246.93 | 658.08 | 295.58 | 660.83 |
| $M_3$ | 5462.43 | 25,725.13 | 7011 | 19,658 |

## TABLE 13.11
### Parameter Estimates for the Tsallis Entropy-Based PDF of 30-Day Minimum Discharge

| | Sabine River | | Trinity River | |
|---|---|---|---|---|
| Parameters | Predam | Postdam | Predam | Postdam |
| $a_0$ | 5.56100 | 114.96400 | 4.51690 | 9.46530 |
| $a_1$ | 0.02450 | 0.30660 | 0.00960 | 0.00050 |
| $c_1$ | 1.46513 | 2.28405 | 1.52875 | 2.64966 |
| $c_2$ | 1.84430 | 12.7035 | 1.3345 | 2.53960 |
| $m$ | 0.9 | 0.9 | 0.9 | 0.9 |
| $M_1$ | 20.60 | 16.23 | 23.35 | 21.93 |
| $M_2$ | 623.15 | 286.18 | 869.65 | 538.28 |
| $M_3$ | 25,112.3 | 5516.125 | 44,250 | 14,438 |

The value $a_0$ is estimated from the total probability and is therefore not a parameter. Parameter $k = (1 - m)/m$, $m$ is the Tsallis parameter. For the solution of Equation 13.25, it is required that $k = (1 - m)/m$ be positive, meaning that $m$ must be less than 1. In this discussion, we consider $m = 0.9$ for solving Equation 13.25. In order to have an idea about the parameter values and their impact on the distribution shape, the sensitivity of the PDF $f(x)$ to each of its parameters ($a_1$, $c_1$, and $c_2$) is evaluated, as represented in Figure 13.3a through c.

Parameters $c_1$, $c_2$, and $k$ can be estimated using the method of moments expressed as

$$\int_{-\infty}^{\infty} f(x)dx = 1 \tag{13.26}$$

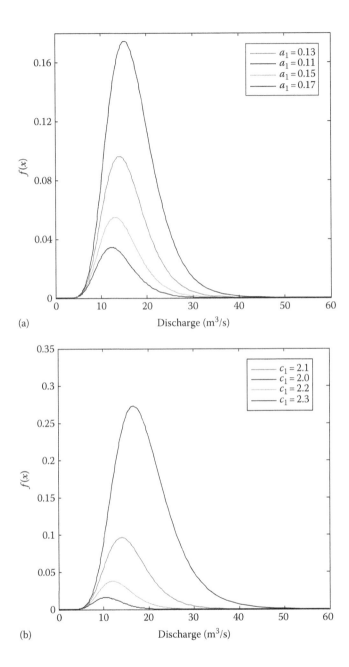

**FIGURE 13.3** Sensitivity of the entropy-based density function $f(x)$ (a) to the scale parameter $a_1$ (the remaining parameters are fixed: $a_0 = 47.94$, $c_1 = 2.11$, $c_2 = 8.86$); (b) to the shape parameter $c_1$ (the remaining parameters are fixed: $a_0 = 47.94$, $a_1 = 0.13$, $c_2 = 8.86$); and

*(Continued)*

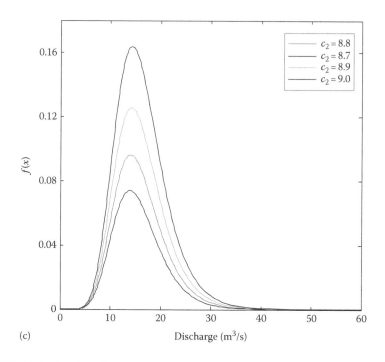

(c)

FIGURE 13.3 (*Continued*)   (c) to the shape parameter $c_2$ (the remaining parameters are fixed: $a_0 = 47.94$, $a_1 = 0.13$, $c_1 = 2.11$).

$$\int_{-\infty}^{\infty} xf(x)dx = \bar{x} = M_1 \tag{13.27}$$

$$\int_{-\infty}^{\infty} x^2 f(x)dx = \overline{x^2} = M_2 \tag{13.28}$$

$$\int_{-\infty}^{\infty} x^3 f(x)dx = \overline{x^3} = M_3 \tag{13.29}$$

However, it is difficult to find the parameters analytically (Koutsoyiannis, 2005a,b) by solving Equations 13.26 through 13.29. Therefore, they are determined numerically. The procedure for the estimation of parameters is suggested by Koutsoyiannis (2005a,b) (https://www.itia.ntua.gr/en/docinfo/641/), National Technical University of Athens, Department of Civil Engineering, Athens, Greece, Accessed June 2014. Explicitly, the procedure leading to the parameter estimation can be summarized in two steps. Given a time series of streamflow, the first step consists in computation of the first, second, and third moments based on the observed data by using Equations 13.27 through 13.29. The second step is iterative and aims to solve numerically the system of equations defined by Equation 13.30 given Equations 13.31 through 13.33 as follows:

$$\mu_q = \xi^{-q/c_1} \frac{B[(c_2+q)/c_1, 1+1/k-(c_2+q)/c_1]}{B(c_2/c_1, 1+1/k-c_2/c_1)} \tag{13.30}$$

where the beta function $B$ is defined by

$$B(\theta, \tau) = \int_0^1 t^{\theta-1}(1-t)^{\tau-1} dt \tag{13.31}$$

and

$$\xi = \frac{k*a_1}{(1+k*a_0)} \tag{13.32}$$

Note that constant $a_0$ is determined by the relation:

$$\left(\frac{1}{c_1}\right)(1+k*a_0)^{-1-1/k}\xi^{-c_1/c_2}B\left(\frac{c_2}{c_1}, 1+\frac{1}{k}-\frac{c_2}{c_1}\right) = 1 \tag{13.33}$$

The objective is to retrieve the triplet $(a_1, c_1, c_2)$ which respects the system of equations. However, in the computation process, the use of computational software (e.g., MATLAB® or R) requires an input of an initial guessed triplet for parameters $a_1$, $c_1$, and $c_2$. Following the procedure, the results are presented in the Tables 13.8 through 13.11. Using the parameter values so determined, the probability distribution is determined, as shown in Figures 13.4 through 13.7.

Computed probability values, based on the estimated parameters for pre- and postdam periods, are presented in Tables 13.12 and 13.15.

Results in Tables 13.12 through 13.15 are employed for comparative analysis presented in Figure 13.8a and b.

### Example 13.3

Compute the OWA operator for the data in Example 13.1.

#### Solution

The values of nonsatisfaction level of biological parameters are aggregated based on Yager's (1999) method in which the OWA operator is computed using Equation 13.6. The OWA operators are computed by maximizing entropy as shown in Tables 13.16 and 13.17 and Figures 13.9 and 13.10. The maximum entropy is 15.61 for the Sabine River basin compared to 13.96 for the Trinity River basin.

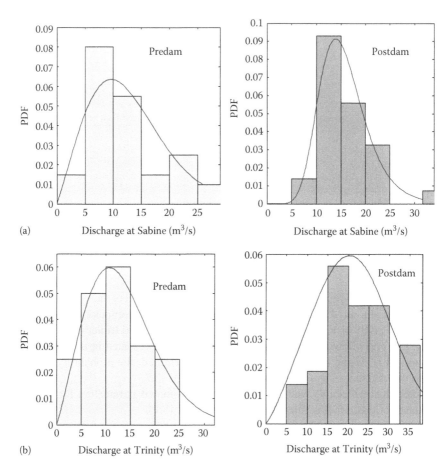

**FIGURE 13.4** PDF of annual minimum 1-day mean discharge corresponding to predam period and postdam period for (a) the Sabine River and (b) the Trinity River.

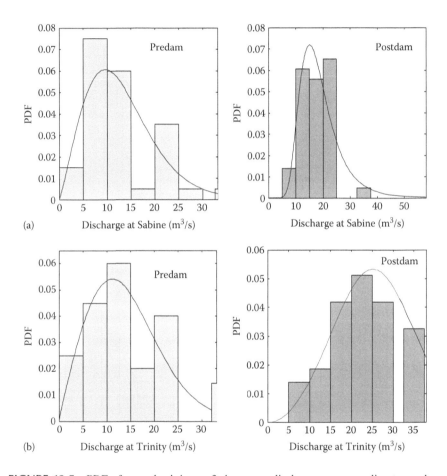

**FIGURE 13.5** PDF of annual minimum 3-day mean discharge corresponding to predam period and postdam period for (a) Sabine River and (b) Trinity River.

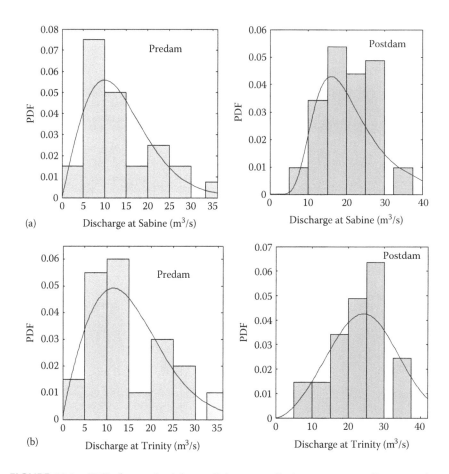

**FIGURE 13.6** PDF of annual minimum 7-day mean discharge corresponding to predam period and postdam period for (a) Sabine River and (b) Trinity River.

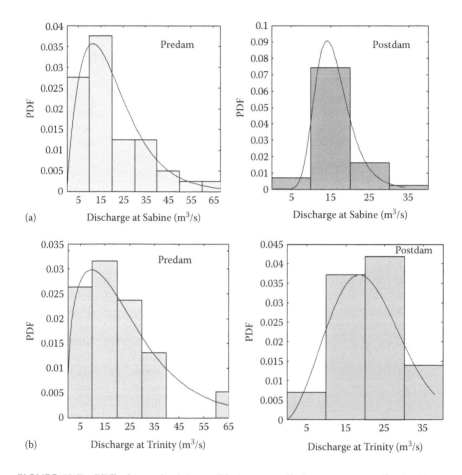

**FIGURE 13.7** PDF of annual minimum 30-day mean discharge corresponding to (a) predam period and postdam period and (b) preperiod and postperiod.

**TABLE 13.12**

**Probability Estimates of Annual Minimum 1-Day Mean Discharge at the Sabine and Trinity Stream Gauge Stations**

| 1-Day Minimum Discharge (m³/s) | Sabine/Toledo Dam Predam PDF | Sabine/Toledo Dam Postdam PDF | Trinity/Livingston Dam Predam PDF | Trinity/Livingston Dam Postdam PDF |
|---|---|---|---|---|
| 2 | 0.0190 | 0.0000 | 0.0147 | 0.0049 |
| 4 | 0.0370 | 0.0003 | 0.0310 | 0.0115 |
| 6 | 0.0498 | 0.0049 | 0.0451 | 0.0189 |
| 8 | 0.0556 | 0.0226 | 0.0550 | 0.0268 |
| 10 | 0.0549 | 0.0535 | 0.0595 | 0.0347 |
| 12 | 0.0494 | 0.0814 | 0.0588 | 0.0422 |
| 14 | 0.0411 | 0.0913 | 0.0539 | 0.0489 |
| 16 | 0.0321 | 0.0826 | 0.0464 | 0.0544 |
| 18 | 0.0238 | 0.0643 | 0.0378 | 0.0580 |
| 20 | 0.0169 | 0.0452 | 0.0293 | 0.0595 |
| 22 | 0.0115 | 0.0295 | 0.0218 | 0.0586 |
| 24 | 0.0076 | 0.0184 | 0.0156 | 0.0554 |
| 26 | 0.0049 | 0.0111 | 0.0109 | 0.0502 |
| 28 | 0.0031 | 0.0066 | 0.0073 | 0.0437 |
| 30 | 0.0019 | 0.0039 | 0.0048 | 0.0363 |
| 32 | 0.0012 | 0.0023 | 0.0031 | 0.0289 |
| 34 | 0.0007 | 0.0014 | 0.0020 | 0.0220 |
| 36 | 0.0004 | 0.0008 | 0.0013 | 0.0160 |
| 38 | 0.0003 | 0.0005 | 0.0008 | 0.0112 |
| 40 | 0.0002 | 0.0003 | 0.0005 | 0.0075 |
| 42 | 0.0001 | 0.0002 | 0.0003 | 0.0049 |
| 44 | 0.0001 | 0.0001 | 0.0002 | 0.0030 |
| 46 | 0.0000 | 0.0001 | 0.0001 | 0.0018 |
| 48 | 0.0000 | 0.0000 | 0.0001 | 0.0011 |
| 50 | 0.0000 | 0.0000 | 0.0000 | 0.0006 |
| 52 | 0.0000 | 0.0000 | 0.0000 | 0.0003 |
| 54 | 0.0000 | 0.0000 | 0.0000 | 0.0002 |
| 56 | 0.0000 | 0.0000 | 0.0000 | 0.0001 |
| 58 | 0.0000 | 0.0000 | 0.0000 | 0.0001 |
| 60 | 0.0000 | 0.0000 | 0.0000 | 0.0000 |

TABLE 13.13

## Probability Estimates of Annual Minimum 3-Day Mean Discharge at the Sabine and Trinity Stream Gauge Stations

| 3-Day Minimum Discharge (m³/s) | Sabine/Toledo Dam | | Trinity/Livingston Dam | |
|---|---|---|---|---|
| | Predam PDF | Postdam PDF | Predam PDF | Postdam PDF |
| 2 | 0.0179 | 0.0000 | 0.0167 | 0.0006 |
| 4 | 0.0365 | 0.0000 | 0.0305 | 0.0023 |
| 6 | 0.0507 | 0.0016 | 0.0416 | 0.0052 |
| 8 | 0.0587 | 0.0114 | 0.0494 | 0.0094 |
| 10 | 0.0605 | 0.0332 | 0.0535 | 0.0146 |
| 12 | 0.0571 | 0.0568 | 0.0538 | 0.0206 |
| 14 | 0.0505 | 0.0701 | 0.0508 | 0.0273 |
| 16 | 0.0423 | 0.0707 | 0.0454 | 0.0342 |
| 18 | 0.0338 | 0.0628 | 0.0385 | 0.0408 |
| 20 | 0.0260 | 0.0515 | 0.0313 | 0.0464 |
| 22 | 0.0194 | 0.0403 | 0.0244 | 0.0507 |
| 24 | 0.0141 | 0.0306 | 0.0183 | 0.0530 |
| 26 | 0.0100 | 0.0228 | 0.0132 | 0.0532 |
| 28 | 0.0070 | 0.0169 | 0.0093 | 0.0511 |
| 30 | 0.0048 | 0.0125 | 0.0063 | 0.0471 |
| 32 | 0.0033 | 0.0092 | 0.0042 | 0.0417 |
| 34 | 0.0022 | 0.0068 | 0.0028 | 0.0353 |
| 36 | 0.0015 | 0.0051 | 0.0018 | 0.0287 |
| 38 | 0.0010 | 0.0038 | 0.0011 | 0.0224 |
| 40 | 0.0007 | 0.0029 | 0.0007 | 0.0167 |
| 42 | 0.0004 | 0.0022 | 0.0004 | 0.0121 |
| 44 | 0.0003 | 0.0017 | 0.0003 | 0.0084 |
| 46 | 0.0002 | 0.0013 | 0.0002 | 0.0056 |
| 48 | 0.0001 | 0.0010 | 0.0001 | 0.0037 |
| 50 | 0.0001 | 0.0008 | 0.0001 | 0.0023 |
| 52 | 0.0001 | 0.0006 | 0.0000 | 0.0014 |
| 54 | 0.0000 | 0.0005 | 0.0000 | 0.0008 |
| 56 | 0.0000 | 0.0004 | 0.0000 | 0.0005 |
| 58 | 0.0000 | 0.0003 | 0.0000 | 0.0003 |
| 60 | 0.0000 | 0.0003 | 0.0000 | 0.0002 |

**TABLE 13.14**

**Probability Estimate of Annual Minimum 7-Day Mean Discharge at the Sabine and Trinity River Gauge Stations**

| 7-Day Minimum Discharge (m³/s) | Sabine/Toledo Dam | | Trinity/Livingston Dam | |
| | Predam PDF | Postdam PDF | Predam PDF | Postdam PDF |
|---|---|---|---|---|
| 2 | 0.0180 | 0.0000 | 0.0164 | 0.0010 |
| 4 | 0.0342 | 0.0001 | 0.0287 | 0.0032 |
| 6 | 0.0464 | 0.0022 | 0.0383 | 0.0063 |
| 8 | 0.0536 | 0.0096 | 0.0449 | 0.0103 |
| 10 | 0.0558 | 0.0216 | 0.0485 | 0.0150 |
| 12 | 0.0539 | 0.0333 | 0.0491 | 0.0200 |
| 14 | 0.0490 | 0.0407 | 0.0471 | 0.0253 |
| 16 | 0.0424 | 0.0429 | 0.0432 | 0.0305 |
| 18 | 0.0351 | 0.0412 | 0.0380 | 0.0352 |
| 20 | 0.0281 | 0.0371 | 0.0322 | 0.0390 |
| 22 | 0.0218 | 0.0320 | 0.0264 | 0.0416 |
| 24 | 0.0165 | 0.0270 | 0.0210 | 0.0426 |
| 26 | 0.0122 | 0.0223 | 0.0162 | 0.0419 |
| 28 | 0.0089 | 0.0182 | 0.0123 | 0.0395 |
| 30 | 0.0063 | 0.0148 | 0.0090 | 0.0358 |
| 32 | 0.0045 | 0.0120 | 0.0065 | 0.0310 |
| 34 | 0.0031 | 0.0097 | 0.0047 | 0.0257 |
| 36 | 0.0022 | 0.0079 | 0.0033 | 0.0204 |
| 38 | 0.0015 | 0.0064 | 0.0023 | 0.0155 |
| 40 | 0.0010 | 0.0052 | 0.0015 | 0.0113 |
| 42 | 0.0007 | 0.0043 | 0.0010 | 0.0079 |
| 44 | 0.0005 | 0.0035 | 0.0007 | 0.0053 |
| 46 | 0.0003 | 0.0029 | 0.0005 | 0.0034 |
| 48 | 0.0002 | 0.0024 | 0.0003 | 0.0021 |
| 50 | 0.0001 | 0.0020 | 0.0002 | 0.0013 |
| 52 | 0.0001 | 0.0016 | 0.0001 | 0.0008 |
| 54 | 0.0001 | 0.0014 | 0.0001 | 0.0004 |
| 56 | 0.0000 | 0.0012 | 0.0001 | 0.0002 |
| 58 | 0.0000 | 0.0010 | 0.0000 | 0.0001 |
| 60 | 0.0000 | 0.0008 | 0.0000 | 0.0001 |

TABLE 13.15
Probability Estimate of Annual Minimum 30-Day Mean
Discharge at the Sabine and Trinity River Gauge Stations

| | Sabine/Toledo Dam | | Trinity/Livingston Dam | |
|---|---|---|---|---|
| | Predam | Postdam | Predam | Postdam |
| 30-Day Minimum Discharge (m³/s) | PDF | PDF | PDF | PDF |
| 2 | 0.0139 | 0.0000 | 0.0212 | 0.0022 |
| 4 | 0.0231 | 0.0000 | 0.0257 | 0.0063 |
| 6 | 0.0294 | 0.0013 | 0.0280 | 0.0116 |
| 8 | 0.0333 | 0.0122 | 0.0290 | 0.0174 |
| 10 | 0.0352 | 0.0420 | 0.0292 | 0.0232 |
| 12 | 0.0356 | 0.0760 | 0.0288 | 0.0286 |
| 14 | 0.0349 | 0.0906 | 0.0279 | 0.0329 |
| 16 | 0.0333 | 0.0822 | 0.0268 | 0.0358 |
| 18 | 0.0311 | 0.0625 | 0.0254 | 0.0371 |
| 20 | 0.0286 | 0.0425 | 0.0239 | 0.0367 |
| 22 | 0.0260 | 0.0270 | 0.0223 | 0.0348 |
| 24 | 0.0233 | 0.0165 | 0.0207 | 0.0316 |
| 26 | 0.0207 | 0.0099 | 0.0190 | 0.0277 |
| 28 | 0.0183 | 0.0059 | 0.0175 | 0.0234 |
| 30 | 0.0160 | 0.0035 | 0.0159 | 0.0191 |
| 32 | 0.0139 | 0.0021 | 0.0145 | 0.0152 |
| 34 | 0.0120 | 0.0012 | 0.0131 | 0.0116 |
| 36 | 0.0104 | 0.0008 | 0.0118 | 0.0087 |
| 38 | 0.0089 | 0.0005 | 0.0106 | 0.0063 |
| 40 | 0.0076 | 0.0003 | 0.0095 | 0.0045 |
| 42 | 0.0065 | 0.0002 | 0.0085 | 0.0032 |
| 44 | 0.0055 | 0.0001 | 0.0076 | 0.0022 |
| 46 | 0.0047 | 0.0001 | 0.0068 | 0.0015 |
| 48 | 0.0040 | 0.0001 | 0.0060 | 0.0010 |
| 50 | 0.0034 | 0.0000 | 0.0053 | 0.0006 |
| 52 | 0.0029 | 0.0000 | 0.0047 | 0.0004 |
| 54 | 0.0024 | 0.0000 | 0.0042 | 0.0003 |
| 56 | 0.0020 | 0.0000 | 0.0037 | 0.0002 |
| 58 | 0.0017 | 0.0000 | 0.0033 | 0.0001 |
| 60 | 0.0015 | 0.0000 | 0.0029 | 0.0001 |

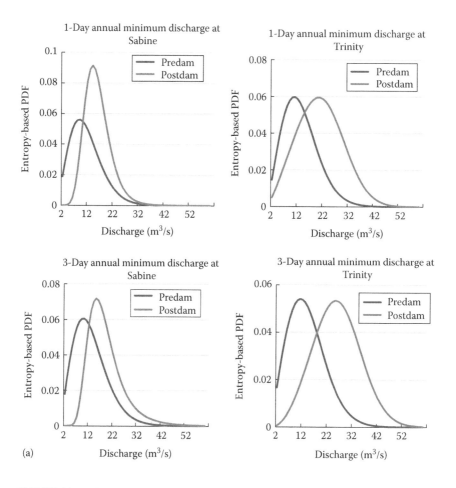

**FIGURE 13.8**  Comparing changes in (a) 1-day minimum and 3-day minimum streamflow and

*(Continued)*

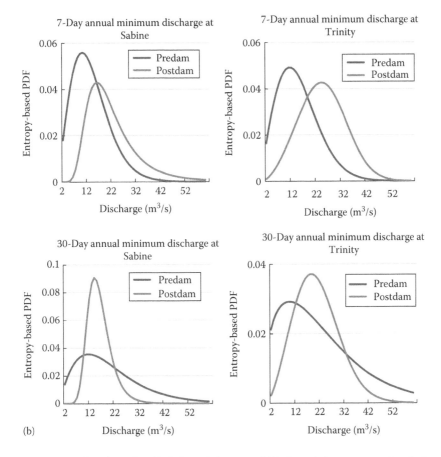

**FIGURE 13.8 (*Continued*)** (b) 7-day minimum and 30-day minimum streamflow before and after the dams on Sabine and Trinity Rivers, using the Tsallis entropy-based PDF estimates.

**TABLE 13.16**
**OWA Operator and Entropy for the Sabine River**

| $j$ | $w_j$ | Entropy | Cumulative Entropy | $j$ | $w_j$ | Entropy | Cumulative Entropy |
|---|---|---|---|---|---|---|---|
| 1 | 0.0504 | 1.26 | 1.26 | 17 | 0.031 | 0.31 | 10.45 |
| 2 | 0.049 | 0.89 | 2.15 | 18 | 0.030 | 0.69 | 11.13 |
| 3 | 0.048 | 0.67 | 2.82 | 19 | 0.029 | 0.57 | 11.71 |
| 4 | 0.047 | 0.23 | 3.05 | 20 | 0.027 | 0.71 | 12.42 |
| 5 | 0.046 | 0.41 | 3.46 | 21 | 0.026 | 0.29 | 12.71 |
| 6 | 0.044 | 1.06 | 4.53 | 22 | 0.025 | 0.75 | 13.46 |
| 7 | 0.043 | 0.30 | 4.83 | 23 | 0.024 | 0.31 | 13.77 |
| 8 | 0.042 | 0.04 | 4.87 | 24 | 0.023 | 0.27 | 14.04 |
| 9 | 0.041 | 0.16 | 5.03 | 25 | 0.021 | 0.32 | 14.36 |
| 10 | 0.040 | 1.11 | 6.14 | 26 | 0.020 | 0.32 | 14.68 |
| 11 | 0.038 | 1.19 | 7.33 | 27 | 0.019 | 0.06 | 14.74 |
| 12 | 0.037 | 0.30 | 7.62 | 28 | 0.018 | 0.04 | 14.77 |
| 13 | 0.036 | 0.68 | 8.30 | 29 | 0.017 | 0.28 | 15.05 |
| 14 | 0.035 | 0.76 | 9.07 | 30 | 0.015 | 0.44 | 15.50 |
| 15 | 0.033 | 0.20 | 9.27 | 31 | 0.014 | 0.30 | 15.79 |
| 16 | 0.032 | 0.87 | 10.14 | | | | |

**TABLE 13.17**
**OWA Operator and Entropy for the Trinity River**

| $j$ | $w_j$ | Entropy | Cumulative Entropy | $j$ | $w_j$ | Entropy | Cumulative Entropy |
|---|---|---|---|---|---|---|---|
| 1 | 0.050 | 0.30 | 0.30 | 17 | 0.031 | 0.96 | 8.82 |
| 2 | 0.049 | 0.74 | 1.04 | 18 | 0.030 | 0.60 | 9.42 |
| 3 | 0.048 | 1.20 | 2.24 | 19 | 0.029 | 0.83 | 10.25 |
| 4 | 0.047 | 0.05 | 2.29 | 20 | 0.027 | 0.82 | 11.07 |
| 5 | 0.046 | 1.23 | 3.52 | 21 | 0.026 | 0.73 | 11.80 |
| 6 | 0.044 | 0.40 | 3.92 | 22 | 0.025 | 0.32 | 12.13 |
| 7 | 0.043 | 0.09 | 4.00 | 23 | 0.024 | 0.26 | 12.39 |
| 8 | 0.042 | 0.50 | 4.51 | 24 | 0.023 | 0.18 | 12.57 |
| 9 | 0.041 | 0.16 | 4.67 | 25 | 0.021 | 0.56 | 13.12 |
| 10 | 0.040 | 0.40 | 5.06 | 26 | 0.020 | 0.46 | 13.59 |
| 11 | 0.038 | 0.19 | 5.25 | 27 | 0.019 | 0.13 | 13.72 |
| 12 | 0.037 | 0.11 | 5.37 | 28 | 0.018 | 0.25 | 13.97 |
| 13 | 0.036 | 0.61 | 5.98 | 29 | 0.017 | 0.31 | 14.28 |
| 14 | 0.035 | 0.76 | 6.74 | 30 | 0.015 | 0.37 | 14.65 |
| 15 | 0.033 | 0.60 | 7.34 | 31 | 0.014 | 0.30 | 14.95 |
| 16 | 0.032 | 0.52 | 7.86 | | | | |

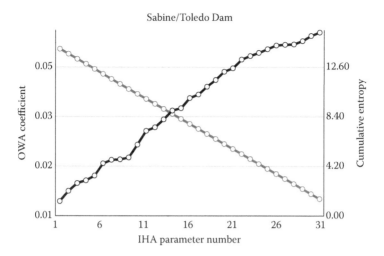

**FIGURE 13.9**   OWA operator and entropy for the Sabine River.

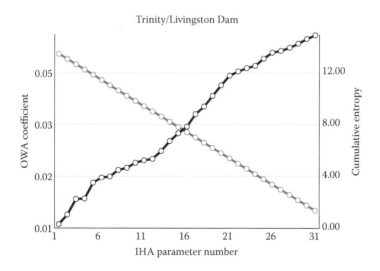

**FIGURE 13.10**   OWA operator and entropy for the Trinity River.

## Example 13.4

Compute the eco-index for the data from Example of 13.1.

**Solution**

The eco-index can be computed with the computed value of *NSL* for each of the parameters represented by the argument $(a_j)$ and can be written as

$$\text{Eco-index} = F(a_1, a_2, ..., a_{33}) = F(NSL_1, NSL_2, ..., NSL_{32}) = \sum_{j=1}^{32} w_j b_j \qquad (13.34)$$

---

**TABLE 13.18**

**NSL Values for the Sabine River Station**

| Parameter | H_pre | H_post | NSL |
|---|---|---|---|
| January | 0.69 | 0.70 | 0.018 |
| February | 0.69 | 0.73 | 0.043 |
| March | 0.73 | 0.68 | 0.057 |
| April | 0.58 | 0.73 | 0.150 |
| May | 0.54 | 0.64 | 0.097 |
| June | 0.63 | 0.60 | 0.023 |
| July | 0.60 | 0.48 | 0.123 |
| August | 0.35 | 0.74 | 0.390 |
| September | 0.47 | 0.68 | 0.219 |
| October | 0.54 | 0.55 | 0.007 |
| November | 0.49 | 0.48 | 0.003 |
| December | 0.57 | 0.67 | 0.097 |
| Min1D | 0.71 | 0.67 | 0.042 |
| Max1D | 0.68 | 0.71 | 0.029 |
| Min3D | 0.70 | 0.56 | 0.142 |
| Max3D | 0.69 | 0.70 | 0.014 |
| Min7D | 0.69 | 0.60 | 0.097 |
| Max7D | 0.72 | 0.69 | 0.023 |
| Min30D | 0.66 | 0.63 | 0.037 |
| Max30D | 0.71 | 0.69 | 0.018 |
| Min90D | 0.60 | 0.68 | 0.073 |
| Max90D | 0.73 | 0.74 | 0.006 |
| JulianMin1D | 0.60 | 0.66 | 0.067 |
| JulianMax1D | 0.61 | 0.68 | 0.068 |
| NLowPls | 0.61 | 0.67 | 0.056 |
| NHighPls | 0.61 | 0.67 | 0.056 |
| DurLowPls | 0.53 | 0.28 | 0.253 |
| DurHighPls | 0.66 | 0.40 | 0.253 |
| RiseRate | 0.75 | 0.70 | 0.045 |
| FallRate | 0.70 | 0.71 | 0.007 |
| NumHydrRev | 0.68 | 0.66 | 0.029 |

Results of eco-index at the Sabine River and Trinity River gauge stations are shown in Tables 13.18 through 13.21 and the eco-index of the two gage stations is presented in Figure 13.11.

Figure 13.12a and b shows the magnitude of the alteration induced by the constructed dams on the natural flow regime at the Sabine and Trinity gaging stations. Figure 13.12a and b justifies the alteration on the natural flow regime through the measure of dispersion, which is the spread between the 25th and 75th percentiles, divided by the median.

## TABLE 13.19
## OWA Operator Estimated Based on the Sorted NSL Values for the Sabine River Station

| Rank | NSL_Sorted | OWA | Eco-Index | Cumulative Eco-Index |
|---|---|---|---|---|
| 1 | 0.390 | 0.061 | 0.0239 | 0.0239 |
| 2 | 0.253 | 0.059 | 0.0150 | 0.0389 |
| 3 | 0.253 | 0.057 | 0.0145 | 0.0534 |
| 4 | 0.219 | 0.055 | 0.0121 | 0.0655 |
| 5 | 0.150 | 0.053 | 0.0080 | 0.0735 |
| 6 | 0.142 | 0.052 | 0.0073 | 0.0808 |
| 7 | 0.123 | 0.050 | 0.0061 | 0.0869 |
| 8 | 0.097 | 0.048 | 0.0046 | 0.0915 |
| 9 | 0.097 | 0.046 | 0.0044 | 0.0959 |
| 10 | 0.097 | 0.044 | 0.0042 | 0.1001 |
| 11 | 0.073 | 0.042 | 0.0031 | 0.1032 |
| 12 | 0.068 | 0.040 | 0.0027 | 0.1059 |
| 13 | 0.067 | 0.038 | 0.0025 | 0.1084 |
| 14 | 0.057 | 0.036 | 0.0020 | 0.1104 |
| 15 | 0.056 | 0.034 | 0.0019 | 0.1123 |
| 16 | 0.056 | 0.032 | 0.0018 | 0.1141 |
| 17 | 0.045 | 0.030 | 0.0014 | 0.1155 |
| 18 | 0.043 | 0.028 | 0.0012 | 0.1167 |
| 19 | 0.042 | 0.026 | 0.0011 | 0.1178 |
| 20 | 0.037 | 0.024 | 0.0009 | 0.1187 |
| 21 | 0.029 | 0.023 | 0.0007 | 0.1194 |
| 22 | 0.029 | 0.021 | 0.0006 | 0.1200 |
| 23 | 0.023 | 0.019 | 0.0004 | 0.1204 |
| 24 | 0.023 | 0.017 | 0.0004 | 0.1208 |
| 25 | 0.018 | 0.015 | 0.0003 | 0.1211 |
| 26 | 0.018 | 0.013 | 0.0002 | 0.1213 |
| 27 | 0.014 | 0.011 | 0.0002 | 0.1215 |
| 28 | 0.007 | 0.009 | 0.0001 | 0.1216 |
| 29 | 0.007 | 0.007 | 0.0000 | 0.1216 |
| 30 | 0.006 | 0.005 | 0.0000 | 0.1216 |
| 31 | 0.003 | 0.003 | 0.0000 | 0.1216 |

**TABLE 13.20**
**NSL Values for Trinity River Station**

| Parameter | H_pre | H_post | NSL |
|---|---|---|---|
| January | 0.64 | 0.56 | 0.079 |
| February | 0.66 | 0.68 | 0.026 |
| March | 0.65 | 0.65 | 0.005 |
| April | 0.53 | 0.72 | 0.191 |
| May | 0.64 | 0.65 | 0.001 |
| June | 0.63 | 0.68 | 0.049 |
| July | 0.52 | 0.38 | 0.144 |
| August | 0.46 | 0.50 | 0.041 |
| September | 0.64 | 0.52 | 0.118 |
| October | 0.55 | 0.51 | 0.043 |
| November | 0.49 | 0.57 | 0.082 |
| December | 0.53 | 0.67 | 0.136 |
| Min1D | 0.72 | 0.74 | 0.023 |
| Max1D | 0.72 | 0.73 | 0.012 |
| Min3D | 0.72 | 0.75 | 0.022 |
| Max3D | 0.71 | 0.74 | 0.025 |
| Min7D | 0.72 | 0.72 | 0.000 |
| Max7D | 0.71 | 0.70 | 0.016 |
| Min30D | 0.68 | 0.68 | 0.001 |
| Max30D | 0.72 | 0.72 | 0.000 |
| Min90D | 0.61 | 0.61 | 0.001 |
| Max90D | 0.72 | 0.69 | 0.028 |
| JulianMin1D | 0.64 | 0.68 | 0.041 |
| JulianMax1D | 0.62 | 0.68 | 0.063 |
| NLowPls | 0.62 | 0.62 | 0.005 |
| NHighPls | 0.61 | 0.62 | 0.007 |
| DurLowPls | 0.55 | 0.48 | 0.068 |
| DurHighPls | 0.65 | 0.68 | 0.026 |
| RiseRate | 0.71 | 0.73 | 0.020 |
| FallRate | 0.72 | 0.72 | 0.005 |
| NumHydrRev | 0.73 | 0.72 | 0.015 |

**TABLE 13.21**

**OWA Operator Estimated Based on the Sorted NSL Values
for the Trinity River Station**

| Parameter | NSL_Sorted | OWA | Eco-Index | Cumulative Eco-Index |
|---|---|---|---|---|
| 1 | 0.191 | 0.061 | 0.0117 | 0.0117 |
| 2 | 0.144 | 0.059 | 0.0085 | 0.0202 |
| 3 | 0.136 | 0.057 | 0.0078 | 0.0280 |
| 4 | 0.118 | 0.055 | 0.0065 | 0.0345 |
| 5 | 0.082 | 0.053 | 0.0044 | 0.0389 |
| 6 | 0.079 | 0.052 | 0.0041 | 0.0430 |
| 7 | 0.068 | 0.050 | 0.0034 | 0.0464 |
| 8 | 0.063 | 0.048 | 0.0030 | 0.0494 |
| 9 | 0.049 | 0.046 | 0.0022 | 0.0516 |
| 10 | 0.043 | 0.044 | 0.0019 | 0.0535 |
| 11 | 0.041 | 0.042 | 0.0017 | 0.0552 |
| 12 | 0.041 | 0.040 | 0.0016 | 0.0568 |
| 13 | 0.028 | 0.038 | 0.0011 | 0.0579 |
| 14 | 0.026 | 0.036 | 0.0010 | 0.0589 |
| 15 | 0.026 | 0.034 | 0.0009 | 0.0598 |
| 16 | 0.025 | 0.032 | 0.0008 | 0.0606 |
| 17 | 0.023 | 0.030 | 0.0007 | 0.0613 |
| 18 | 0.022 | 0.028 | 0.0006 | 0.0619 |
| 19 | 0.020 | 0.026 | 0.0005 | 0.0624 |
| 20 | 0.016 | 0.024 | 0.0004 | 0.0628 |
| 21 | 0.015 | 0.023 | 0.0003 | 0.0631 |
| 22 | 0.012 | 0.021 | 0.0002 | 0.0633 |
| 23 | 0.007 | 0.019 | 0.0001 | 0.0634 |
| 24 | 0.005 | 0.017 | 0.0001 | 0.0635 |
| 25 | 0.005 | 0.015 | 0.0001 | 0.0636 |
| 26 | 0.005 | 0.013 | 0.0001 | 0.0637 |
| 27 | 0.001 | 0.011 | 0.0000 | 0.0637 |
| 28 | 0.001 | 0.009 | 0.0000 | 0.0637 |
| 29 | 0.001 | 0.007 | 0.0000 | 0.0637 |
| 30 | 0.000 | 0.005 | 0.0000 | 0.0637 |
| 31 | 0.000 | 0.003 | 0.0000 | 0.0637 |

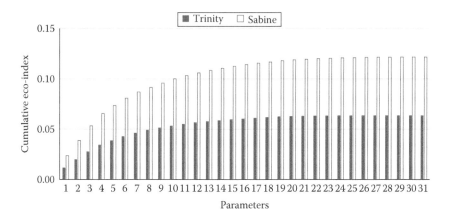

**FIGURE 13.11**   Eco-Index comparison between the Trinity and the Sabine Rivers.

The computed values of eco-index, presented in Figure 13.11, show that the Trinity River's ecosystem is more vulnerable compared to the Sabine River. The eco-index of the Trinity River is almost two times lower than the value of the Sabine River. The point may be related to the differences in hydrological characteristics of the dams and the morphology of the two watersheds. In fact the storage capacity of the Toledo reservoir (Sabine River) is two times greater than the capacity of the Livingston reservoir (Trinity). This difference is quite critical, as it allows a better outflow control at the Sabine downstream. This can be understood by comparing the monthly median flow in the two rivers (Figure 13.13a and b), which show a moderate decreasing trend of the postdam median flow rate of the Sabine River during the period May–November, while this trend looks more abrupt in the Trinity river. It can be inferred that the postdam conditions on the Trinity River may not have been regulated wisely to meet the need of the ecosystem.

This exercise shows that an information-based eco-index, which reflects the non-satisfaction level flow regime, can guide eco-managers of Sabine and Trinity Rivers in a proper direction, while allocating water resources among potential users, and where to concentrate their attention, while mitigating the dam-induced effects and future alteration on sustained flow regime. In addition, these values of eco-index indirectly portray eventual alteration occurring in the local ecosystem particularly downstream. At the regional watershed scale, the relative values of this eco-index can show where regional eco-managers need to pay attention. Many times the paucity of hydrological data hinders our understanding of the state of a system. The entropy-based eco-index uses observed flows. Often times water resource development activities within a river basin rely on spatial homogeneity or heterogeneity. Therefore, having spatial information about the ecosystem alterations and their influence at the subbasin level or grid scale needs attention when developing water resources.

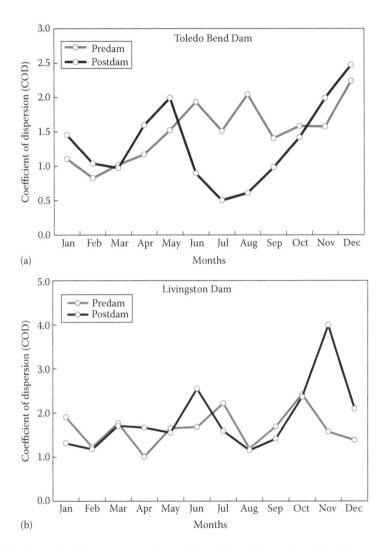

**FIGURE 13.12**    Coefficient of dispersion at (a) Toledo Bend Reservoir and (b) Livingston dam.

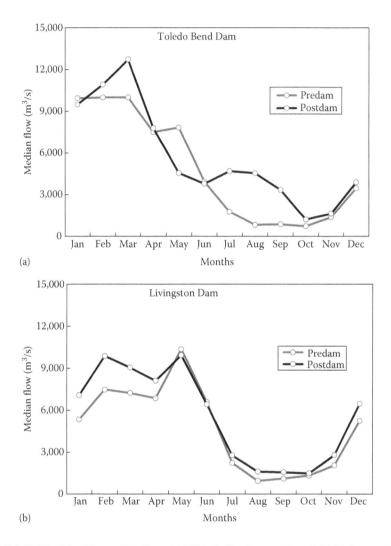

(a)

(b)

**FIGURE 13.13**    Monthly median flow at (a) Toledo Bend reservoir and (b) Livingston dam.

# REFERENCES

Arthington, A.H. (1998). Comparative evaluation of environmental flow assessment techniques: Review of holistic methodologies. Land and Water Resources Research and Development Corporation Occasional Paper No. 26/98, Canberra, Australian Capital Territory, Australia.

Dyson, M., Berkamp, G., and Scanlon, J. (2003). *Flow: The Essentials of Environmental Flows*. IUCN, Cambridge, U.K.

Fullér, R. and Majlender, P. (2001). An analytic approach for obtaining maximal entropy OWA operator weights. *Fuzzy Sets and Systems*, 124, 53–57.

Kim, Z. and Singh, V.P. (2014). Assessment of environmental flow requirements by entropy-based multi-criteria decision. *Water Resources Management*, 28, 459–474.

Koutsoyiannis, D. (2005a). Uncertainty, entropy, scaling and hydrological stochastics. 1. Marginal distributional properties of hydrological processes and state scaling. *Hydrological Sciences Journal*, 50(3), 381–404.

Koutsoyiannis, D. (2005b). Uncertainty, entropy, scaling and hydrological stochastics. 2. Time dependence of hydrological processes and state scaling. *Hydrological Sciences Journal*, 50(3), 405–426.

Lamata, M.T. (2004). Ranking of alternatives with ordered weighted averaging operators. *International Journal of Intelligent Systems*, 19, 473–482.

Lytle, D.H. and Poff, N.L. (2004). Adaptation to natural flow regimes. *Trends in Ecology and Evolution*, 19, 94–100.

Naiman, R.J., Bunn, S.E., Nilsson, C., Petts, G.E., Pinay, G., and Thompson, L.C. (2002). Legitimizing fluvial ecosystems as users of water: An overview. *Environmental Management*, 30, 455–467.

O'Hagan, M. (1988). Aggregating template or rule antecedents in real-time expert systems with fuzzy set logic. *Proceedings of the 22nd Annual IEEE Asilomar Conference on Signals, Systems and Computers*, Pacific Grove, CA, pp. 681–689.

Papalexiou, S.M. and Koutsoyiannis, D. (2012). Entropy based derivation of probability distributions: A case study to daily rainfall. *Advances in Water Resources*, 45, 51–57.

Poff, N. L., Allan, J.D., Bain, M.B., Karr, J.R., Prestegaard, K.L., Richter, B.D., Sparks, R.E. and Stromberg, J.C. (1997). The natural flow regime. BioScience, 47(11), 769–784.

Postel, S., and Richter, B.D. (2003). *Rivers for Life: Managing Water for People and Nature*. Island Press, Washington, DC.

Richter, B.D., Baumgartner, J.V., Powel, J., and Braun, D.P. (1996). A method for assessing hydrologic alteration within ecosystems. *Conservation Biology*, 10, 1163–1174.

Singh, V.P. (1998). The use of entropy in hydrology and water resources. *Hydrological Processes*, 11, 587–626.

Smakhtin, V., Revenga, C., and Doll, P. (2004). A pilot global assessment of environmental water requirements and scarcity. *Water International*, 29, 307–320.

Stromberg, J.C. (1997). The natural flow regime: A paradigm for river conservation and restoration. *BioScience*, 47, 769–784.

Tharme, R.E. (2003). A global perspective on environmental flow assessment: Emerging trends in the development and application of environmental flow methodologies for rivers. *River Research and Applications*, 19, 397–442.

TNC (The Nature Conservancy). (2009). User's manual for the Indicators of Hydrologic Alteration (IHA) Software, Ver 7.1.

Tsallis, C. (1988). Possible generalization of Boltzmann–Gibbs statistics. *Journal of Statistical Physics*, 52(1–2), 479–487.

Tsallis, C. and Brigatti, E. (2004). Nonextensive statistical mechanics: A brief introduction. *Continuum Mechanics and Thermodynamics*, 16(3), 223–235.

Vorosmarty, C.J., Green, P., Salisbury, J., and Lammers, R.B. (2000). Global water resources vulnerability from climate change and population growth. *Science*, 289, 284–288.

Yager, R.R. (1999). Induced ordered weighted averaging operators. *IEEE Transactions on Systems, Man and Cybernetics*, 29, 141–150.

# 14 Redundancy Measures for Water Distribution Networks

Water distribution systems are vital for our way of life, for energy generation, for industries, and for waste disposal. These networks should be designed in such a manner that they are reliable and optimal. Fundamental to either type of optimization is reliability, while others deal with the reliability of within water distribution networks. The objective of this study is to present measures of reliability of water distribution networks based on the Tsallis entropy.

## 14.1 OPTIMIZATION OF WATER DISTRIBUTION NETWORKS

Design of a water distribution system entails competing objectives, including the minimization of head losses, cost, risk, and departures from specified values of water quantity, pressure, and quality; and the maximization of reliability (Perelman et al., 2008); and is hence a multiobjective optimization problem. However, the design problem can be formulated as a single objective optimization problem, where the system capital and operational costs are minimized and at the same time the laws of hydraulics are satisfied and the targets of water quantity and pressure at demand nodes are met. Approaches to the optimization of water distribution networks have been either deterministic or stochastic.

### 14.1.1 Deterministic Optimization

Approaches to deterministic optimization are exemplified by the studies of Goulter and Coals (1986), Su et al. (1987), Lansey et al. (1989), Goulter and Bouchart (1990) among others that focused on reliability within an optimization framework with respect to the hydraulic performance of the network under a range of mechanical failures and demands. In a similar vein, Ekinci and Konak (2009) developed an optimization strategy for water distribution networks by minimizing head losses for least cost design, whereas Eiger et al. (1994) presented a two-stage decomposition model for optimization of water distribution networks. Todini (2000) used a resilience index for developing a technique for looped water distribution network design.

### 14.1.2 STOCHASTIC OPTIMIZATION

Incorporating the uncertainty of nodal water demands and pipe roughness, Giustolisi et al. (2009) developed a multiobjective optimization scheme with the objective of minimizing costs and maximizing hydraulic reliability. Kwon and Lee (2008) analyzed the reliability of pipe networks using probability concepts focusing on transient flow that can cause failure of the water distribution system. On the other hand, entropy theory has been applied to develop measures for water distribution network reliability (Singh and Oh, 2014). In a review of explorative uses of entropy, Templeman (1992) discussed application of entropy to water supply network analysis.

### 14.1.3 ENTROPY-BASED OPTIMIZATION

Awumah (1990) and Awumah et al. (1990, 1991) used the Shannon entropy (Shannon, 1948) to develop redundancy measures for water distribution systems. Xu and Jowitt (1992) remarked in a discussion of the study by Awumah et al. (1991) and their entropy-based measure needed further investigation. Redundancy in a water distribution network fundamentally means that demand points or nodes have alternative supply paths for water in the event that some links go out of service. In a redundant network, there is sufficient residual capacity to meet water flow requirements. Being a characteristic of the water distribution system, redundancy is related to its reliability. Therefore, to ensure reliability the water distribution network design must incorporate some amount of redundancy.

Tanyimboh and Templeman (1993a) described methods using entropy for computing the most likely flows in the links of the networks with incomplete data. Tanyimboh and Templeman (1993b) developed an algorithm for computing the maximum entropy flows for single source networks. Chen and Templeman (1995) developed entropy-based methods for mathematical planning. Perelman and Ostfeld (2007) developed a cross-entropy-based algorithm for optimal design of water distribution systems. Subsequently, Perelman et al. (2008) extended the cross-entropy based algorithm to multiobjective optimization for water distribution systems design. The extended algorithm coupled the cross-entropy algorithm (Rubinstein, 1997) and some features of multiobjective evolutionary techniques (Fonseca and Fleming, 1996). Shibu and Janga Reddy (2013) applied cross-entropy (Kullback and Leibler, 1951) for optimal design of water distribution networks. Many studies have dealt with the reliability of overall water supply systems (Germanopoulos et al., 1986; Goulter and Coals, 1986; Su et al., 1987; Beim and Hobbs, 1988; Hobbs and Beim, 1988; Wagner et al., 1988a,b; Goulter and Bouchart, 1990).

## 14.2　RELIABILITY MEASURE

The need for reliability stems from uncertainties in consumer demand, fire flow requirements and their locations, pumping systems failure, inefficient storage, pipe failures and their locations, valve leakages and their locations, and reduced capacity due to sedimentation. Goulter (1987, 1988, 1992) argued that the shape or layout of a network determines the level of reliability that can be imposed on the network.

When designing a water distribution network a critically important requirement is that the flow to a demand node must be carried by multiple links instead of just one link, and these links should be connected directly to the node. Although these links may carry equal or unequal proportions of flow to the demand node, Goulter and Coals (1986) and Walters (1988) have shown that from a reliability point of view it is more advantageous to carry equal proportions of flow for two reasons. First, in the case of unequal proportions of flow the network reliability is severely impacted if the link carrying the larger proportion goes out of service. Second, it is hydraulically inefficient. It is known that for a fixed pipe size, discharge $q$ carried by a pipe is approximately proportional to the 0.54 power of the head loss $h_L$ in that pipe, that is, $q \propto h_L^{0.54}$ or approximately $h_L \propto q^2$. When a larger pipe fails and the flow is to be increased in a smaller pipe, then head loss would increase quadratically as a function of discharge. For example, doubling the flow would quadruple the head loss and tripling the flow would increase the head loss nine fold. On the other hand, increasing the flow in a larger pipe would not cause the same order of head loss.

## 14.3 TSALLIS ENTROPY–BASED REDUNDANCY MEASURES

The Tsallis entropy, $S$, (Tsallis, 1988) for a discrete random variable $X$ with probability distribution $P = \{p_i, i = 1, 2,..., N\}$, where $p_i$ are probabilities for $X = x_i, i = 1, 2, ..., N$, can be expressed as

$$S(x) = \frac{1}{m-1}\left(1-\sum_{i=1}^{N} p_i^m\right) = \frac{1}{m-1}\sum_{i=1}^{N} p_i\left(1- p_i^{m-1}\right) \tag{14.1}$$

where
  $m$ is a real number
  $N$ is the number of values of $X$ takes on
  $S$ describes the uncertainty associated with $p_i$ and in turn with $X$

If $(1- p_i^{m-1})/(m-1)$ is considered as a measure of uncertainty, then Equation 14.1 represents the average uncertainty of $X$.

## 14.4 REDUNDANCY MEASURES

Let a network consist of $N$ nodes, as shown in Figure 14.1, and let the number of links incident at node $j$ be $n(j)$. A particular link incident at node $j$ is denoted by $i$; thus $i = 1, 2, 3, ..., n(j)$. Let the flow carried by this $i$th link to node $j$ be denoted by $q_{ij}$, flow in pipe-connecting links incident on node $j$, or the total flow at node $j$ by $Q_j$ and the fraction of flow carried by link $i$ by $W_{ij}$. Then, for a particular flow pattern, fraction $W_{ij}$ can be denoted by

$$W_{ij} = \frac{q_{ij}}{Q_j} \tag{14.2}$$

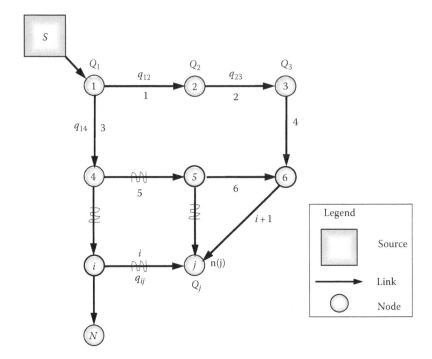

**FIGURE 14.1**    A water distribution network layout with $N$ nodes.

Clearly,

$$\sum_{i=1}^{n(j)} W_{ij} = 1 \tag{14.3}$$

where

$$Q_j = \sum_{i=1}^{n(j)} q_{ij} \tag{14.4}$$

Term $W_{ij}$ defines the relative contribution of link $i$ to flow at node $j$ and is therefore an indicator of relative flow capacity of the link incident at node $j$. Thus, it can be construed as a measure of the potential contribution of the link to the required demand at the node if a link failed. Thus, it enables consideration of relative flow capacities of links in the redundancy measure.

Now, let $Q_0$ be the total flow in the network, which is equal to the sum of flows in all links in the network, that is,

$$Q_0 = \sum_{j=1}^{N} Q_j \tag{14.5}$$

where $N$ is the number of nodes in the network. It should be emphasized that $Q_0$ is not the total demand in the network or the total flow supply to the network; it is usually greater than the total demand in the network.

To develop a Tsallis entropy–based redundancy measure of the network with $N$ nodes, where the nodes may be considered to constitute subsystems, the Tsallis entropy of a node $j$ can now be expressed in terms of $W_{ij}$ as

$$S_j = \frac{K}{m-1}\left[\sum_{i=1}^{n(j)} W_{ij} - W_{ij}^m\right] = \frac{K}{m-1}\left\{\sum_{i=1}^{n(j)}\left[\frac{q_{ij}}{Q_j} - \left(\frac{q_{ij}}{Q_j}\right)^m\right]\right\} \tag{14.6}$$

where
  $K$ is a positive constant and is often taken as one
  $m$ is the entropy index and is a real number
  $S_j$ is an entropic measure of redundancy at node $j$ and is local redundancy

Maximizing $S_j$ would maximize redundancy of node $j$ and is equivalent to maximizing entropy at node $j$. The maximum value of $S_j$ is achieved when all $W_{ij}$s or $q_{ij}/Q_j$s are equal. This occurs when all $q_{ij}$s are equal.

For the entire water distribution network, redundancy is a function of redundancies $S_j$s of individual nodes in the network. The overall network redundancy can be assessed using two approaches:

**Approach 1:** The network redundancy can be assessed by the relative importance of a link to its node and its importance is recognized by $q_{ij}/Q_j$. In this case, the redundancy is maximized at each node. It may, however, be noted that the network redundancy is not a sum of nodal redundancies.

**Approach 2:** The network redundancy can be assessed by the relative importance of a link to the total flow and its importance is recognized by $q_{ij}/Q_0$. Here the proposition is that the importance of a link relative to the local flow is not as important as it is to the total flow. In this case also, the network redundancy is not a sum of nodal redundancies. In order to acknowledge the relative importance of a link to the entire network, Awumah et al. (1990) suggested that $q_{ij}/Q_j$ should be replaced by $q_{ij}/Q_0$ in Equation 14.6. Then, the nodal redundancy $S_{j*}$ can be expressed as

$$S_{j*} = \frac{1}{m-1}\sum_{i=1}^{n(j)}\left[\frac{q_{ij}}{Q_0} - \left(\frac{q_{ij}}{Q_0}\right)^m\right] = \frac{1}{m-1}\left[\frac{Q_j}{Q_0} - \sum_{i=1}^{n(j)}\left(\frac{q_{ij}}{Q_0}\right)^m\right] \tag{14.7}$$

It may be noted that $S_{j*}$ given by Equation 14.7 is similar to Equation 14.6. In this case also, the maximum value of $S_{j*}$ will occur when the $q_{ij}$ values are equal at the $j$th each node. It can also be shown that the maximum network redundancy will be achieved when all the $q_{ij}$ values are equal. It may, however, be noted that

$$S_{j*} = \frac{1}{m-1}\sum_{i=1}^{n(j)}\left[\frac{q_{ij}}{Q_0} - \left(\frac{q_{ij}}{Q_0}\right)^m\right] \neq \frac{1}{m-1}\left[1 - \sum_{i=1}^{n(j)}\left(\frac{q_{ij}}{Q_0}\right)^{m-1}\right] \tag{14.8}$$

This is because

$$\sum_{i=1}^{n(j)} q_{ij} = Q_j \neq Q_0 \quad \text{and} \quad \sum_{i=1}^{n(j)} \frac{q_{ij}}{Q_0} \neq 1 \tag{14.9}$$

Therefore, in the second approach Equation 14.7 can be used in the spirit of Tsallis entropy or considering it via partial Tsallis entropy (Niven, 2004).

The network redundancy or reliability, $S_N$, cannot, however, be expressed as the sum of local or nodal redundancies:

$$S_N \neq \sum_{j=1}^{N} S_j \quad \text{or} \quad S_{N*} \neq \sum_{j=N}^{N} S_{j*} \tag{14.10}$$

where $S_{j*}$ is the network redundancy using the second approach. What is important here is the consideration of the relative significance of links incident upon a node as opposed to the simple consideration of individual redundancies in the network. In this chapter both approaches are employed for assessing the overall network redundancy. From now onward, approach two will be denoted by subscript *.

The network redundancy for $N$ nodes is a function of redundancies of individual nodes, $S_j$s, in the network but will not be a simple summation of these nodal redundancies because of the nonextensive property of the Tsallis entropy. Therefore, it is now important to discuss the additivity property.

## 14.5  ADDITIVITY PROPERTY FOR INDEPENDENT SYSTEMS FOR FIRST APPROACH

In order to illustrate the additivity property of the Tsallis entropy for independent systems, let there be three independent systems, $A$, $B$, and $C$, with ensembles of configurational possibilities $E^A = \{1, 2,..., i,..., M\}$, $E^B = \{1, 2,..., j,..., N\}$, and $E^C = \{1, 2,..., k,..., R\}$, respectively, and the corresponding probabilities as $P^A = \{p_i^A, i = 1,2,...,M\}$, $P^B = \{p_j^B, j = 1,2,...,N\}$, and $P^C = \{p_k^C, k = 1,2,...,R\}$.

Then, one needs to deal with $A \cup B \cup C$ and the corresponding ensembles of possibilities $E^{A \cup B \cup C} = \{(1, 1, 1), (1, 2, 2),..., (i, j, k),..., (M, N, R)\}$. Let the corresponding probabilities be denoted as $p_{ijk}^{A \cup B \cup C}$. Because systems $A$, $B$, and $C$ are independent, one can write

$$p_{ijk}^{A \cup B \cup C} = p_i^A p_j^B p_k^C, \quad \forall (i, j, k) \tag{14.11}$$

Therefore,

$$\sum_{i,j,k}^{M,N,R} \left( p_{ijk}^{A \cup B \cup C} \right)^m = \left[ \sum_{i=1}^{M} \left( p_i^A \right)^m \right]\left[ \sum_{j=1}^{N} \left( p_j^B \right)^m \right]\left[ \sum_{k=1}^{R} \left( p_k^C \right)^m \right] \tag{14.12}$$

Taking the logarithm of Equation 14.12, one gets

$$\log\left[\sum_{i,j,k}^{M,N,R}\left(p_{ijk}^{A\cup B\cup C}\right)^m\right] = \log\left[\sum_{i=1}^{M}\left(p_i^A\right)^m\right] + \log\left[\sum_{j=1}^{N}\left(p_j^B\right)^m\right] + \log\left[\sum_{k=1}^{R}\left(p_k^C\right)^m\right]$$

(14.13)

Each term of Equation 14.13 is now considered:

$$\log\left[\sum_{i,j,k}^{M,N,R}\left(p_{ijk}^{A\cup B\cup C}\right)^m\right] = \log\left\{1 - \frac{(m-1)\left[1-\sum_{i=1}^{M}\left(p_{ijk}^{A\cup B\cup C}\right)^m\right]}{(m-1)}\right\}$$

$$= \log[1-(m-1)S_{A\cup B\cup C}]$$

(14.14)

Similarly,

$$\log\left[\sum_{i=1}^{M}\left(p_i^A\right)^m\right] = \log[1-(m-1)S_A]$$

(14.15)

$$\log\left[\sum_{j=1}^{M}\left(p_j^B\right)^m\right] = \log[1-(m-1)S_B]$$

(14.16)

$$\log\left[\sum_{k=1}^{R}\left(p_k^C\right)^m\right] = \log[1-(m-1)S_C]$$

(14.17)

Therefore,

$$\log[1-(m-1)S_{A\cup B\cup C}] = \log[1-(m-1)S_A] + \log[1-(m-1)S_B] + \log[1-(m-1)S_C]$$

(14.18)

Equation 14.18 can be recast as

$$1-(m-1)S_{A\cup B\cup C} = [1-(m-1)S_A][1-(m-1)S_B][1-(m-1)S_C]$$ (14.19)

This can be simplified as

$$1-(m-1)S_{A\cup B\cup C} = [1-(m-1)S_A-(m-1)S_B+(m-1)^2S_AS_B][1-(m-1)S_C]$$ (14.20)

or

$$1 - (m-1)S_{A \cup B \cup C} = 1 - (m-1)S_A - (m-1)S_B - (m-1)S_C + (m-1)^2 S_A S_B$$
$$\quad + (m-1)^2 S_A S_C + (m-1)^2 S_B S_C - (m-1)^3 S_A S_B S_C \qquad (14.21)$$

Equation 14.21 exhibits a pattern and can be written as

$$S_{A \cup B \cup C} = S_A + S_B + S_C - (m-1)[S_A S_B + S_A S_C + S_B S_C] + (m-1)^2 S_A S_B S_C \qquad (14.22)$$

Denoting $A$, $B$, and $C$ by 1, 2, and 3, respectively, Equation 14.22 can be written in compact form as

$$S_{1 \cup 2 \cup 3} = \sum_{1 \le j \le 3} S_j + (1-m) \sum_{1 \le j1 < j2 \le 3} S_{j1} S_{j2} + (1-m)^2 \sum_{1 \le j1 < j2 < j3 \le 3} S_{j1} S_{j2} S_{j3} \qquad (14.23)$$

Equation 14.22 can be generalized to any number of independent systems. First, consider two cases. If there are two systems 1 and 2 with Tsallis entropy $S^1$ and $S^2$, then the joint entropy can be written as

$$S_{1 \cup 2} = S_1 + S_2 - (m-1)S_1 S_2 \qquad (14.24)$$

or

$$S_{1 \cup 2} = \sum_{1 \le j \le 2} S_j + (1-m) \sum_{1 \le j1 < j2 \le 2} S_{j1} S_{j2} \qquad (14.25)$$

If there are four independent systems with Tsallis entropies as $S_1$, $S_2$, $S_3$, and $S_4$, then the joint entropy can be expressed as

$$S_{1 \cup 2 \cup 3 \cup 4} = S_1 + S_2 + S_3 + S_4 - (m-1)[S_1 S_2 + S_1 S_3 + S_1 S_4 + S_2 S_3 + S_2 S_4 + S_3 S_4]$$
$$\quad + (m-1)^2 [S_1 S_2 S_3 + S_1 S_3 S_4 + S_1 S_2 S_4 + S_2 S_3 S_4] - (m-1)^3 S_1 S_2 S_3 S_4 \qquad (14.26)$$

Equation 14.22 can be generalized to any number of independent systems. Let these systems be represented as $A_i, i = 1, 2, \ldots, n$. Then, the joint entropy can be written in compact form as

$$S_{1 \cup 2 \cup \cdots \cup n} = \sum_{1 \le j \le n} S_j + (1-m) \sum_{1 \le j1 < j2 \le n} S_{j1} S_{j2} + (1-m)^2 \sum_{1 \le j1 < j2 < j3 \le n} S_{j1} S_{j2} S_3 + \cdots$$
$$\quad + (1-m)^{n-1} \sum_{1 \le j1 < j2 < \cdots < jn \le n} S_{j1} S_{j2} \cdots S_{jn} \qquad (14.27)$$

It can be shown that the network redundancy (with $N$ nodes) can be expressed as

$$S_{1 \cup 2 \cup \cdots \cup N} = \sum_{1 \le j \le N} S_j + (1-m) \sum_{1 \le j1 < j2 \le N} S_{j1} S_{j2} + (1-m)^2 \sum_{1 \le j1 < j2 < j3 \le N} S_{j1} S_{j2} S_{j3} + \cdots$$

$$+ (1-m)^{N-1} \sum_{1 \le j1 < j2 \cdots < jN \le N} S_{j1} S_{j2} \cdots S_{jN} \qquad (14.28)$$

where
  $S_j$ is the redundancy of node $j$
  $N$ is number of nodes

In order to develop an appreciation for Equation 14.28, it will be instructive to expand Equation 14.28 in terms of flow quantities. The first term on the right side can be expressed as

$$\sum_{1 \le j \le N} S_j = \sum_{j=1}^{N} \frac{1}{m-1} \left\{ 1 - \sum_{i=1}^{n(j)} \left( \frac{q_{ij}}{Q_j} \right)^m \right\} \qquad (14.29)$$

Note that the nodal redundancy is modified by the factor $(Q_j/Q_0)^m$.

The second term on the right side of Equation 14.28 will be the sum of combinations of two node entropies as

$$-(m-1) \sum_{1 \le j1 < j2 \le N}^{N} S_{j1} S_{j2} = -(m-1)[S_1 S_2 + S_1 S_3 + \cdots + S_1 S_N + S_2 S_3 + \cdots + S_{N-1} S_N] \qquad (14.30)$$

Consider only one term: $-(m-1)S_1 S_2$, which can be written as

$$-(m-1)S_1 S_2 = -\left\{ 1 - \sum_{i=1}^{n(1)} \left( \frac{q_{ij}}{Q_j} \right)^m \right\} \left\{ 1 - \sum_{i=1}^{n(2)} \left( \frac{q_{ij}}{Q_j} \right)^m \right\} \qquad (14.31)$$

In a similar manner, other terms can be expanded.

## 14.6 ADDITIVITY PROPERTY OF TSALLIS ENTROPY FOR INDEPENDENT SYSTEMS FOR APPROACH TWO

In this approach, Equations 14.13 through 14.15 are valid. It may be noted that $p_{ij} = q_{ij}/Q_0$ and unlike in approach one, $\sum_{i=1}^{n(j)} p_i \ne 1$. First, we consider three independent systems, $A$, $B$, and $C$, as before. Each term of Equation 14.15 is now considered:

$$\log\left[\sum_{i=1}^{M}\left(p_i^A\right)^m\right] = \log\left[\frac{Q_A}{Q_0} - (m-1)S_{A*}\right] \qquad (14.32)$$

$$\log\left[\sum_{i=1}^{N}\left(p_i^B\right)^m\right] = \log\left[\frac{Q_B}{Q_0} - (m-1)S_{B*}\right] \qquad (14.33)$$

$$\log\left[\sum_{i=1}^{R}\left(p_i^C\right)^m\right] = \log\left[\frac{Q_C}{Q_0} - (m-1)S_{C*}\right] \qquad (14.34)$$

$$\log\left[\sum_{i,j,k}^{M,N,R}\left(p_{ijk}^{A\cup B\cup C}\right)^m\right] = \log\left[\frac{Q_A Q_B Q_C}{Q_0^{\;3}} - (m-1)S_{*A\cup B\cup C}\right] \qquad (14.35)$$

where $Q_A$, $Q_B$, and $Q_C$ denote, respectively, flow of systems $A$, $B$, and $C$; $S_{A*}$, $S_{B*}$, and $S_{C*}$ denote, respectively, redundancies of systems $A$, $B$, and $C$. Therefore,

$$\left[\frac{Q_A Q_B Q_C}{Q_0^{\;3}} - (m-1)S_{*A\cup B\cup C}\right] = \left[\frac{Q_A}{Q_0} - (m-1)S_{A*}\right]\left[\frac{Q_B}{Q_0} - (m-1)S_{B*}\right]\left[\frac{Q_C}{Q_0} - (m-1)S_{C*}\right]$$

$$(14.36)$$

Equation 14.36 can be recast as

$$S_{*A\cup B\cup C} = \frac{1}{Q_0^2}[Q_B Q_C S_{A*} + Q_A Q_C S_{B*} + Q_A Q_B S_C] - \frac{(m-1)}{Q_0^2}(m-1)[Q_C S_{A*} S_{B*}$$

$$+ Q_B S_{A*} S_{C*} + Q_A S_{B*} S_{C*}] + (m-1)^2 S_{A*} S_{B*} S_{C*} \qquad (14.37)$$

One can see in Equation 14.37 a pattern emerging and hence it can be generalized pot any number of systems.

Now we consider two cases. If there are two systems A and B with Tsallis entropy $S_{A*}$ and $S_{B*}$, then the joint Tsallis entropy–based redundancy can be written as

$$S_{*A\cup B} = \frac{1}{Q_0}(Q_B S_{A*} + Q_A S_{B*}) - (m-1)S_{A*}S_{B*} \qquad (14.38)$$

For generalization, let there be $N$ systems, denoted as 1, 2, …, $N$. Then, Equation 14.37 can be generalized as

$$S_{1 \cup 2 \cup 3 \cup 4 \cdots \cup N}$$

$$= (m-1)^{N-1} \prod_{j=1}^{N} S_{j*} - \frac{(m-1)^{N-2}}{Q_0} [Q_1 S_{2*} \cdots S_{N*} + Q_2 S_{1*} S_{3*} \cdots Q_N + \cdots Q_N S_{1*} S_{2*}]$$

$$+ \cdots + \frac{(m-1)}{Q_0^{N-1}} [Q_1 Q_2 \cdots Q_{N-1} S_N + Q_1 Q_2 \cdots Q_{N-2} S_{N-1} + \cdots + Q_2 Q_2 \cdots Q_N Q_1]$$

(14.39)

It can be seen that Equation 14.39 for network redundancy for approach two is significantly different from Equation 14.28 for approach one.

## 14.7 TRANSMISSION OF REDUNDANCY THROUGH NETWORK

In a water distribution network, nodes are connected to one another. Failure of one link affects not only the node it is incident upon, but also the downstream nodes, because the links upstream of the node service the downstream nodes via redistribution of flows. Consider, for example Figure 14.2, which shows a simple network in which one node receives flow from another. In this network, node 1 has three independent paths or links and hence has some degree of redundancy. Node 2 receives flow from node 1 and an independent path or link and hence has some redundancy due to both sources of flow. Because there is some redundancy in node 1, it may be transferred to node 2. Likewise, node 3 will have some redundancy due to the redundancy in node 2 and hence indirectly due to that in node 1. Intuitively, one can estimate the redundancy at node 2 by the proportion of flow coming from node 1 to the total flow coming into node 1. The implication here is that any remainder of flow at node 1 will be transmitted to the downstream node 2 and hence some to node 3. Thus, redundancy from node 2 will be transmitted to node 3 directly in proportion to the ratio of the total flow entering node 3 from node 2 to the total flow entering node 2. This ratio of flows defines what Awumah et al. (1991) called transmissivity. More precisely, transmissivity can be defined empirically as the ratio of flow through

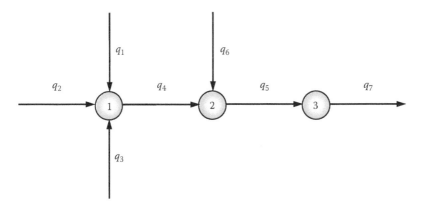

**FIGURE 14.2** A simple network.

the link to the total flow into node at the upstream end of the link. In this sense, redundancy in one area of the network impacts the redundancy in another.

In order to determine the propagation of redundancy in one node upstream to another node downstream, one can define the percentage of the redundancy at the upstream node that is transmitted to the downstream node and approximate it by the ratio of flow coming from that upstream node into the downstream node to the total flow entering the upstream node. Consider two nodes: upstream node $k$ and downstream node $j$. Then the transmissivity of the connection between these two nodes from node $k$ to node $j$ can be expressed as

$$t_{jk} = \frac{q_{kj}}{\sum_{i=1}^{n_k} q_{ik}} \tag{14.40}$$

where
$t_{jk}$ is the transmissivity from node $k$ to node $j$
$q_{kj}$ is the flow in link $k$ incident on node $j$
the denominator is the sum of flows into node $k$

Note that flow is positive toward node $k$. The entropy-based measure can now be extended to include this consideration of transmissivity.

For approach one, the measure defined by Equation 14.6 can be expressed as

$$S'_j = S_j + \sum t_{jk} S'_k \tag{14.41}$$

where
$S'_j$ is the measure of the total (global) redundancy at node $j$
$S'_k$ is the redundancy at node $k$

Similarly, for approach two the measure defined by Equation 14.7 can expressed as

$$S'_{j*} = S_{j*} + \sum_{*} t_{jk} S'_{k*} \tag{14.42}$$

where
$S'_{j*}$ is the global redundancy at $j$
$S'_{k*}$ is the redundancy at node $k$ for partial Tsallis entropy

Equations 14.41 and 14.42 show that the global redundancy at node $j$ is the sum of local redundancy at the node and the contribution from upstream supplies to the redundancy performance of that node. It should be noted that the global redundancy at each node depends on the global redundancy of all upstream nodes. Therefore, it may be necessary to apply Equation 14.40 recursively with distance or with the number of nodes distant from the source. In this recursive manner, redundancy of a particular node due to the redundancy at all upstream nodes is included in the redundancy measure of that node.

To illustrate the transmissivity concept for both approaches, consider the network in Figure 14.2. For approach one, the global redundancy at node 1 (using simple notation of the figure) can be expressed as

$$S_1' = S_1 = \frac{1}{(m-1)} \left[ 1 - \sum_{i=1}^{3} \left( \frac{q_i}{Q_1} \right)^m \right], \quad Q_1 = \sum_{i=1}^{3} q_i \tag{14.43}$$

For node 2,

$$S_2' = S_2 + t_{21} S_1' \tag{14.44}$$

where $t_{21}$ is the transmissivity from node 1 to node 2 and is given by

$$t_{21} = \frac{q_4}{q_1 + q_2 + q_3} \tag{14.45}$$

Therefore,

$$S_2' = \frac{1}{(m-1)} \left[ \frac{q_4}{q_4 + q_6} - \left( \frac{q_4}{q_4 + q_6} \right)^m + \left( \frac{q_6}{q_4 + q_6} \right) - \left( \frac{q_6}{q_4 + q_6} \right)^m \right] + \frac{q_4}{q_1 + q_2 + q_3} S_1' \tag{14.46}$$

Similarly, for node 3

$$S_3' = S_3 + t_{32} S_2' \tag{14.47}$$

where $t_{32}$ is the redundancy between nodes 2 and 3 and can be written as

$$t_{32} = \frac{q_5}{q_4 + q_6} \tag{14.48}$$

Note that transmissivities will always be less than or equal to unity.

Now for approach two, the global redundancy at node 1 (using simple notation of the figure) can be expressed as

$$S_{1*}' = S_{1*} = \frac{1}{(m-1)} \left[ \sum_{i=1}^{3} \frac{q_i}{Q_0} - \left( \frac{q_i}{Q_0} \right)^m \right], \quad Q_0 = \sum_{j=1}^{N} Q_j \tag{14.49}$$

For node 2,

$$S_{2*}' = S_{2*} + t_{21} S_{1*}' \tag{14.50}$$

where $t_{21}$ is the transmissivity from node 1 to node 2. The global redundancy at node 2 is given as

$$S'_{2*} = \frac{1}{(m-1)}\left[\sum_{i=1}^{2} \frac{q_i}{Q_0} - \left(\frac{q_i}{Q_0}\right)^m\right] + \frac{q_4}{Q_0} S'_{1*} \qquad (14.51)$$

Similarly, for node 3

$$S'_{3*} = S_{3*} + t_{32}S'_{2*} \quad t_{32} = \frac{q_5}{Q_0} \qquad (14.52)$$

where $t_{32}$ is the redundancy between nodes 2 and 3. Note that transmissivities will always be less than or equal to unity.

## 14.8   CONSIDERATION OF PATH DEPENDENCY

When a link connecting to a node fails, alternative paths that supply water to the node may originate some distance away in the immediate vicinity of the failed link. The number of these alternative paths significantly affects the network redundancy and reliability. Thus far, it has been implicitly assumed that the number of alternative paths from the source to a demand node (point) is the same as the number of links incident on the node. This assumption is not always realistic. The contribution to the redundancy at a node by one of its incident links depends therefore on both the percentage of flow it brings to the node but also the number of paths between the supply source and the node via that link. A network has redundancy because even if one of the links fails, some nodes may continue to operate uninterrupted. Of course, the redundancy would depend on the failure of a particular link and its contribution to flow at a given node.

The redundancy measure may incorporate the contribution of links to the node through what Awumah (1990) called path parameter. Let the path parameter for node $j$ be $a_j$, which is considered to be equal to the number of alternative independent paths between the source and node $j$. The number of these paths depends on the degree of overlap between paths. It is essential to know the number of links used by different paths. Thus, the total number of independent paths may be less than the total number of all paths. In the case of dependent paths, the effective alternative independent number of paths from the given number of dependent paths can be derived.

Following Awumah (1990), let the number of paths to which a link belongs define the degree of that link. If different paths have no common links then each link in these independent paths has one degree. If a link is common between two paths, then it has a degree of dependency of one unit. If the link is shared by three paths then it has a degree of dependency of two units. If the degree of link is denoted by $d_l$ then its degree of dependency $D_l$ is given as

$$D_l = d_l - 1 \qquad (14.53)$$

If the number of alternative dependent paths from the source to the given node is $n_d$, then the effective number of independent paths can be obtained by removing dependencies from the links. The required path parameter $a_j$, which is the adjusted number of independent paths can be expressed for node $j$ as

$$a_j = n_d \left[ \frac{\sum_{i=1}^{M} d_i - \sum_{i=1}^{M} D_i}{\sum_{i=1}^{M} d_i} \right] \tag{14.54}$$

where $M$ is the number of links in the $n_d$ number of paths. Equation 14.54 can be written as

$$a_j = n_d \left[ 1 - \frac{\sum_{i=1}^{M} D_i}{\sum_{i=1}^{M} d_i} \right] \tag{14.55}$$

The term within brackets is the factor that reduces the number of dependent paths to the equivalent number of independent paths. For independent paths, term $\sum_{l=1}^{M} D_i = 0$, and hence $a_j = n_d$.

Equation 14.54 shows that the value of $a_j$ decreases with increasing dependency. In the case of network in Figure 14.2, node 2 has three independent paths from the source wherein link between nodes 1 and 2 is common to these independent paths. The degree of this link is three and its degree of dependency is two. For node 2, the value of path parameter then becomes 3. Similarly, for node 3, there are three independent paths. The degree of link connecting nodes 2 and 3 is 3 and the degree of dependence is 2. For this node, the value of path parameter then becomes 3.18.

## 14.9  MODIFICATION OF REDUNDANCY MEASURE WITH PATH PARAMETER

The nodal redundancy measure given by Equation 14.6 can now be modified by including the path parameter as

$$S_j = \frac{1}{m-1} \sum_{i=1}^{n(j)} \left[ \frac{q_{ij}}{a_{ij} Q_j} - \left( \frac{q_{ij}}{a_{ij} Q_j} \right)^{m-1} \right] \tag{14.56}$$

Note that the objective is to increase the basic redundancy measure if the number of independent paths between the source and the node is greater than one. Likewise, for approach two, the nodal redundancy measure can be modified as

$$S_{j*} = \frac{1}{m-1} \sum_{i=1}^{n(j)} \left[ \frac{q_{ij}}{a_{ij} Q_0} - \left( \frac{q_{ij}}{a_{ij} Q_0} \right)^{m-1} \right] \tag{14.57}$$

It may be interesting to note the usefulness of Equations 14.56 and 14.57. First, in the case of nodes with one incident links but having several paths through the network upstream of the single incident link, if the equivalent number of paths exceeds one, then $a_{ij} > 0$ and the second term will make a positive contribution to the redundancy of the node. Second, for nodes with two or more incident links where each link is equal to one path from the source to the node, the term will cease to have a significance. Third, for nodes with several incident links so that equivalent paths through some of these links are less than one, $a_{ij}$ would be less than one and the second term would be negative. Then the redundancy measure would less than that given by Equation 14.56 for approach one and less than that given by Equation 14.57 for approach two. The path parameters would still be less than one, since it measures total equivalent paths, not the value for a particular link.

## 14.10 MODIFICATION OF REDUNDANCY MEASURE BY AGE FACTOR

Let $u_{ij}$ be the age factor parameter for the pipe material in link $ij$. This reflects the degree of deterioration of the pipe with age and the consequent reduction in carrying capacity and its contribution to redundancy. If the Hazen–Williams formula for flow through pipes is used then its friction coefficient $C_{ij}$ reflects the characteristic of the pipe material as well as its age. Its value ranges from 100 to 150. For example, for steel and plastic pipes, it is between 140 and 150, and for bricks it is around 100. For cast iron pipes the $C$ values can degrade from about 130 to 75 over a period of 50 years. Awumah (1990) expressed

$$u_{ij} = 0.2 \ln C_{ij}(t) \tag{14.58}$$

where $t$ is time after installation of pipes in years.

Awumah (1990) used $C = 150$ as the upper reference limit and scaled down all values therefrom. The reference point value for the age factor parameter from Equation 14.58 is $\ln(150) = 5.0$. Dividing the parameter by 5.0 so that the age factor parameter for pipes with the Hazen–Williams friction coefficient $C_{ij} = 150$ becomes unity leads to the scale factor of 0.2.

The Tsallis entropy–based redundancy Equation 14.6 can be modified for approach one, as

$$S_j = \frac{1}{m-1} \sum_{i=1}^{n(j)} u_{ij} \left[ \frac{q_{ij}}{Q_j} - \left( \frac{q_{ij}}{Q_j} \right)^{m-1} \right] \tag{14.59}$$

and Equation 14.7 for approach two as

$$S_j = \frac{1}{m-1} \sum_{i=1}^{n(j)} u_{ij} \left[ \frac{q_{ij}}{Q_0} - \left( \frac{q_{ij}}{Q_0} \right)^{m-1} \right] \tag{14.60}$$

where $u_{ij}$ is the age factor for link $ij$ incident on node $j$.

The redundancy measure incorporating both the path parameter and the age factor can be expressed by modifying Equation 14.59 as

$$S_j = \frac{1}{m-1} \sum_{i=1}^{n(j)} u_{ij} \left[ \frac{q_{ij}}{a_{ij}Q_j} - \left( \frac{q_{ij}}{a_{ij}Q_j} \right)^{m-1} \right] \tag{14.61}$$

and modifying Equation 14.60

$$S_j = \frac{1}{m-1} \sum_{i=1}^{n(j)} u_{ij} \left[ \frac{q_{ij}}{a_{ij}Q_0} - \left( \frac{q_{ij}}{a_{ij}Q_0} \right)^{m-1} \right] \tag{14.62}$$

## 14.11 MODIFICATION OF OVERALL NETWORK REDUNDANCY

The overall network redundancy, given by Equation 14.27 for approach one, can be modified by incorporating the age factor $u_{ij}$ and path parameter $a_{ij}$ as

$$
\begin{aligned}
\hat{S} = &\sum_{1 \le j \le N} \frac{1}{m-1} \sum_{j=1}^{N} \left[ 1 - \sum_{i=1}^{n(j)} u_{ij} \left( \frac{q_{ij}}{a_{ij}Q_j} \right)^{m} \right] \\
&- \frac{1}{(m-1)} \sum_{1 \le j1 < j2 \le N} \sum_{j1=1}^{N} \left[ 1 - \sum_{i=1}^{n(j)} u_{ij1} \left( \frac{q_{ij1}}{a_{ij1}Q_j} \right)^{m} \right] \sum_{j2=1}^{N} \left[ 1 - \sum_{i=1}^{n(j)} u_{ij2} \left( \frac{q_{ij2}}{a_{ij2}Q_j} \right)^{m} \right] \\
&+ \frac{1}{m-1} \sum_{1 \le j1 < j2 < j3 \le N} \sum_{j1=1}^{N} \left[ 1 - \sum_{i=1}^{n(j)} u_{ij1} \left( \frac{q_{ij1}}{a_{ij1}Q_j} \right)^{m} \right] \sum_{j2=1}^{N} \left[ 1 - \sum_{i=1}^{n(j)} u_{ij2} \left( \frac{q_{ij2}}{a_{ij2}Q_j} \right)^{m} \right] \\
&\times \sum_{j3=1}^{N} \left[ 1 - \sum_{i=1}^{n(j)} u_{ij3} \left( \frac{q_{ij3}}{a_{ij3}Q_j} \right)^{m} \right] + \cdots \\
&+ (-1)^{N-1} \frac{1}{m-1} \sum_{1 \le j1 < j2 < \cdots jN \le N} \sum_{j1=1}^{N} \left[ 1 - \sum_{i=1}^{n(j)} u_{ij1} \left( \frac{q_{ij1}}{a_{ij1}Q_j} \right)^{m} \right] \\
&\times \sum_{j2=1}^{N} \left[ 1 - \sum_{i=1}^{n(j)} u_{ij2} \left( \frac{q_{ij2}}{a_{ij2}Q_j} \right)^{m} \right] \cdots \sum_{jN=1}^{N} \left[ 1 - \sum_{i=1}^{n(j)} u_{ijN} \left( \frac{q_{ijN}}{a_{ijN}Q_j} \right)^{m} \right]
\end{aligned}
\tag{14.63}
$$

The term within brackets represents the contribution from node $j$ to the network redundancy:

$$\tilde{S}_j = \frac{1}{m-1} \sum_{i=1}^{n(j)} u_{ij} \left[ \frac{q_{ij}}{a_{ij}Q_j} - \left( \frac{q_{ij}}{a_{ij}Q_j} \right)^{m-1} \right] \quad (14.64)$$

The contribution from a node can be decomposed as

$$\tilde{S}_j = \frac{1}{m-1} \left[ 1 - \sum_{i=1}^{n(j)} u_{ij} \left( \frac{q_{ij}}{a_{ij}Q_j} \right)^m \right] \quad (14.65)$$

Let the age factor be accommodated on a nodal basis, that is, $u_{ij} = u_j$ and

$$n(j)u_j = U_j \quad (14.66)$$

Equation 14.65 can be modified as

$$\tilde{S}_j = \frac{1}{m-1} \left[ 1 - U_j \sum_{i=1}^{n(j)} \left( \frac{q_{ij}}{a_{ij}Q_j} \right)^m \right] \quad (14.67)$$

Summing Equation 14.67 over the $N$ nodes yields the overall network redundancy:

$$\hat{S} = \sum_{1 \le j \le N} \frac{1}{m-1} \sum_{j=1}^{N} \left[ 1 - U_j \sum_{i=1}^{n(j)} \left( \frac{q_{ij}}{a_{ij}Q_j} \right)^m \right] - \frac{1}{(m-1)} \sum_{1 \le j1 < j2 \le N} \sum_{j1=1}^{N} \left[ 1 - U_{j1} \sum_{i=1}^{n(j)} \left( \frac{q_{ij1}}{a_{ij1}Q_j} \right)^m \right]$$

$$\times \sum_{j2=1}^{N} \left[ 1 - U_{j2} \sum_{i=1}^{n(j)} \left( \frac{q_{ij2}}{a_{ij2}Q_j} \right)^m \right] + \frac{1}{(m-1)} \sum_{1 \le j1 < j2 < j3 \le N} \sum_{j1=1}^{N} \left[ 1 - U_{j1} \sum_{i=1}^{n(j)} \left( \frac{q_{ij1}}{a_{ij1}Q_j} \right)^m \right]$$

$$\times \sum_{j2=1}^{N} \left[ 1 - U_{j2} \sum_{i=1}^{n(j)} \left( \frac{q_{ij2}}{a_{ij2}Q_j} \right)^m \right] \sum_{j3=1}^{N} \left[ 1 - U_{j3} \sum_{i=1}^{n(j)} \left( \frac{q_{ij3}}{a_{ij3}Q_j} \right)^m \right] + \cdots$$

$$+ (-1)^{N-1} \frac{1}{(m-1)} \sum_{1 \le j1 < j2 < \cdots < jN \le N} \sum_{j1=1}^{N} \left[ 1 - U_{j1} \sum_{i=1}^{n(j)} \left( \frac{q_{ij1}}{a_{ij1}Q_j} \right)^m \right]$$

$$\times \sum_{j2=1}^{N} \left[ 1 - U_{j2} \sum_{i=1}^{n(j)} \left( \frac{q_{ij2}}{a_{ij2}Q_j} \right)^m \right] \cdots \sum_{jN=1}^{N} \left[ 1 - U_{jN} \sum_{i=1}^{n(j)} \left( \frac{q_{ijN}}{a_{ijN}Q_j} \right)^m \right] \quad (14.68)$$

Equation 14.68 is similar to Equation 14.63, except for the second term that includes $U_j$—the sum of age factor parameters of the links incident on node $j$. The second term is similar to the expression for useful entropy. Thus, the overall network redundancy is the sum of weighted nodal useful entropies and the useful entropies among nodes.

For approach two, the overall network redundancy, given by Equation 14.37, is modified by incorporating the transmission of redundancy expressed by Equation 14.42, path dependency by Equation 14.58 and age factor by Equation 14.60. For brevity, the network redundancy is expressed in terms of nodal redundancies rather than flows as

$$\tilde{S}_* = (m-1)^{N-1} \prod_{j=1}^{N} S_{j*}$$

$$- \frac{(m-1)^{N-2}}{Q_0} [Q_1 S_{2*} \cdots S_{N*} + Q_2 S_{1*} S_{3*} \cdots S_{N*} + \cdots + Q_N S_{1*} S_{2*} \cdots S_{N-1*}]$$

$$+ \cdots + \frac{(m-1)}{Q_0^{N-1}} [Q_1 Q_2 \cdots Q_{N-1} S_{N*} + Q_1 Q_2 \cdots Q_{N-2} S_{N-1*} + \cdots + Q_2 Q_2 \cdots Q_N Q_1]$$

$$(14.69)$$

In real world, there are distribution networks that have several source nodes to serve the demand nodes. The procedure for redundancy calculation will, however, remain the same. Since all nodes are interconnected, each node upstream of another node is a source node to the node downstream to it. Incident links are counted as those links connected to the upstream nodes, which are source nodes to the downstream node for which redundancy is determined. In case of multiple sources connected to a demand node by a single link, then this node may have multiple incident links and will have nonzero redundancy measure. For approach one, Equations 14.64 or 14.63 for single or multiple links can be used for redundancy measure. Likewise, Equation 14.68 can be employed for approach two.

Another aspect that occurs in real world is that flows in links are not fixed but vary with time in response to changing demands at nodes. The question arises: which flow pattern yields the redundancy measure. To answer this question entails defining the flow pattern. One may compute the redundancy measure for peak demand pattern or average flow pattern. The method of computation, however, remains unaltered.

### Example 14.1

Following Xu and Jowitt (1992), consider three simple distribution network layouts, as shown in Figure 14.3. The demand at point (node) $A$ is one unit and that at node $B$ is 10 units. In configuration 2, the demand at point $B$ is supplied via node $A$. Compute the redundancy of the three layouts.

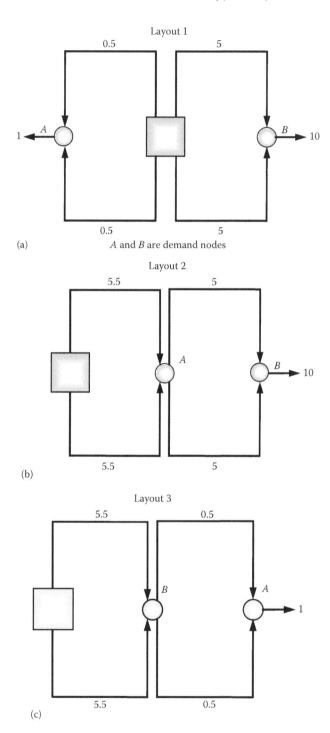

**FIGURE 14.3**  Three simple distribution network layouts a, b, and c.

## Solution

Only approach one is employed here. First, layout 1 is considered. For the $j$th node

(1)
$$S_j = \frac{1}{m-1}\sum_{i=1}^{n(j)}\left[\frac{q_{ij}}{Q_j} - \left(\frac{q_{ij}}{Q_j}\right)^m\right]$$

It is assumed that $m = 3$.

$$S_A = \frac{1}{3-1}\times 2\times\left[\frac{0.5}{1} - \left(\frac{0.5}{1}\right)^3\right] = 0.38$$

$$S_B = \frac{1}{3-1}\times 2\times\left[\frac{5}{10} - \left(\frac{5}{10}\right)^3\right] = 0.38$$

(2)
$$S_{j^*} = \frac{1}{m-1}\left[1 - \sum_{i=1}^{n(j)}\left(\frac{q_{ij}}{Q_0}\right)^m\right]$$

$$S_{A^*} = \frac{1}{3-1}\times\left[1 - 2\times\left(\frac{0.5}{11}\right)^3\right] = 0.50$$

$$S_{B^*} = \frac{1}{3-1}\times\left[1 - 2\times\left(\frac{5}{11}\right)^3\right] = 0.41$$

(3)
$$S_N = S^{A\cup B} = \sum_{1\le j\le n} S^j + (1-m)\sum_{1\le j1\le j2\le n} S^{jA}S^{jB}$$

$$S^{jA} = \frac{1}{m-1} - \frac{1}{m-1}\sum_{j=1}^{N}\left(\frac{Q_j}{Q_0}\right)^m + \sum_{j=1}^{N}\left(\frac{Q_j}{Q_0}\right)^m$$

$$S_{A^*} = \frac{1}{3-1} - \frac{1}{3-1}\times\left(\frac{1}{11}\right)^3 + \left(\frac{1}{11}\right)^3\times 0.50 = 0.50$$

$$S^{jB} = \frac{1}{m-1} - \frac{1}{m-1}\sum_{j=1}^{N}\left(\frac{Q_j}{Q_0}\right)^m + \sum_{j=1}^{N}\left(\frac{Q_j}{Q_0}\right)^m$$

$$S_{B^*} = \frac{1}{3-1} - \frac{1}{3-1}\times\left(\frac{10}{11}\right)^3 + \left(\frac{10}{11}\right)^3\times 0.41 = 0.43$$

$$\sum_{1\le j\le N} S^j = S^{jA} + S^{jB} = 0.5 + 0.43 = 0.93$$

$$\sum_{1\le j1 < j2\le N} S^{jA}S^{jB} = S^{jA} \times S^{jB} = 0.5 \times 0.43 = 0.21$$

$$S_N = S^{jA\cup jB} = \sum_{1\le j\le n} S^j + (1-m)\sum_{1\le j1 < j2\le n} S^{jA}S^{jB} = 0.93 + (1-3)\times 0.21 = 0.50$$

For layout 2,

$$S_j = \frac{1}{m-1}\sum_{i=1}^{n(j)}\left[\frac{q_{ij}}{Q_j} - \left(\frac{q_{ij}}{Q_j}\right)^m\right]$$

(1)    $$S_A = \frac{1}{3-1}\times 2 \times\left[\frac{5.5}{11} - \left(\frac{5.5}{11}\right)^3\right] = 0.38$$

$$S_B = \frac{1}{3-1}\times 2 \times\left[\frac{5}{10} - \left(\frac{5}{10}\right)^3\right] = 0.38$$

(2)    $$S_{j^*} = \frac{1}{m-1}\left[1-\sum_{i=1}^{n(j)}\left(\frac{q_{ij}}{Q_0}\right)^m\right]$$

$$S_{A^*} = \frac{1}{3-1}\times\left[1-2\times\left(\frac{5.5}{21}\right)^3\right] = 0.48$$

$$S_{B^*} = \frac{1}{3-1}\times\left[1-2\times\left(\frac{5}{21}\right)^3\right] = 0.49$$

(3)    $$S_N = S^{jA\cup jB} = \sum_{1\le j\le n} S^j + (1-m)\sum_{1\le j1 < j2\le n} S^{jA}S^{jB}$$

$$S^{jA} = \frac{1}{m-1} - \frac{1}{m-1}\sum_{j=1}^{N}\left(\frac{Q_j}{Q_0}\right)^m + \sum_{j=1}^{N}\left(\frac{Q_j}{Q_0}\right)^m$$

$$S_{A^*} = \frac{1}{3-1} - \frac{1}{3-1}\times\left(\frac{11}{21}\right)^3 + \left(\frac{11}{21}\right)^3 \times 0.48 = 0.50$$

$$S^{jB} = \frac{1}{m-1} - \frac{1}{m-1} \sum_{j=1}^{N} \left(\frac{Q_j}{Q_0}\right)^m + \sum_{j=1}^{N} \left(\frac{Q_j}{Q_0}\right)^m$$

$$S_{B^*} = \frac{1}{3-1} - \frac{1}{3-1} \times \left(\frac{10}{21}\right)^3 + \left(\frac{10}{21}\right)^3 \times 0.41 = 0.50$$

$$\sum_{1 \leq j \leq N} S^j = S^{jA} + S^{jB} = 0.5 + 0.50 = 1.00$$

$$\sum_{1 \leq j1 < j2 \leq N} S^{jA} S^{jB} = S^{jA} \times S^{jB} = 0.5 \times 0.50 = 0.25$$

$$S_N = S^{jA \cup jB} = \sum_{1 \leq j \leq n} S^j + (1-m) \sum_{1 \leq j1 < j2 \leq n} S^{jA} S^{jB} = 1.00 + (1-3) \times 0.25 = 0.50$$

For layout 3,

(1)
$$S_j = \frac{1}{m-1} \sum_{i=1}^{n(j)} \left[\frac{q_{ij}}{Q_j} - \left(\frac{q_{ij}}{Q_j}\right)^m\right]$$

$$S_A = \frac{1}{3-1} \times 2 \times \left[\frac{0.5}{1} - \left(\frac{0.5}{1}\right)^3\right] = 0.38$$

$$S_B = \frac{1}{3-1} \times 2 \times \left[\frac{5.5}{11} - \left(\frac{5.5}{11}\right)^3\right] = 0.38$$

(2)
$$S_{j^*} = \frac{1}{m-1} \left[1 - \sum_{i=1}^{n(j)} \left(\frac{q_{ij}}{Q_0}\right)^m\right]$$

$$S_{A^*} = \frac{1}{3-1} \times \left[1 - 2 \times \left(\frac{5.5}{12}\right)^3\right] = 0.50$$

$$S_{B^*} = \frac{1}{3-1} \times \left[1 - 2 \times \left(\frac{5.5}{12}\right)^3\right] = 0.40$$

(3)
$$S_N = S^{jA \cup jB} = \sum_{1 \leq j \leq n} S^j + (1-m) \sum_{1 \leq j1 < j2 \leq n} S^{jA} S^{jB}$$

$$S^{jA} = \frac{1}{m-1} - \frac{1}{m-1}\sum_{j=1}^{N}\left(\frac{Q_j}{Q_0}\right)^m + \sum_{j=1}^{N}\left(\frac{Q_j}{Q_0}\right)^m$$

$$S_{A^*} = \frac{1}{3-1} - \frac{1}{3-1}\times\left(\frac{1}{12}\right)^3 + \left(\frac{1}{12}\right)^3 \times 0.50 = 0.50$$

$$S^{jB} = \frac{1}{m-1} - \frac{1}{m-1}\sum_{j=1}^{N}\left(\frac{Q_j}{Q_0}\right)^m + \sum_{j=1}^{N}\left(\frac{Q_j}{Q_0}\right)^m$$

$$S_{B^*} = \frac{1}{3-1} - \frac{1}{3-1}\times\left(\frac{11}{21}\right)^3 + \left(\frac{11}{21}\right)^3 \times 0.40 = 0.43$$

$$\sum_{1\leq j\leq N} S^j = S^{jA} + S^{jB} = 0.5 + 0.43 = 0.93$$

$$\sum_{1\leq j1 < j2\leq N} S^{jA}S^{jB} = S^{jA}\times S^{jB} = 0.5\times 0.43 = 0.21$$

$$S_N = S^{jA\cup jB} = \sum_{1\leq j\leq n} S^j + (1-m)\sum_{1\leq j1 < j2\leq n} S^{jA}S^{jB} = 0.93 + (1-3)\times 0.21 = 0.50$$

## Example 14.2

Consider five simple single node layouts as shown in Figure 14.4. Compute the redundancy of each layout.

**Solution**

For case 1: A

$$S_j = \frac{1}{m-1}\sum_{i=1}^{n(j)}\left[\frac{q_{ij}}{Q_j} - \left(\frac{q_{ij}}{Q_j}\right)^m\right] = \frac{1}{3-1}\times\left[\frac{240}{240} - \left(\frac{240}{240}\right)^3\right] = 0$$

Case 2: B

$$S_j = \frac{1}{m-1}\sum_{i=1}^{n(j)}\left[\frac{q_{ij}}{Q_j} - \left(\frac{q_{ij}}{Q_j}\right)^m\right] = \frac{1}{3-1}\times 2\times\left[\frac{120}{240} - \left(\frac{120}{240}\right)^3\right] = 0.38$$

Case 3: C

$$S_j = \frac{1}{m-1}\sum_{i=1}^{n(j)}\left[\frac{q_{ij}}{Q_j} - \left(\frac{q_{ij}}{Q_j}\right)^m\right] = \frac{1}{3-1}\times\left\{\left[\frac{160}{240} - \left(\frac{160}{240}\right)^3\right] + \left[\frac{80}{240} - \left(\frac{80}{240}\right)^3\right]\right\} = 0.33$$

Case 1: A

Case 2: B

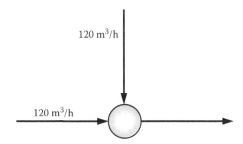

Case 3: C

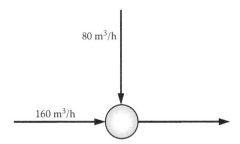

Case 4: D

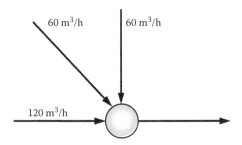

**FIGURE 14.4** Five simple single-node water distribution layouts. *(Continued)*

Case 5: E

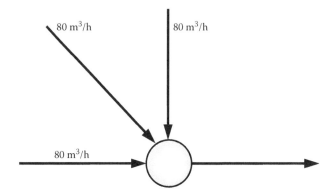

**FIGURE 14.4 (*Continued*)** Five simple single-node water distribution layouts.

Case 4: D

$$S_j = \frac{1}{m-1}\sum_{i=1}^{n(j)}\left[\frac{q_{ij}}{Q_j}-\left(\frac{q_{ij}}{Q_j}\right)^m\right]$$

$$= \frac{1}{3-1}\times\left\{\left[\frac{120}{240}-\left(\frac{120}{240}\right)^3\right]+\left[\frac{60}{240}-\left(\frac{60}{240}\right)^3\right]+\left[\frac{60}{240}-\left(\frac{60}{240}\right)^3\right]\right\}=0.42$$

Case 5: E

$$S_j = \frac{1}{m-1}\sum_{i=1}^{n(j)}\left[\frac{q_{ij}}{Q_j}-\left(\frac{q_{ij}}{Q_j}\right)^m\right]=\frac{1}{3-1}\times 3\times\left[\frac{80}{240}-\left(\frac{80}{240}\right)^3\right]=0.44$$

## Example 14.3

Consider a layout as shown in Figure 14.5. The demand at the source is 2000 m³/h. The demand at each node of the layout is specified as follows:

| Node | Demand (m³/h) | Node | Demand (m³/h) | Node | Demand (m³/h) |
|------|---------------|------|---------------|------|---------------|
| 1 | 2000 | 5 | 175 | 9 | 225 |
| 2 | 150 | 6 | 150 | 10 | 150 |
| 3 | 175 | 7 | 225 | 11 | 175 |
| 4 | 175 | 8 | 225 | 12 | 175 |

Note the sum of demands at nodes equals the demand at the source. Compute the redundancy of each layout, taking into consideration transmissivity, path parameter, and age factor. Consider friction factor as 115.

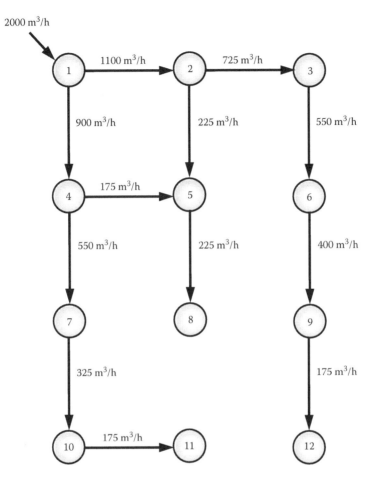

**FIGURE 14.5**   A water distribution network layout.

## Solution

The network redundancy can be expressed as

(1)
$$S_N = \frac{1}{m-1}\left\{\sum_{j=1}^{n(j)}\left[\frac{q_{ij}}{Q_i} - \left(\frac{q_{ij}}{Q_i}\right)^m\right]\right\}$$

Since the number of link incidences equals one for nodes 1, 2, 3, 4, 6, 7, 8, 9, 10, 11, and 12, $S_j = 0$ in this layout. The only node with redundancy is node 5 for which the entropy is computed as

$$S_5 = \frac{1}{m-1}\left\{\sum_{j=1}^{n(j)}\left[\frac{q_{ij}}{Q_j}-\left(\frac{q_{ij}}{Q_j}\right)^m\right]\right\}$$

$$= \frac{1}{3-1}\times\left\{\left[\left(\frac{225}{400}\right)-\left(\frac{225}{400}\right)^3\right]+\left[\left(\frac{175}{400}\right)-\left(\frac{175}{400}\right)^3\right]\right\}=0.3691$$

(2)                    $$S_{j^*} = \frac{1}{m-1}\left[1-\sum_{j=1}^{n(j)}\left(\frac{q_{ij}}{Q_0}\right)^m\right]$$

$$S_{1^*} = 0$$

$$S_{2^*} = \frac{1}{3-1}\times\left[1-\left(\frac{1100}{5525}\right)^3\right]=0.4961$$

$$S_{3^*} = \frac{1}{3-1}\times\left[1-\left(\frac{725}{5525}\right)^3\right]=0.4989$$

$$S_{4^*} = \frac{1}{3-1}\times\left[1-\left(\frac{900}{5525}\right)^3\right]=0.4978$$

$$S_{5^*} = \frac{1}{3-1}\times\left[1-\left(\frac{225}{5525}\right)^3-\left(\frac{175}{5525}\right)^3\right]=0.50$$

$$S_{6^*} = \frac{1}{3-1}\times\left[1-\left(\frac{550}{5525}\right)^3\right]=0.4995$$

$$S_{7^*} = \frac{1}{3-1}\times\left[1-\left(\frac{550}{5525}\right)^3\right]=0.4995$$

$$S_{8^*} = \frac{1}{3-1}\times\left[1-\left(\frac{225}{5525}\right)^3\right]=0.50$$

$$S_{9^*} = \frac{1}{3-1}\times\left[1-\left(\frac{400}{5525}\right)^3\right]=0.4998$$

$$S_{10^*} = \frac{1}{3-1} \times \left[1 - \left(\frac{325}{5525}\right)^3\right] = 0.4999$$

$$S_{11^*} = \frac{1}{3-1} \times \left[1 - \left(\frac{175}{5525}\right)^3\right] = 0.50$$

$$S_{12^*} = \frac{1}{3-1} \times \left[1 - \left(\frac{175}{5525}\right)^3\right] = 0.50$$

(3) Determine $S_N$

$$S_N = S^{1 \cup 2 \cup \cdots \cup 12} = \sum_{1 \le j \le 12} S^j + (1-m) \sum_{1 \le j1 < j2 \le 12} S^{j1} S^{j2} + (1-m)^2 \sum_{1 \le j1 < j2 < j3 \le 12} S^{j1} S^{j2} S^{j3}$$

$$+ (1-m)^{11} \sum_{1 \le j1 < j2 < \cdots < jN \le 12} S^{j1} S^{j2} \cdots S^{j12}$$

Find

$$\sum_{1 \le j \le N} S^j = \frac{N}{m-1} - \frac{1}{m-1} \sum_{j=1}^{N} \left(\frac{Q_j}{Q_0}\right)^m + \sum_{j=1}^{N} \left(\frac{Q_j}{Q_0}\right)^m S_{j^*}$$

$$S^1 = 0$$

$$S^2 = \frac{1}{3-1} - \frac{1}{3-1} \times \left(\frac{1100}{5525}\right)^3 + \left(\frac{1100}{5525}\right)^3 \times 0.4961 = 0.5$$

$$S^3 = \frac{1}{3-1} - \frac{1}{3-1} \times \left(\frac{725}{5525}\right)^3 + \left(\frac{725}{5525}\right)^3 \times 0.8258 = 0.5007$$

$$S^4 = \frac{1}{3-1} - \frac{1}{3-1} \times \left(\frac{900}{5525}\right)^3 + \left(\frac{900}{5525}\right)^3 \times 0.4978 = 0.5$$

$$S^5 = \frac{1}{3-1} - \frac{1}{3-1} \times \left(\frac{400}{5525}\right)^3 + \left(\frac{400}{5525}\right)^3 \times 0.6982 = 0.5001$$

$$S^6 = \frac{1}{3-1} - \frac{1}{3-1} \times \left(\frac{550}{5525}\right)^3 + \left(\frac{550}{5525}\right)^3 \times 1.1260 = 0.5006$$

$$S^7 = \frac{1}{3-1} - \frac{1}{3-1} \times \left(\frac{550}{5525}\right)^3 + \left(\frac{550}{5525}\right)^3 \times 0.8037 = 0.5003$$

$$S^8 = \frac{1}{3-1} - \frac{1}{3-1} \times \left(\frac{225}{5525}\right)^3 + \left(\frac{225}{5525}\right)^3 \times 0.8927 = 0.5$$

$$S^9 = \frac{1}{3-1} - \frac{1}{3-1} \times \left(\frac{400}{5525}\right)^3 + \left(\frac{400}{5525}\right)^3 \times 1.3187 = 0.5003$$

$$S^{10} = \frac{1}{3-1} - \frac{1}{3-1} \times \left(\frac{325}{5525}\right)^3 + \left(\frac{325}{5525}\right)^3 \times 0.9748 = 0.5001$$

$$S^{11} = \frac{1}{3-1} - \frac{1}{3-1} \times \left(\frac{175}{5525}\right)^3 + \left(\frac{175}{5525}\right)^3 \times 1.0249 = 0.5$$

$$S^{12} = \frac{1}{3-1} - \frac{1}{3-1} \times \left(\frac{175}{5525}\right)^3 + \left(\frac{175}{5525}\right)^3 \times 1.0769 = 0.5$$

$$\sum_{1 \leq j \leq N} S^j = S^1 + S^2 + \cdots + S^{12} = 0 + 0.5 + \cdots + 0.5 = 5.5$$

$$\sum_{1 \leq j1 < j2 \leq N} S^{j1}S^{j2} = S^1 S^2 + S^1 S^3 + \cdots + S^{11} S^{12} = 13.7498$$

$$\sum_{1 \leq j1 < j2 < j3 \leq N} S^{j1}S^{j2}S^{j3} = S^1 S^2 S^3 + S^1 S^2 S^4 + \cdots + S^{10} S^{11} S^{12} = 20.6245$$

$$\sum_{1 \leq j1 < j2 < j3 < j4 \leq N} S^{j1}S^{j2}S^{j3}S^{j4} = S^1 S^2 S^3 S^4 + S^1 S^2 S^3 S^5 + \cdots + S^9 S^{10} S^{11} S^{12} = 20.6244$$

$$\sum_{1 \leq j1 < j2 \cdots < j5 \leq N} S^{j1}S^{j2} \cdots S^{j5} = S^1 S^2 S^3 S^4 S^5 + S^1 S^2 S^3 S^4 S^6 + \cdots + S^8 S^9 S^{10} S^{11} S^{12} = 14.4369$$

$$\sum_{1 \leq j1 < j2 \cdots < j6 \leq N} S^{j1}S^{j2} \cdots S^{j6} = S^1 S^2 S^3 S^4 S^5 S^6 + S^1 S^2 S^3 S^4 S^5 S^7 + \cdots + S^7 S^8 S^9 S^{10} S^{11} S^{12} = 7.2184$$

$$\sum_{1 \leq j1 < j2 \cdots < j7 \leq N} S^{j1}S^{j2} \cdots S^{j7} = S^1 S^2 S^3 S^4 S^5 S^6 S^7 + S^1 S^2 S^3 S^4 S^5 S^6 S^8 + \cdots + S^6 S^7 S^8 S^9 S^{10} S^{11} S^{12}$$

$$= 2.5780$$

$$\sum_{1 \le j1 < j2 \cdots < j8 \le N} S^{j1}S^{j2} \cdots S^{j8} = S^1 S^2 S^3 S^4 S^5 S^6 S^7 S^8 + \cdots + S^5 S^6 S^7 S^8 S^9 S^{10} S^{11} S^{12} = 0.6445$$

$$\sum_{1 \le j1 < j2 \cdots < j9 \le N} S^{j1}S^{j2} \cdots S^{j9} = S^1 S^2 S^3 S^4 S^5 S^6 S^7 S^8 S^9 + \cdots + S^4 S^5 S^6 S^7 S^8 S^9 S^{10} S^{11} S^{12} = 0.1074$$

$$\sum_{1 \le j1 < j2 \cdots < j10 \le N} S^{j1}S^{j2} \cdots S^{j10} = S^1 S^2 S^3 S^4 S^5 S^6 S^7 S^8 S^9 S^{10} + \cdots + S^3 S^4 S^5 S^6 S^7 S^8 S^9 S^{10} S^{11} S^{12}$$

$$= 0.0107$$

$$\sum_{1 \le j1 < j2 \cdots < j11 \le N} S^{j1}S^{j2} \cdots S^{j11} = S^1 S^2 S^3 S^4 S^5 S^6 S^7 S^8 S^9 S^{10} S^{11} + \cdots + S^2 S^3 S^4 S^5 S^6 S^7 S^8 S^9 S^{10} S^{11} S^{12}$$

$$= 0.0005$$

$$\sum_{1 \le j1 < j2 \cdots < j12 \le N} S^{j1}S^{j2} \cdots S^{j11} = S^1 S^2 S^3 S^4 S^5 S^6 S^7 S^8 S^9 S^{10} S^{11} S^{12} = 0.00$$

Therefore,

$$S_N = S^{1 \cup 2 \cup \cdots \cup 12} = \sum_{1 \le j \le 12} S^j + (1-m) \sum_{1 \le j1 < j2 \le 12} S^{j1}S^{j2} + (1-m)^2 \sum_{1 \le j1 < j2 < j3 \le 12} S^{j1}S^{j2}S^{j3}$$

$$+ (1-m)^{11} \sum_{1 \le j1 < j2 < \cdots < jN \le 12} S^{j1}S^{j2} \cdots S^{j12}$$

$$= 5.5 + (1-3) \times 13.7498 + \cdots + (1-3)^{11} \times 0 = 0.49995$$

Transmission of redundancy: There is one loop for this layout. For all nodes, except node 5, the total redundancy is 0.

(1)
$$S_5 = \frac{1}{m-1} \left\{ \sum_{i=1}^{n(j)} \left[ \frac{q_{ij}}{Q_j} - \left( \frac{q_{ij}}{Q_j} \right)^m \right] \right\}$$

$$= \frac{1}{3-1} \times \left[ \left( \frac{225}{400} \right) - \left( \frac{225}{400} \right)^3 + \left( \frac{175}{400} \right) - \left( \frac{175}{400} \right)^3 \right] = 0.3691$$

(2)
$$S_{j^*} = \frac{1}{m-1} \left[ 1 - \sum_{i=1}^{n(j)} \left( \frac{q_{ij}}{Q_0} \right)^m \right]$$

$$S_{1^*} = 0$$

$$S_{2^*} = \frac{1}{3-1} \times \left[ 1 - \left( \frac{1100}{5525} \right)^3 \right] = 0.4961$$

$$S_{3^*} = \frac{1}{3-1} \times \left[ 1 - \left( \frac{725}{5525} \right)^3 \right] = 0.4989$$

$$S_{4^*} = \frac{1}{3-1} \times \left[ 1 - \left( \frac{900}{5525} \right)^3 \right] = 0.4978$$

$$S_{5^*} = \frac{1}{3-1} \times \left[ 1 - \left( \frac{225}{5525} \right)^3 - \left( \frac{175}{5525} \right)^3 \right] = 0.50$$

$$S_{6^*} = \frac{1}{3-1} \times \left[ 1 - \left( \frac{550}{5525} \right)^3 \right] = 0.4995$$

$$S_{7^*} = \frac{1}{3-1} \times \left[ 1 - \left( \frac{550}{5525} \right)^3 \right] = 0.4995$$

$$S_{8^*} = \frac{1}{3-1} \times \left[ 1 - \left( \frac{225}{5525} \right)^3 \right] = 0.50$$

$$S_{9^*} = \frac{1}{3-1} \times \left[ 1 \quad \left( \frac{400}{5525} \right)^3 \right] = 0.4998$$

$$S_{10^*} = \frac{1}{3-1} \times \left[ 1 - \left( \frac{325}{5525} \right)^3 \right] = 0.4999$$

$$S_{11^*} = \frac{1}{3-1} \times \left[ 1 - \left( \frac{175}{5525} \right)^3 \right] = 0.50$$

$$S_{12^*} = \frac{1}{3-1} \times \left[ 1 - \left( \frac{175}{5525} \right)^3 \right] = 0.50$$

(3) Transmissivity

$$t_{21} = \frac{1100}{2000} = 0.55, \quad t_{32} = \frac{725}{1100} = 0.6591,$$

$$t_{41} = \frac{400}{2000} = 0.45, \quad t_{52} = \frac{225}{1100} = 0.2045$$

$$t_{54} = \frac{175}{900} = 0.1944, \quad t_{63} = \frac{550}{725} = 0.7586,$$

$$t_{74} = \frac{550}{900} = 0.6111, \quad t_{85} = \frac{225}{400} = 0.5625$$

$$t_{96} = \frac{400}{550} = 0.7273, \quad t_{10-7} = \frac{325}{550} = 0.5909,$$

$$t_{10-11} = \frac{175}{325} = 0.5385, \quad t_{12-9} = \frac{175}{400} = 0.4375$$

(4) Determine $S'_j$.

Node 1: $S'_{1^*} = S_{1^*} = 0$

Node 2: $S'_{2^*} = S_{2^*} + t_{21}S'_{1^*} = 0.4961 + 0.55 \times 0 = 0.4961$

Node 3: $S'_{3^*} = S_{3^*} + t_{32}S'_{2^*} = 0.4989 + 0.6591 \times 0.4961 = 0.8258$

Node 4: $S'_{4^*} = S_{4^*} + t_{41}S'_{1^*} = 0.4978 + 0.45 \times 0 = 0.4978$

Node 5: $S'_{5^*} = S_{5^*} + t_{52}S'_{2^*} + t_{54}S'_{4^*}$

$$= 0.5 + 0.2045 \times 0.4961 + 0.1944 \times 0.4987 = 0.6981$$

Node 6: $S'_{6^*} = S_{6^*} + t_{63}S'_{3^*} = 0.4995 + 0.7586 \times 0.8258 = 1.1260$

Node 7: $S'_{7^*} = S_{7^*} + t_{74}S'_{4^*} = 0.4995 + 0.6111 \times 0.4978 = 0.8037$

Node 8: $S'_{8^*} = S_{8^*} + t_{85}S'_{5^*} = 0.5 + 0.5625 \times 0.6981 = 0.8927$

Node 9: $S'_{9^*} = S_{9^*} + t_{96}S'_{6^*} = 0.4998 + 0.7273 \times 1.1260 = 1.3187$

Node 10: $S'_{10^*} = S_{10^*} + t_{10-7}S'_{7^*} = 0.4999 + 0.5909 \times 0.8037 = 0.9748$

Node 11: $S'_{11^*} = S_{11^*} + t_{11-10}S'_{10^*} = 0.5 + 0.5385 \times 0.9748 = 1.0249$

Node 12: $S'_{12^*} = S_{12^*} + t_{12-9}S'_{9^*} = 0.5 + 0.4375 \times 1.3187 = 1.0769$

(5) Determine $S_N$

$$S_N = S^{1 \cup 2 \cup \cdots \cup 12} = \sum_{1 \le j \le 12} S^j + (1-m) \sum_{1 \le j1 < j2 \le 12} S^{j1}S^{j2} + (1-m)^2 \sum_{1 \le j1 < j2 < j3 \le 12} S^{j1}S^{j2}S^{j3}$$

$$+ (1-m)^{11} \sum_{1 \le j1 < j2 < \cdots < jN \le 12} S^{j1}S^{j2} \cdots S^{j12}$$

Determine $S_j$

$$\sum_{1 \le j \le N} S^j = \frac{N}{m-1} - \frac{1}{m-1} \sum_{j=1}^{N} \left(\frac{Q_j}{Q_0}\right)^m + \sum_{j=1}^{N} \left(\frac{Q_j}{Q_0}\right)^m S_{j^*}$$

$$S^1 = 0$$

$$S^2 = \frac{1}{3-1} - \frac{1}{3-1} \times \left(\frac{1100}{5525}\right)^3 + \left(\frac{1100}{5525}\right)^3 \times 0.4961 = 0.5$$

$$S^3 = \frac{1}{3-1} - \frac{1}{3-1} \times \left(\frac{725}{5525}\right)^3 + \left(\frac{725}{5525}\right)^3 \times 0.4989 = 0.5007$$

$$S^4 = \frac{1}{3-1} - \frac{1}{3-1} \times \left(\frac{900}{5525}\right)^3 + \left(\frac{900}{5525}\right)^3 \times 0.4978 = 0.5$$

$$S^5 = \frac{1}{3-1} - \frac{1}{3-1} \times \left(\frac{400}{5525}\right)^3 + \left(\frac{400}{5525}\right)^3 \times 0.5 = 0.5001$$

$$S^6 = \frac{1}{3-1} - \frac{1}{3-1} \times \left(\frac{550}{5525}\right)^3 + \left(\frac{550}{5525}\right)^3 \times 0.4995 = 0.5006$$

$$S^7 = \frac{1}{3-1} - \frac{1}{3-1} \times \left(\frac{550}{5525}\right)^3 + \left(\frac{550}{5525}\right)^3 \times 0.4995 = 0.5006$$

$$S^8 = \frac{1}{3-1} - \frac{1}{3-1} \times \left(\frac{225}{5525}\right)^3 + \left(\frac{225}{5525}\right)^3 \times 0.50 = 0.5$$

$$S^9 = \frac{1}{3-1} - \frac{1}{3-1} \times \left(\frac{400}{5525}\right)^3 + \left(\frac{400}{5525}\right)^3 \times 0.4998 = 0.5003$$

$$S^{10} = \frac{1}{3-1} - \frac{1}{3-1} \times \left(\frac{325}{5525}\right)^3 + \left(\frac{325}{5525}\right)^3 \times 0.4999 = 0.5001$$

$$S^{11} = \frac{1}{3-1} - \frac{1}{3-1} \times \left(\frac{175}{5525}\right)^3 + \left(\frac{175}{5525}\right)^3 \times 0.50 = 0.5$$

$$S^{12} = \frac{1}{3-1} - \frac{1}{3-1} \times \left(\frac{175}{5525}\right)^3 + \left(\frac{175}{5525}\right)^3 \times 0.50 = 0.5$$

$$\sum_{1 \leq j \leq N} S^j = S^1 + S^2 + \cdots + S^{12} = 0 + 0.5 + \cdots + 0.5 = 5.2022$$

$$\sum_{1 \leq j1 < j2 \leq N} S^{j1}S^{j2} = S^1S^2 + S^1S^3 + \cdots + S^{11}S^{12} = 13.7608$$

$$\sum_{1\le j1<j2<j3\le N} S^{j1}S^{j2}S^{j3} = S^1S^2S^3 + S^1S^2S^4 + \cdots + S^{10}S^{11}S^{12} = 20.6493$$

$$\sum_{1\le j1<j2<j3<j4\le N} S^{j1}S^{j2}S^{j3}S^{j4} = S^1S^2S^3S^4 + S^1S^2S^3S^5 + \cdots + S^9S^{10}S^{11}S^{12} = 20.6574$$

$$\sum_{1\le j1<j2\cdots<j5\le N} S^{j1}S^{j2}\cdots S^{j5} = S^1S^2S^3S^4S^5 + S^1S^2S^3S^4S^6 + \cdots + S^8S^9S^{10}S^{11}S^{12} = 14.4658$$

$$\sum_{1\le j1<j2\cdots<j6\le N} S^{j1}S^{j2}\cdots S^{j6} = S^1S^2S^3S^4S^5S^6 + S^1S^2S^3S^4S^5S^7 + \cdots + S^7S^8S^9S^{10}S^{11}S^{12} = 7.2358$$

$$\sum_{1\le j1<j2\cdots<j7\le N} S^{j1}S^{j2}\cdots S^{j7} = S^1S^2S^3S^4S^5S^6S^7 + S^1S^2S^3S^4S^5S^6S^8 + \cdots + S^6S^7S^8S^9S^{10}S^{11}S^{12}$$

$$= 2.2852$$

$$\sum_{1\le j1<j2\cdots<j8\le N} S^{j1}S^{j2}\cdots S^{j8} = S^1S^2S^3S^4S^5S^6S^7S^8 + \cdots + S^5S^6S^7S^8S^9S^{10}S^{11}S^{12} = 0.6466$$

$$\sum_{1\le j1<j2\cdots<j9\le N} S^{j1}S^{j2}\cdots S^{j9} = S^1S^2S^3S^4S^5S^6S^7S^8S^9 + \cdots + S^4S^5S^6S^7S^8S^9S^{10}S^{11}S^{12} = 0.1078$$

$$\sum_{1\le j1<j2\ldots<j10\le N} S^{j1}S^{j2}\cdots S^{j10} = S^1S^2S^3S^4S^5S^6S^7S^8S^9S^{10} + \cdots + S^3S^4S^5S^6S^7S^8S^9S^{10}S^{11}S^{12}$$

$$= 0.0108$$

$$\sum_{1\le j1<j2\cdots<j11\le N} S^{j1}S^{j2}\cdots S^{j11} = S^1S^2S^3S^4S^5S^6S^7S^8S^9S^{10}S^{11} + \cdots + S^2S^3S^4S^5S^6S^7S^8S^9S^{10}S^{11}S^{12}$$

$$= 0.0005$$

$$\sum_{1\le j1<j2\cdots<j12\le N} S^{j1}S^{j2}\cdots S^{j11} = S^1S^2S^3S^4S^5S^6S^7S^8S^9S^{10}S^{11}S^{12} = 0.00$$

Therefore,

$$S_N = S^{1\cup2\cup\cdots\cup12} = \sum_{1\le j\le12} S^j + (1-m)\sum_{1\le j1<j2\le12} S^{j1}S^{j2} + (1-m)^2 \sum_{1\le j1<j2<j3\le12} S^{j1}S^{j2}S^{j3}$$

$$+ (1-m)^{11} \sum_{1\le j1<j2<\cdots<jN\le12} S^{j1}S^{j2}\cdots S^{j12}$$

$$= 5.5022 + (1-3)\times13.7608 + \cdots + (1-3)^{11}\times0 = 0.4999$$

Path parameter redundancy:

(1) Now compute path parameter using $a_{ij} = n_{d_{ij}} \left[ 1 - \dfrac{\sum_{l=1}^{M_{ij}} D_l}{\sum_{l=1}^{M_{ij}} d_l} \right].$

$a_1 = a_{s1} = 1 \times \left[ 1 - \dfrac{0}{1} \right] = 1$   (Path: S-1)

$a_2 = a_{12} = 1 \times \left[ 1 - \dfrac{0}{2} \right] = 1$   (Path: S-1-2)

$a_3 = a_{23} = 1 \times \left[ 1 - \dfrac{0}{2} \right] = 1$   (Path: S-1-2-3)

$a_4 = a_{14} = 1 \times \left[ 1 - \dfrac{0}{2} \right] = 1$   (Path: S-1-4)

$a_5 = a_{25} + a_{45} = 1 \times \left[ 1 - \dfrac{1}{4} \right] + 1 \times \left[ 1 - \dfrac{1}{4} \right]$

$\qquad = \dfrac{3}{4} + \dfrac{3}{4} = 1.5$   (Path: S-1-2-5, S-1-4-5)

$a_6 = a_{36} = 1 \times \left[ 1 - \dfrac{0}{4} \right] = 1$   (Path: S-1-2-3-6)

$a_7 = a_{47} = 1 \times \left[ 1 - \dfrac{0}{3} \right] = 1$   (Path: S-1-4-7)

$a_8 = a_{58} = 2 \times \left[ 1 - \dfrac{2}{8} \right] = \dfrac{3}{2} = 1.5$   (Paths: S-1-2-5-8, S-1-4-5-8)

$a_9 = a_{69} = 1 \times \left[ 1 - \dfrac{0}{5} \right] = 1$   (Path: S-1-2-3-6-9)

$a_{10} = a_{710} = 1 \times \left[ 1 - \dfrac{0}{4} \right] = 1$   (Path: S-1-4-7-10)

$a_{11} = a_{1011} = 1 \times \left[ 1 - \dfrac{0}{5} \right] = 1$   (Path: S-1-4-7-10-11)

$a_{12} = a_{912} = 1 \times \left[ 1 - \dfrac{0}{6} \right] = 1$   (Path: S-1-2-3-6-9-12)

$$(2) \qquad S_j = \frac{1}{m-1}\left[1 - \sum_{i=1}^{n(j)}\left(\frac{q_{ij}}{a_{ij}Q_j}\right)^3\right]$$

$$S_1 = 0$$

$$S_2 = \frac{1}{3-1}\times\left[1 - \left(\frac{1100}{1\times1100}\right)^3\right] = 0$$

$$S_3 = \frac{1}{3-1}\times\left[1 - \left(\frac{725}{1\times725}\right)^3\right] = 0$$

$$S_4 = \frac{1}{3-1}\times\left[1 - \left(\frac{900}{1\times900}\right)^3\right] = 0$$

$$S_5 = \frac{1}{3-1}\times\left[1 - \left(\frac{225}{0.75\times400}\right)^3 - \left(\frac{175}{0.75\times400}\right)^3\right] = 0.1898$$

$$S_6 = \frac{1}{3-1}\times\left[1 - \left(\frac{550}{1\times550}\right)^3\right] = 0$$

$$S_7 = \frac{1}{3-1}\times\left[1 - \left(\frac{550}{1\times550}\right)^3\right] = 0$$

$$S_8 = \frac{1}{3-1}\times\left[1 - \left(\frac{225}{1.5\times225}\right)^3\right] = 0.3519$$

$$S_9 = \frac{1}{3-1}\times\left[1 - \left(\frac{400}{1\times400}\right)^3\right] = 0$$

$$S_{11} = \frac{1}{3-1}\times\left[1 - \left(\frac{325}{1\times325}\right)^3\right] = 0$$

$$S_{12} = \frac{1}{3-1}\times\left[1 - \left(\frac{175}{1\times175}\right)^3\right] = 0$$

(3) $\quad S_{j^*} = \dfrac{1}{m-1}\left[1 - \displaystyle\sum_{i=1}^{n(j)}\left(\dfrac{q_{ij}}{a_{ij}Q_0}\right)^3\right]$

$S_{1^*} = 0$

$S_{2^*} = \dfrac{1}{3-1}\times\left[1 - \left(\dfrac{1100}{1\times 5525}\right)^3\right] = 0.4961$

$S_{3^*} = \dfrac{1}{3-1}\times\left[1 - \left(\dfrac{725}{1\times 5525}\right)^3\right] = 0.4989$

$S_{4^*} = \dfrac{1}{3-1}\times\left[1 - \left(\dfrac{900}{1\times 5525}\right)^3\right] = 0.4978$

$S_{5^*} = \dfrac{1}{3-1}\times\left[1 - \left(\dfrac{225}{0.75\times 5525}\right)^3 - \left(\dfrac{175}{0.75\times 5525}\right)^3\right] = 0.4999$

$S_{6^*} = \dfrac{1}{3-1}\times\left[1 - \left(\dfrac{550}{1\times 5525}\right)^3\right] = 0.4995$

$S_{7^*} - \dfrac{1}{3-1}\times\left[1 - \left(\dfrac{550}{1\times 5525}\right)^3\right] = 0.4995$

$S_{8^*} = \dfrac{1}{3-1}\times\left[1 - \left(\dfrac{225}{1.5\times 5525}\right)^3\right] = 0.50$

$S_{9^*} = \dfrac{1}{3-1}\times\left[1 - \left(\dfrac{400}{1\times 5525}\right)^3\right] = 0.4998$

$S_{10^*} = \dfrac{1}{3-1}\times\left[1 - \left(\dfrac{325}{1\times 5525}\right)^3\right] = 0.4999$

$S_{11^*} = \dfrac{1}{3-1}\times\left[1 - \left(\dfrac{175}{1\times 5525}\right)^3\right] = 0.5$

$S_{12^*} = \dfrac{1}{3-1}\times\left[1 - \left(\dfrac{175}{1\times 5525}\right)^3\right] = 0.5$

(4) Determine $S_N$

$$S_N = S^{1 \cup 2 \cup \cdots \cup 12} = \sum_{1 \le j \le 12} S^j + (1-m) \sum_{1 \le j1 < j2 \le 12} S^{j1}S^{j2} + (1-m)^2 \sum_{1 \le j1 < j2 < j3 \le 12} S^{j1}S^{j2}S^{j3}$$

$$+ (1-m)^{11} \sum_{1 \le j1 < j2 < \cdots < jN \le 12} S^{j1}S^{j2} \cdots S^{j12}$$

Determine $S_j$

$$\sum_{1 \le j \le N} S^j = \frac{N}{m-1} - \frac{1}{m-1} \sum_{j=1}^{N} \left(\frac{Q_j}{Q_0}\right)^m + \sum_{j=1}^{N} \left(\frac{Q_j}{Q_0}\right)^m S_{j^*}$$

$$S^1 = 0$$

$$S^2 = \frac{1}{3-1} - \frac{1}{3-1} \times \left(\frac{1100}{5525}\right)^3 + \left(\frac{1100}{5525}\right)^3 \times 0.4961 = 0.5$$

$$S^3 = \frac{1}{3-1} - \frac{1}{3-1} \times \left(\frac{725}{5525}\right)^3 + \left(\frac{725}{5525}\right)^3 \times 0.4989 = 0.50$$

$$S^4 = \frac{1}{3-1} - \frac{1}{3-1} \times \left(\frac{900}{5525}\right)^3 + \left(\frac{900}{5525}\right)^3 \times 0.4978 = 0.5$$

$$S^5 = \frac{1}{3-1} - \frac{1}{3-1} \times \left(\frac{400}{5525}\right)^3 + \left(\frac{400}{5525}\right)^3 \times 0.4999 = 0.50$$

$$S^6 = \frac{1}{3-1} - \frac{1}{3-1} \times \left(\frac{550}{5525}\right)^3 + \left(\frac{550}{5525}\right)^3 \times 0.4995 = 0.50$$

$$S^7 = \frac{1}{3-1} - \frac{1}{3-1} \times \left(\frac{550}{5525}\right)^3 + \left(\frac{550}{5525}\right)^3 \times 0.4995 = 0.50$$

$$S^8 = \frac{1}{3-1} - \frac{1}{3-1} \times \left(\frac{225}{5525}\right)^3 + \left(\frac{225}{5525}\right)^3 \times 0.50 = 0.50$$

$$S^9 = \frac{1}{3-1} - \frac{1}{3-1} \times \left(\frac{400}{5525}\right)^3 + \left(\frac{400}{5525}\right)^3 \times 0.4998 = 0.50$$

$$S^{10} = \frac{1}{3-1} - \frac{1}{3-1} \times \left(\frac{325}{5525}\right)^3 + \left(\frac{325}{5525}\right)^3 \times 0.4999 = 0.50$$

$$S^{11} = \frac{1}{3-1} - \frac{1}{3-1} \times \left(\frac{175}{5525}\right)^3 + \left(\frac{175}{5525}\right)^3 \times 0.50 = 0.50$$

$$S^{12} = \frac{1}{3-1} - \frac{1}{3-1} \times \left(\frac{175}{5525}\right)^3 + \left(\frac{175}{5525}\right)^3 \times 0.50 = 0.50$$

$$\sum_{1 \le j \le N} S^j = S^1 + S^2 + \cdots + S^{12} = 0 + 0.5 + \cdots + 0.5 = 5.50$$

$$\sum_{1 \le j1 < j2 \le N} S^{j1}S^{j2} = S^1 S^2 + S^1 S^3 + \cdots + S^{11}S^{12} = 13.7498$$

$$\sum_{1 \le j1 < j2 < j3 \le N} S^{j1}S^{j2}S^{j3} = S^1 S^2 S^3 + S^1 S^2 S^4 + \cdots + S^{10}S^{11}S^{12} = 20.6245$$

$$\sum_{1 \le j1 < j2 < j3 < j4 \le N} S^{j1}S^{j2}S^{j3}S^{j4} = S^1 S^2 S^3 S^4 + S^1 S^2 S^3 S^5 + \cdots + S^9 S^{10}S^{11}S^{12} = 20.6244$$

$$\sum_{1 \le j1 < j2 \ldots < j5 \le N} S^{j1}S^{j2} \cdots S^{j5} = S^1 S^2 S^3 S^4 S^5 + S^1 S^2 S^3 S^4 S^6 + \cdots + S^8 S^9 S^{10}S^{11}S^{12} = 14.4369$$

$$\sum_{1 \le j1 < j2 \cdots < j6 \le N} S^{j1}S^{j2} \cdots S^{j6} = S^1 S^2 S^3 S^4 S^5 S^6 + S^1 S^2 S^3 S^4 S^5 S^7 + \cdots + S^7 S^8 S^9 S^{10}S^{11}S^{12}$$

$$= 7.2184$$

$$\sum_{1 \le j1 < j2 \ldots < j7 \le N} S^{j1}S^{j2} \cdots S^{j7} = S^1 S^2 S^3 S^4 S^5 S^6 S^7 + S^1 S^2 S^3 S^4 S^5 S^6 S^8 + \cdots + S^6 S^7 S^8 S^9 S^{10}S^{11}S^{12}$$

$$= 2.578$$

$$\sum_{1 \le j1 < j2 \cdots < j8 \le N} S^{j1}S^{j2} \cdots S^{j8} = S^1 S^2 S^3 S^4 S^5 S^6 S^7 S^8 + \cdots + S^5 S^6 S^7 S^8 S^9 S^{10}S^{11}S^{12} = 0.6445$$

$$\sum_{1 \le j1 < j2 \cdots < j9 \le N} S^{j1}S^{j2} \cdots S^{j9} = S^1 S^2 S^3 S^4 S^5 S^6 S^7 S^8 S^9 + \cdots + S^4 S^5 S^6 S^7 S^8 S^9 S^{10}S^{11}S^{12} = 0.1074$$

$$\sum_{1 \le j1 < j2 \ldots < j10 \le N} S^{j1}S^{j2} \cdots S^{j10} = S^1 S^2 S^3 S^4 S^5 S^6 S^7 S^8 S^9 S^{10} + \cdots + S^3 S^4 S^5 S^6 S^7 S^8 S^9 S^{10}S^{11}S^{12}$$

$$= 0.0107$$

$$\sum_{1 \le j1 < j2 \cdots < j11 \le N} S^{j1} S^{j2} \cdots S^{j11} = S^1 S^2 S^3 S^4 S^5 S^6 S^7 S^8 S^9 S^{10} S^{11} + \cdots + S^2 S^3 S^4 S^5 S^6 S^7 S^8 S^9 S^{10} S^{11} S^{12}$$

$$= 0.0005$$

$$\sum_{1 \le j1 < j2 \cdots < j12 \le N} S^{j1} S^{j2} \cdots S^{j11} = S^1 S^2 S^3 S^4 S^5 S^6 S^7 S^8 S^9 S^{10} S^{11} S^{12} = 0.00$$

Therefore,

$$S_N = S^{1 \cup 2 \cup \cdots \cup 12} = \sum_{1 \le j \le 12} S^j + (1-m) \sum_{1 \le j1 < j2 \le 12} S^{j1} S^{j2} + (1-m)^2 \sum_{1 \le j1 < j2 < j3 \le 12} S^{j1} S^{j2} S^{j3}$$

$$+ (1-m)^{11} \sum_{1 \le j1 < j2 < \cdots < jN \le 12} S^{j1} S^{j2} \cdots S^{j12}$$

$$= 5.5022 + (1-3) \times 13.7498 + \cdots + (1-3)^{11} \times 0 = 0.4995$$

Age factor and path parameter of redundancy are as follows:

Now, consider the age factor since the friction factor for all links is 115. Therefore, In 115 = 4.7449 and $u_{ij} = 0.2$ In $C_{ij}(t) = 0.9490$.

(1) Now compute path parameter using $a_{ij} = n_{d_{ij}} \left[ 1 - \dfrac{\sum_{l=1}^{M_{ij}} \dfrac{D_l}{d_l}}{\sum_{l=1}^{M_{ij}} d_l} \right].$

$$a_1 = a_{s1} = 1 \times \left[ 1 - \frac{0}{1} \right] = 1 \quad \text{(Path: S-1)}$$

$$a_2 = a_{12} = 1 \times \left[ 1 - \frac{0}{2} \right] = 1 \quad \text{(Path: S-1-2)}$$

$$a_3 = a_{23} = 1 \times \left[ 1 - \frac{0}{3} \right] = 1 \quad \text{(Path: S-1-2-3)}$$

$$a_4 = a_{14} = 1 \times \left[ 1 - \frac{0}{2} \right] = 1 \quad \text{(Path: S-1-4)}$$

$$a_5 = a_{25} + a_{45} = 1 \times \left[ 1 - \frac{1}{4} \right] + 1 \times \left[ 1 - \frac{1}{4} \right] = \frac{3}{4} + \frac{3}{4} = 1.5 \quad \text{(Paths: S-1-2-5, S-1-4-5)}$$

$$a_6 = a_{36} = 1 \times \left[ 1 - \frac{0}{4} \right] = 1 \quad \text{(Path: S-1-2-3-6)}$$

$$a_7 = a_{47} = 1 \times \left[1 - \frac{0}{3}\right] = 1 \quad \text{(Path: S-1-4-7)}$$

$$a_8 = a_{58} = 2 \times \left[1 - \frac{2}{8}\right] = \frac{3}{2} = 1.5 \quad \text{(Paths: S-1-2-5-8, S-1-4-5-8)}$$

$$a_9 = a_{69} = 1 \times \left[1 - \frac{0}{5}\right] = 1 \quad \text{(Path: S-1-2-3-6-9)}$$

$$a_{10} = a_{710} = 1 \times \left[1 - \frac{0}{4}\right] = 1 \quad \text{(Path: S-1-4-7-10)}$$

$$a_{11} = a_{1011} = 1 \times \left[1 - \frac{0}{5}\right] = 1 \quad \text{(Path: S-1-4-7-10-11)}$$

$$a_{12} = a_{912} = 1 \times \left[1 - \frac{0}{6}\right] = 1 \quad \text{(Path: S-1-2-3-6-9-12)}$$

(2) $S_j = \dfrac{1}{m-1}\left[1 - \displaystyle\sum_{i=1}^{n(j)} u_{ij}\left(\dfrac{q_{ij}}{a_{ij}Q_j}\right)^3\right]$

$$S_1 = 0$$

$$S_2 = \frac{1}{3-1} \times \left[1 - 0.9490 \times \left(\frac{1100}{1 \times 1100}\right)^3\right] = 0.0255$$

$$S_3 = \frac{1}{3-1} \times \left[1 - 0.9490 \times \left(\frac{725}{1 \times 725}\right)^3\right] = 0.0255$$

$$S_4 = \frac{1}{3-1} \times \left[1 - 0.949 \times \left(\frac{900}{1 \times 900}\right)^3\right] = 0.0255$$

$$S_5 = \frac{1}{3-1} \times \left[1 - 0.9490 \times \left(\frac{225}{0.75 \times 400}\right)^3 - 0.9490 \times \left(\frac{175}{0.75 \times 400}\right)^3\right] = 0.2056$$

$$S_6 = \frac{1}{3-1} \times \left[1 - 0.9490 \times \left(\frac{550}{1 \times 550}\right)^3\right] = 0.0255$$

$$S_7 = \frac{1}{3-1} \times \left[ 1 - 0.9490 \times \left( \frac{550}{1 \times 550} \right)^3 \right] = 0.0255$$

$$S_8 = \frac{1}{3-1} \times \left[ 1 - 0.9490 \times \left( \frac{225}{1.5 \times 225} \right)^3 \right] = 0.3594$$

$$S_9 = \frac{1}{3-1} \times \left[ 1 - 0.9490 \times \left( \frac{400}{1 \times 400} \right)^3 \right] = 0.0255$$

$$S_{10} = \frac{1}{3-1} \times \left[ 1 - 0.9490 \times \left( \frac{325}{1 \times 325} \right)^3 \right] = 0.0255$$

$$S_{11} = \frac{1}{3-1} \times \left[ 1 - 0.9490 \times \left( \frac{175}{1 \times 175} \right)^3 \right] = 0.0255$$

$$S_{12} = \frac{1}{3-1} \times \left[ 1 - 0.9490 \times \left( \frac{175}{1 \times 175} \right)^3 \right] = 0.0255$$

(3)  $$\tilde{S}_{j*} = \frac{1}{m-1} \left[ 1 - \left( \frac{Q_j}{Q_0} \right)^m \sum_{j=1}^{n(j)} u_{ij} \left( \frac{q_{ij}}{a_{ij} Q_j} \right)^3 \right]$$

$$\tilde{S}_1 = 0$$

$$\tilde{S}_2 = \frac{1}{3-1} \times \left[ 1 - \left( \frac{1100}{1 \times 5525} \right)^3 \left( 0.9490 \times \left( \frac{1100}{1 \times 1100} \right)^3 \right) \right] = 0.4963$$

$$\tilde{S}_3 = \frac{1}{3-1} \times \left[ 1 - \left( \frac{725}{1 \times 5525} \right)^3 \left( 0.9490 \times \left( \frac{725}{1 \times 725} \right)^3 \right) \right] = 0.4989$$

$$\tilde{S}_4 = \frac{1}{3-1} \times \left[ 1 - \left( \frac{900}{1 \times 5525} \right)^3 \left( 0.9490 \times \left( \frac{900}{1 \times 900} \right)^3 \right) \right] = 0.4979$$

$$\tilde{S}_5 = \frac{1}{3-1} \times \left[ 1 - \left( \frac{400}{1 \times 5525} \right)^3 \left( 0.9490 \times \left( \frac{225}{1 \times 400} \right)^3 - 0.9490 \times \left( \frac{175}{1 \times 400} \right)^3 \right) \right]$$

$$= 0.4999$$

$$\tilde{S}_6 = \frac{1}{3-1} \times \left[ 1 - \left( \frac{550}{1 \times 5525} \right)^3 \left( 0.9490 \times \left( \frac{550}{1 \times 550} \right)^3 \right) \right] = 0.4995$$

$$\tilde{S}_7 = \frac{1}{3-1} \times \left[ 1 - \left( \frac{550}{1 \times 5525} \right)^3 \left( 0.9490 \times \left( \frac{550}{1 \times 550} \right)^3 \right) \right] = 0.4995$$

$$\tilde{S}_8 = \frac{1}{3-1} \times \left[ 1 - \left( \frac{225}{1 \times 5525} \right)^3 \left( 0.9490 \times \left( \frac{225}{1.5 \times 225} \right)^3 \right) \right] = 0.50$$

$$\tilde{S}_9 = \frac{1}{3-1} \times \left[ 1 - \left( \frac{400}{1 \times 5525} \right)^3 \left( 0.9490 \times \left( \frac{400}{1 \times 400} \right)^3 \right) \right] = 0.4998$$

$$\tilde{S}_{10} = \frac{1}{3-1} \times \left[ 1 - \left( \frac{325}{1 \times 5525} \right)^3 \left( 0.9490 \times \left( \frac{325}{1 \times 325} \right)^3 \right) \right] = 0.4999$$

$$\tilde{S}_2 = \frac{1}{3-1} \times \left[ 1 - \left( \frac{175}{1 \times 5525} \right)^3 \left( 0.9490 \times \left( \frac{175}{1 \times 175} \right)^3 \right) \right] = 0.50$$

(4) Determine $S_N$

$$S_N = S^{1 \cup 2 \cup \cdots \cup 12} = \sum_{1 \le j \le 12} S^j + (1-m) \sum_{1 \le j1 < j2 \le 12} S^{j1} S^{j2} + (1-m)^2 \sum_{1 \le j1 < j2 < j3 \le 12} S^{j1} S^{j2} S^{j3}$$

$$+ (1-m)^{11} \sum_{1 \le j1 < j2 < \cdots < jN \le 12} S^{j1} S^{j2} \cdots S^{j12}$$

Determine $S_j$

$$\sum_{1 \le j \le N} S^j = \frac{N}{m-1} - \frac{1}{m-1} \sum_{j=1}^{N} \left( \frac{Q_j}{Q_0} \right)^m + \sum_{j=1}^{N} \left( \frac{Q_j}{Q_0} \right)^m S_{j^*}$$

$$S^1 = 0$$

$$S^2 = \frac{1}{3-1} - \frac{1}{3-1} \times \left( \frac{1100}{5525} \right)^3 + \left( \frac{1100}{5525} \right)^3 \times 0.4963 = 0.50$$

$$S^3 = \frac{1}{3-1} - \frac{1}{3-1} \times \left(\frac{725}{5525}\right)^3 + \left(\frac{725}{5525}\right)^3 \times 0.4989 = 0.50$$

$$S^4 = \frac{1}{3-1} - \frac{1}{3-1} \times \left(\frac{900}{5525}\right)^3 + \left(\frac{900}{5525}\right)^3 \times 0.4978 = 0.50$$

$$S^5 = \frac{1}{3-1} - \frac{1}{3-1} \times \left(\frac{400}{5525}\right)^3 + \left(\frac{400}{5525}\right)^3 \times 0.4999 = 0.50$$

$$S^6 = \frac{1}{3-1} - \frac{1}{3-1} \times \left(\frac{550}{5525}\right)^3 + \left(\frac{550}{5525}\right)^3 \times 0.4995 = 0.50$$

$$S^7 = \frac{1}{3-1} - \frac{1}{3-1} \times \left(\frac{550}{5525}\right)^3 + \left(\frac{550}{5525}\right)^3 \times 0.4995 = 0.50$$

$$S^8 = \frac{1}{3-1} - \frac{1}{3-1} \times \left(\frac{225}{5525}\right)^3 + \left(\frac{225}{5525}\right)^3 \times 0.50 = 0.50$$

$$S^9 = \frac{1}{3-1} - \frac{1}{3-1} \times \left(\frac{400}{5525}\right)^3 + \left(\frac{400}{5525}\right)^3 \times 0.4998 = 0.50$$

$$S^{10} = \frac{1}{3-1} - \frac{1}{3-1} \times \left(\frac{325}{5525}\right)^3 + \left(\frac{325}{5525}\right)^3 \times 0.4999 = 0.50$$

$$S^{11} = \frac{1}{3-1} - \frac{1}{3-1} \times \left(\frac{175}{5525}\right)^3 + \left(\frac{175}{5525}\right)^3 \times 0.50 = 0.50$$

$$S^{12} = \frac{1}{3-1} - \frac{1}{3-1} \times \left(\frac{175}{5525}\right)^3 + \left(\frac{175}{5525}\right)^3 \times 0.50 = 0.50$$

$$\sum_{1 \le j \le N} S^j = S^1 + S^2 + \cdots + S^{12} = 0 + 0.5 + \cdots + 0.5 = 5.50$$

$$\sum_{1 \le j1 < j2 \le N} S^{j1}S^{j2} = S^1S^2 + S^1S^3 + \cdots + S^{11}S^{12} = 13.7498$$

$$\sum_{1 \le j1 < j2 < j3 \le N} S^{j1}S^{j2}S^{j3} = S^1S^2S^3 + S^1S^2S^4 + \cdots + S^{10}S^{11}S^{12} = 20.6245$$

$$\sum_{1 \leq j1 < j2 < j3 < j4 \leq N} S^{j1}S^{j2}S^{j3}S^{j4} = S^1S^2S^3S^4 + S^1S^2S^3S^5 + \cdots + S^9S^{10}S^{11}S^{12} = 20.6244$$

$$\sum_{1 \leq j1 < j2 \ldots < j5 \leq N} S^{j1}S^{j2}\cdots S^{j5} = S^1S^2S^3S^4S^5 + S^1S^2S^3S^4S^6 + \cdots + S^8S^9S^{10}S^{11}S^{12} = 14.4369$$

$$\sum_{1 \leq j1 < j2 \ldots < j6 \leq N} S^{j1}S^{j2}\cdots S^{j6} = S^1S^2S^3S^4S^5S^6 + S^1S^2S^3S^4S^5S^7 + \cdots + S^7S^8S^9S^{10}S^{11}S^{12}$$

$$= 7.2184$$

$$\sum_{1 \leq j1 < j2 \cdots < j7 \leq N} S^{j1}S^{j2}\cdots S^{j7} = S^1S^2S^3S^4S^5S^6S^7 + S^1S^2S^3S^4S^5S^6S^8 + \cdots + S^6S^7S^8S^9S^{10}S^{11}S^{12}$$

$$= 2.578$$

$$\sum_{1 \leq j1 < j2 \cdots < j8 \leq N} S^{j1}S^{j2}\cdots S^{j8} = S^1S^2S^3S^4S^5S^6S^7S^8 + \cdots + S^5S^6S^7S^8S^9S^{10}S^{11}S^{12} = 0.6445$$

$$\sum_{1 \leq j1 < j2 \cdots < j9 \leq N} S^{j1}S^{j2}\cdots S^{j9} = S^1S^2S^3S^4S^5S^6S^7S^8S^9 + \cdots + S^4S^5S^6S^7S^8S^9S^{10}S^{11}S^{12} = 0.1074$$

$$\sum_{1 \leq j1 < j2 \cdots < j10 \leq N} S^{j1}S^{j2}\cdots S^{j10} = S^1S^2S^3S^4S^5S^6S^7S^8S^9S^{10} + \cdots + S^3S^4S^5S^6S^7S^8S^9S^{10}S^{11}S^{12}$$

$$= 0.0107$$

$$\sum_{1 \leq j1 < j2 \ldots < j11 \leq N} S^{j1}S^{j2}\cdots S^{j11} = S^1S^2S^3S^4S^5S^6S^7S^8S^9S^{10}S^{11} + \cdots + S^2S^3S^4S^5S^6S^7S^8S^9S^{10}S^{11}S^{12}$$

$$= 0.0005$$

$$\sum_{1 \leq j1 < j2 \cdots < j12 \leq N} S^{j1}S^{j2}\cdots S^{j12} = S^1S^2S^3S^4S^5S^6S^7S^8S^9S^{10}S^{11}S^{12} = 0.00$$

Therefore,

$$S_N = S^{1 \cup 2 \cup \cdots \cup 12} = \sum_{1 \leq j \leq 12} S^j + (1-m)\sum_{1 \leq j1 < j2 \leq 12} S^{j1}S^{j2} + (1-m)^2 \sum_{1 \leq j1 < j2 < j3 \leq 12} S^{j1}S^{j2}S^{j3}$$

$$+ (1-m)^{11} \sum_{1 \leq j1 < j2 < \ldots < jN \leq 12} S^{j1}S^{j2}\cdots S^{j12}$$

$$= 5.50 + (1-3) \times 13.7498 + \cdots + (1-3)^{11} \times 0 = 0.4996$$

## Example 14.4

Consider a layout as shown in Figure 14.6. For the layout, indicate the number of loops and compute the redundancy measure derived from local redundancy and from global redundancy. Show redundancy at each node and redundancy among nodes. Take account of transmissivity, path parameter, and age factor. Assume the friction factor as 140. The demand at each node is 100 (m³/h).

### Solution

Transmissivity $t_{jk}$ from node $k$ to node $j$ is taken into account when computing the global redundancy at node $j$, $S_j$. The overall redundancy $S_N$ and global redundancy are now computed.

(1) $\quad S_j = \frac{1}{m-1} \left\{ \sum_{i=1}^{n(j)} \left[ \frac{q_{ij}}{Q_j} - \left( \frac{q_{ij}}{Q_j} \right)^m \right] \right\}$

$$S_1 = \frac{1}{m-1} \left\{ \sum_{i=1}^{n(j)} \left[ \frac{q_{ij}}{Q_j} - \left( \frac{q_{ij}}{Q_j} \right)^3 \right] \right\}$$

$$= \frac{1}{3-1} \times \left[ \left( \frac{250}{300} \right) - \left( \frac{250}{300} \right)^3 + \left( \frac{50}{300} \right) - \left( \frac{50}{300} \right)^3 \right] = 0.21$$

$$S_2 = \frac{1}{m-1} \left\{ \sum_{i=1}^{n(j)} \left[ \frac{q_{ij}}{Q_j} - \left( \frac{q_{ij}}{Q_j} \right)^3 \right] \right\} = \frac{1}{3-1} \times \left[ \left( \frac{200}{200} \right) - \left( \frac{200}{200} \right)^3 \right] = 0.00$$

$$S_3 = \frac{1}{m-1} \left\{ \sum_{i=1}^{n(j)} \left[ \frac{q_{ij}}{Q_j} - \left( \frac{q_{ij}}{Q_j} \right)^3 \right] \right\} = \frac{1}{3-1} \times \left[ \left( \frac{100}{100} \right) - \left( \frac{100}{100} \right)^3 \right] = 0.00$$

(2) $\quad S_{j^*} = \frac{1}{m-1} \left[ 1 - \sum_{i=1}^{n(j)} \left( \frac{q_{ij}}{Q_0} \right)^3 \right]$

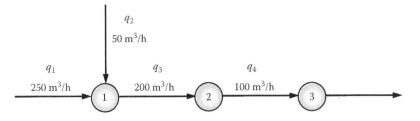

**FIGURE 14.6**   A water distribution layout.

$$S_{1^*} = \frac{1}{3-1} \times \left[ 1 - \left( \frac{250}{300} \right)^3 - \left( \frac{50}{300} \right)^3 \right] = 0.21$$

$$S_{2^*} = \frac{1}{3-1} \times \left[ 1 - \left( \frac{200}{300} \right)^3 \right] = 0.35$$

$$S_{3^*} = \frac{1}{3-1} \times \left[ 1 - \left( \frac{100}{300} \right)^3 \right] = 0.48$$

Determine $S_N$

$$S_N = S^{1 \cup 2 \cup 3} = \sum_{1 \le j \le n} S^j + (1-m) \sum_{1 \le j1 < j2 \le n} S^{j1} S^{j2} + (1-m)^2 \sum_{1 \le j1 < j2 < j3 \le n} S^{j1} S^{j2} S^{j3}$$

Determine $S_j$

$$\sum_{1 \le j \le N} S^j = \frac{N}{m-1} - \frac{1}{m-1} \sum_{j=1}^{N} \left( \frac{Q_j}{Q_0} \right)^m + \sum_{j=1}^{N} \left( \frac{Q_j}{Q_0} \right)^m S_{j^*}$$

$$S^{j1} = \frac{1}{3-1} - \frac{1}{3-1} \times \left( \frac{300}{300} \right)^3 + \left( \frac{300}{300} \right)^3 \times 0.21 = 0.21$$

$$S^{j2} = \frac{1}{3-1} - \frac{1}{3-1} \times \left( \frac{200}{300} \right)^3 + \left( \frac{200}{300} \right)^3 \times 0.35 = 0.46$$

$$S^{j3} = \frac{1}{3-1} - \frac{1}{3-1} \times \left( \frac{100}{300} \right)^3 + \left( \frac{100}{300} \right)^3 \times 0.48 = 0.50$$

$$\sum_{1 \le j \le N} S^j = S^{j1} + S^{j2} + S^{j3} = 0.21 + 0.46 + 0.50 = 1.16$$

Determine $\sum_{1 \le j1 < j2 \le n} S^{j1} S^{j2}$.

$$S^{j1} S^{j2} = 0.21 \times 0.46 = 0.10$$

$$S^{j1} S^{j3} = 0.21 \times 0.50 = 0.10$$

$$S^{j2} S^{j3} = 0.46 \times 0.50 = 0.23$$

$$\sum_{1 \leq j1 < j2 \leq n} S^{j1}S^{j2} = S^{j1}S^{j2} + S^{j1}S^{j3} + S^{j2}S^{j3} = 0.1 + 0.1 + 0.23 = 0.43$$

Determine $\sum_{1 \leq j1 < j2 < j3 \leq n} S^{j1}S^{j2}S^{j3}$.

$$S^{j1}S^{j2}S^{j3} = 0.21 \times 0.46 \times 0.5 = 0.05$$

$$\sum_{1 \leq j1 < j2 < j3 \leq n} S^{j1}S^{j2}S^{j3} = S^{j1}S^{j2}S^{j3} = 0.05$$

Therefore,

$$S_N = S^{1 \cup 2 \cup j} = \sum_{1 \leq j \leq n} S^j + (1-m) \sum_{1 \leq j1 < j2 \leq n} S^{j1}S^{j2} + (1-m)^2 \sum_{1 \leq j1 < j2 < j3 \leq n} S^{j1}S^{j2}S^{j3}$$

$$= 1.16 + (1-3) \times 0.43 + (1-3)^2 \times 0.05 = 0.50$$

Transmission of redundancy

(1) $$S_j = \frac{1}{m-1} \left\{ \sum_{j=1}^{n(j)} \left[ \frac{q_{ij}}{Q_j} - \left( \frac{q_{ij}}{Q_j} \right)^m \right] \right\}$$

$$S_1 = \frac{1}{m-1} \left\{ \sum_{j=1}^{n(j)} \left[ \frac{q_{ij}}{Q_j} - \left( \frac{q_{ij}}{Q_j} \right)^m \right] \right\}$$

$$= \frac{1}{3-1} \times \left\{ \left[ \left( \frac{250}{300} \right) - \left( \frac{225}{300} \right)^3 \right] + \left[ \left( \frac{50}{300} \right) - \left( \frac{50}{300} \right)^3 \right] \right\} = 0.21$$

$$S_2 = \frac{1}{m-1} \left\{ \sum_{j=1}^{n(j)} \left[ \frac{q_{ij}}{Q_j} - \left( \frac{q_{ij}}{Q_j} \right)^m \right] \right\} = \frac{1}{3-1} \times \left[ \left( \frac{200}{200} \right) - \left( \frac{200}{200} \right)^3 \right] = 0.00$$

$$S_3 = \frac{1}{m-1} \left\{ \sum_{j=1}^{n(j)} \left[ \frac{q_{ij}}{Q_j} - \left( \frac{q_{ij}}{Q_j} \right)^m \right] \right\} = \frac{1}{3-1} \times \left[ \left( \frac{100}{100} \right) - \left( \frac{100}{100} \right)^3 \right] = 0.00$$

(2) $$S_{j*} = \frac{1}{m-1} \left[ 1 - \sum_{j=1}^{n(j)} \left( \frac{q_{ij}}{Q_j} \right)^m \right]$$

$$S_{1*} = \frac{1}{3-1} \times \left\{ 1 - \left[ \left( \frac{250}{300} \right)^3 + \left( \frac{50}{300} \right)^3 \right] \right\} = 0.21$$

$$S_{2*} = \frac{1}{3-1} \times \left[ 1 - \left( \frac{200}{300} \right)^3 \right] = 0.35$$

$$S_{3*} = \frac{1}{3-1} \times \left[ 1 - \left( \frac{100}{300} \right)^3 \right] = 0.48$$

(3) Transmissivity

$$t_{21} = \frac{q_3}{q_1 + q_2} = \frac{200}{250 + 50} = \frac{2}{3} = 0.67$$

$$t_{32} = \frac{q_4}{q_3} = \frac{100}{200} = 0.5$$

(4) Determine $S'_{j*}$

Node 1: $S'_{1*} = S_{1*} = 0.21$

Node 2: $S'_{2*} = S_{2*} + t_{21}S'_{1*} = 0.35 + 0.67 \times 0.21 = 0.49$

Node 3: $S'_{3*} = S_{3*} + t_{32}S'_{2*} = 0.48 + 0.50 \times 0.49 = 0.73$

(5) Determine $S_N$

$$S_N = S^{1 \cup 2 \cup 3} = \sum_{1 \leq j \leq n} S^j + (1-m) \sum_{1 \leq j1 < j2 \leq n} S^{j1}S^{j2} + (1-m)^2 \sum_{1 \leq j1 < j2 < j3 \leq n} S^{j1}S^{j2}S^{j3}$$

Determine $\displaystyle\sum_{1 \leq j \leq N} S_j = \frac{N}{m-1} - \frac{1}{m-1} \left[ \sum_{j=1}^{N} \left( \frac{Q_j}{Q_0} \right)^m + \sum_{j=1}^{N} \left( \frac{Q_j}{Q_0} \right)^m S'_{j*} \right].$

$$S^{j1} = \frac{1}{3-1} - \frac{1}{3-1} \times \left( \frac{300}{300} \right) + \left( \frac{300}{300} \right)^3 \times 0.21 = 0.21$$

$$S^{j2} = \frac{1}{3-1} - \frac{1}{3-1} \times \left( \frac{200}{300} \right) + \left( \frac{200}{300} \right)^3 \times 0.49 = 0.50$$

$$S^{j3} = \frac{1}{3-1} - \frac{1}{3-1} \times \left( \frac{100}{300} \right) + \left( \frac{100}{300} \right)^3 \times 0.73 = 0.51$$

$$\sum_{1 \leq j \leq N} S^j = S^{j1} + S^{j2} + S^{j3} = 0.21 + 0.50 + 0.51 = 1.21$$

Determine $\displaystyle\sum_{1\le j1<j2\le n} S^{j1}S^{j2}$.

$$S^{j1}S^{j2} = 0.21\times 0.50 = 0.10$$

$$S^{j1}S^{j3} = 0.21\times 0.51 = 0.11$$

$$S^{j2}S^{j3} = 0.50\times 0.51 = 0.25$$

$$\sum_{1\le j1<j2\le n} S^{j1}S^{j2} = S^{j1}S^{j2}+S^{j1}S^{j3}+S^{j2}S^{j3} = 0.10+0.11+0.25 = 0.46$$

Determine $\displaystyle\sum_{1\le j1<j2\le n} S^{j1}S^{j2}S^{j3}$.

$$S^{j1}S^{j2}S^{j3} = 0.21\times 0.50\times 0.51 = 0.05$$

$$\sum_{1\le j1<j2<j3\le n} S^{j1}S^{j2}S^{j3} = S^{j1}S^{j2}S^{j3} = 0.05$$

Therefore,

$$S_N = S^{1\cup 2\cup 3} = \sum_{1\le j\le N} S^j + (1-m)\sum_{1\le j1<j2\le N} S^{j1}S^{j2} + (1-m)^2 \sum_{1\le j1<j2<j3\le N} S^{j1}S^{j2}S^{j3}$$

$$= 1.21 + (1-3)\times 0.46 + (1-3)^2\times 0.05 = 0.50$$

Path parameter of redundancy is as follows:

(1) Now compute path parameter using $a_{ij} = n_{d_{ij}}\left[1 - \dfrac{\displaystyle\sum_{l=1}^{M_{ij}} D_l}{\displaystyle\sum_{l=1}^{M_{ij}} d_l}\right]$

$$a_1 = a_{s11} + a_{s21} = 1\left[1-\frac{0}{1}\right] + 1\left[1-\frac{0}{1}\right] = 1+1 = 2 \quad \text{(Path: S1-1, S2-1)}$$

$$a_2 = a_{12} = 2\left[1-\frac{1}{4}\right] = 1.50 \quad \text{(Path: S1-1-2, S2-1-2)}$$

$$a_3 = a_{23} = 2\left[1-\frac{2}{6}\right] = 1.33 \quad \text{(Path: S1-1-2-3, S2-1-2-3)}$$

(2)
$$S_j = \frac{1}{m-1}\left[1 - \sum_{j=1}^{n(j)}\left(\frac{q_{ij}}{a_{ij}Q_j}\right)^m\right]$$

$$S_1 = \frac{1}{3-1} \times \left[1 - \left(\frac{250}{1\times300}\right)^3 - \left(\frac{50}{1\times300}\right)^3\right] = 0.21$$

$$S_2 = \frac{1}{3-1} \times \left[1 - \left(\frac{200}{1.5\times200}\right)^3\right] = 0.35$$

$$S_3 = \frac{1}{3-1} \times \left[1 - \left(\frac{100}{1.33\times100}\right)^3\right] = 0.29$$

(3)
$$S_{j^*} = \frac{1}{m-1}\left[1 - \sum_{j=1}^{n(j)}\left(\frac{q_{ij}}{a_{ij}Q_0}\right)^m\right]$$

$$S_{1^*} = \frac{1}{3-1} \times \left[1 - \left(\frac{250}{1\times300}\right)^3 + \left(\frac{50}{1\times300}\right)^3\right] = 0.21$$

$$S_{2^*} = \frac{1}{3-1} \times \left[1 - \left(\frac{200}{1.5\times300}\right)^3\right] = 0.46$$

$$S_{3^*} = \frac{1}{3-1} \times \left[1 - \left(\frac{100}{1.33\times100}\right)^3\right\} = 0.49$$

(4) Determine $S_N$

$$S_N = S^{1\cup2\cup3} = \sum_{1\le j\le N} S^j + (1-m)\sum_{1\le j1<j2\le N} S^{j1}S^{j2} + (1-m)^2 \sum_{1\le j1<j2<j3\le N} S^{j1}S^{j2}S^{j3}$$

Now determine

$$\sum_{1\le j\le N} S^j = \frac{N}{m-1} - \frac{1}{m-1}\sum_{j=1}^{N}\left(\frac{Q_j}{Q_0}\right)^m + \sum_{j=1}^{N}\left(\frac{Q_j}{Q_0}\right)^m S_{j^*}$$

$$S^{j1} = \frac{1}{3-1} - \frac{1}{3-1} \times \left(\frac{300}{300}\right) + \left(\frac{300}{300}\right)^3 \times 0.21 = 0.21$$

$$S^{j2} = \frac{1}{3-1} - \frac{1}{3-1} \times \left(\frac{200}{300}\right) + \left(\frac{200}{300}\right)^3 \times 0.46 = 0.49$$

$$S^{j3} = \frac{1}{3-1} - \frac{1}{3-1} \times \left(\frac{100}{300}\right) + \left(\frac{100}{300}\right)^3 \times 0.49 = 0.50$$

$$\sum_{1 \le j \le N} S^j = S^{j1} + S^{j2} + S^{j3} = 0.21 + 0.49 + 0.50 = 1.20$$

Determine $\displaystyle\sum_{1 \le j1 < j2 \le n} S^{j1} S^{j2}$.

$$S^{j1} S^{j2} = 0.21 \times 0.49 = 0.10$$

$$S^{j1} S^{j3} = 0.21 \times 0.50 = 0.10$$

$$S^{j2} S^{j3} = 0.49 \times 0.50 = 0.24$$

$$\sum_{1 \le j1 < j2 \le N} S^{j1} S^{j2} = S^{j1} S^{j2} + S^{j1} S^{j3} + S^{j2} S^{j3} = 0.10 + 0.10 + 0.24 = 0.45$$

Determine $\displaystyle\sum_{1 \le j1 < j2 \le N} S^{j1} S^{j2} S^{j3}$.

$$S^{j1} S^{j2} S^{j3} = 0.21 \times 0.49 \times 0.50 = 0.05$$

$$\sum_{1 \le j1 < j2 < j3 \le N} S^{j1} S^{j2} S^{j3} = S^{j1} S^{j2} S^{j3} = 0.05$$

Therefore,

$$S_N = S^{1 \cup 2 \cup 3} = \sum_{1 \le j \le N} S^j + (1-m)N \sum_{1 \le j1 < j2 \le N} S^{j1} S^{j2} + (1-m)^2 \sum_{1 \le j1 < j2 < j3 \le N} S^{j1} S^{j2} S^{j3}$$

$$= 1.20 + (1-3) \times 0.45 + (1-3)^2 \times 0.05 = 0.50$$

Age factor and path redundancy are as follows:

Now consider age factor, since the friction factor for all links is 140, then ln 140 = 4.9416. Then,

$$u_{ij} = 0.2 \ln C_{ij}(t) = 0.9883$$

(1) Compute path parameter using $a_{ij} = n_{d_{ij}} \left[ 1 - \dfrac{\sum_{I=1}^{M_{ij}} D_I}{\sum_{I=1}^{M_{ij}} d_I} \right]$

$a_1 = a_{s_11} + a_{s_21} = 1 \times \left[ 1 - \dfrac{0}{1} \right] + 1 \times \left[ 1 - \dfrac{0}{1} \right] = 1 + 1 = 2$   (Path: S1-1, S2-1)

$a_2 = a_{12} = 2 \times \left[ 1 - \dfrac{1}{4} \right] = 1.5$   (Path: S1-1-2, S2-1-2)

$a_3 = a_{23} = 2 \times \left[ 1 - \dfrac{2}{6} \right] = 1.33$   (Path: S1-1-2-3, S2-1-2-3)

(2)   $S_j = \dfrac{1}{m-1} \left[ 1 - \displaystyle\sum_{i=1}^{n(j)} u_{ij} \left( \dfrac{q_{ij}}{a_{ij} Q_j} \right)^3 \right]$

$S_1 = \dfrac{1}{3-1} \times \left[ 1 - 0.9883 \times \left( \dfrac{250}{1 \times 300} \right)^3 - 0.9883 \times \left( \dfrac{50}{1 \times 300} \right)^3 \right] = 0.21$

$S_2 = \dfrac{1}{3-1} \times \left[ 1 - 0.9883 \times \left( \dfrac{200}{1.5 \times 200} \right)^3 \right] = 0.35$

$S_3 = \dfrac{1}{3-1} \times \left[ 1 - 0.9883 \times \left( \dfrac{100}{1.33 \times 100} \right)^3 \right] = 0.29$

(3)   $\tilde{S}_j = \dfrac{1}{m-1} \left[ 1 - \left( \dfrac{Q_j}{Q_0} \right)^m \displaystyle\sum_{i=1}^{n(j)} u_{ij} \left( \dfrac{q_{ij}}{a_{ij} Q_j} \right)^3 \right]$

$\tilde{S}_1 = \dfrac{1}{3-1} \times \left[ 1 - \left( \dfrac{300}{300} \right)^3 \left\{ (0.9883 \times \left( \dfrac{250}{1 \times 300} \right)^3 + 0.9883 \times \left( \dfrac{50}{1 \times 300} \right)^3 \right\} \right] = 0.421$

$\tilde{S}_2 = \dfrac{1}{3-1} \times \left[ 1 - \left( \dfrac{200}{300} \right)^3 (0.9883 \times \left( \dfrac{200}{1.5 \times 200} \right)^3 \right] = 0.46$

$\tilde{S}_3 = \dfrac{1}{3-1} \times \left[ 1 - \left( \dfrac{100}{300} \right)^3 (0.9883 \times \left( \dfrac{100}{1.33 \times 100} \right)^3 \right] = 0.49$

(4) Determine $S_N$

$$S_N = S^{1\cup2\cup3} = \sum_{1\leq j\leq N} S^j + (1-m) \sum_{1\leq j1<j2\leq N} S^{j1}S^{j2} + (1-m)^2 \sum_{1\leq j1<j2<j3\leq N} S^{j1}S^{j2}S^{j3}$$

Now determine

$$\sum_{1\leq j\leq N} S^j = \frac{N}{m-1} - \frac{1}{m-1} \sum_{j=1}^{N}\left(\frac{Q_j}{Q_0}\right)^m + \sum_{j=1}^{N}\left(\frac{Q_j}{Q_0}\right)^m S_j;$$

$$S^{j1} = \frac{1}{3-1} - \frac{1}{3-1} \times \left(\frac{300}{300}\right) + \left(\frac{300}{300}\right)^3 \times 0.21 = 0.21$$

$$S^{j2} = \frac{1}{3-1} - \frac{1}{3-1} \times \left(\frac{200}{300}\right) + \left(\frac{200}{300}\right)^3 \times 0.46 = 0.49$$

$$S^{j3} = \frac{1}{3-1} - \frac{1}{3-1} \times \left(\frac{200}{300}\right) + \left(\frac{200}{300}\right)^3 \times 0.46 = 0.50$$

$$\sum_{1\leq j\leq n} S^j = S^{j1} + S^{j2} + S^{j3} = 0.21 + 0.49 + 0.50 = 1.20$$

Determine $\sum_{1\leq j1<j2\leq n} S^{j1}S^{j2}$.

$$S^{j1}S^{j2} = 0.21 \times 0.49 = 0.10$$

$$S^{j1}S^{j3} = 0.21 \times 0.50 = 0.11$$

$$S^{j2}S^{j3} = 0.49 \times 0.50 = 0.24$$

$$\sum_{1\leq j1<j2\leq n} S^{j1}S^{j2} = S^{j1}S^{j2} + S^{j1}S^{j3} + S^{j2}S^{j3} = 0.10 + 0.11 + 0.24 = 0.45$$

Determine $\sum_{1\leq j1<j2<j3\leq n} S^{j1}S^{j2}S^{j3}$.

$$S^{j1}S^{j2}S^{j3} = 0.21 \times 0.49 \times 0.50 = 0.05$$

$$\sum_{1 \le j1 < j2 < j3 \le n} S^{j1}S^{j2}S^{j3} = S^{j1}S^{j2}S^{j3} = 0.05$$

Therefore,

$$S_N = S^{1 \cup 2 \cup 3} = \sum_{1 \le j \le n} S^{j} + (1-m) \sum_{1 \le j1 < j2 \le n} S^{j1}S^{j2} + (1-m)^2 \sum_{1 \le j1 < j2 < j3 \le n} S^{j1}S^{j2}S^{j3}$$

$$= 1.20 + (1-3) \times 0.45 + (1-3)^2 \times 0.05 = 0.50$$

## 14.12 RELATION BETWEEN REDUNDANCY AND RELIABILITY

Awumah (1990) computed parameters of Nodal Pair Reliability (NPR) and Percentage of Demand Supplied at adequate Pressure (PSPF) for a number of layouts, and compared them with entropy-based redundancy measures. Quimpo and Shamsi (1991) and Wagner et al. (1988a, b) have used NPR to calculate the probability that the source node and each of demand nodes are connected. The resilience of water distribution systems can be assessed by the PSPF and hence one can make a statement of hydraulic redundancy. The relation between redundancy and NPR reliability can be expressed, using the data from Awumah (1990), as

$$NPR = 0.770 \exp(0.0418S) \tag{14.70}$$

where $S$ is the network redundancy. Equation 14.70 has a coefficient of determination of 0.984, explaining more than 96% of the variability. This suggests that with the knowledge of entropy or redundancy, the water distribution network reliability can be determined.

In a similar manner, the relation between PSPF and network redundancy can be expressed as

$$PSPF = 48.282 \exp(0.26S) \tag{14.71}$$

Equation 14.71 has a coefficient of determination of 0.983 and explains more than 96% variability. Since $S$ is a common parameter between Equations 14.70 and 14.71, it is clear that NPR and PSPF should be strongly related, suggesting that if one type of reliability is known then the other type can be determined. The relation between NPR and PSPF can now be written as

$$PSPF = 0.1392 \exp(7.6046 NPR) \tag{14.72}$$

Equation 14.72 has a coefficient of determination of 0.985 and explains more than 97% variability. Thus, water distribution systems can be designed using either of these two types of reliability.

## REFERENCES

Awumah, K. (1990). Entropy based measures in water distribution network design. Unpublished Ph.D. thesis, University of Manitoba, Winnipeg, Manitoba, Canada.

Awumah, K., Goulter, I., and Bhatt, S.K. (1990). Assessment of reliability in water distribution networks using entropy based measures. *Stochastic Hydrology and Hydraulics*, 4(4), 309–320, doi:10.1007/BF01544084.

Awumah, K., Goulter, I., and Bhatt, S.K. (1991). Entropy-based redundancy measures in water distribution networks. *Journal of Hydraulic Engineering*, ASCE, 117(5), 595–614, doi:10.1061/(ASCE)0733-9429(1991)117:5(595).

Beim, K. and Hobbs, B. (1988). Analytical simulation of water system capacity reliability, 2. A Markov chain approach and verification of the models. *Water Resources Research*, 24(9), 1445–1458, doi:10.1029/WR024i009p01445.

Chen, G.J. and Templeman, A.B. (1995). On entropy-based methods for nonlinear programming problems. *Engineering Optimization*, 23(3), 225–238, doi:10.1080/03052159508941355.

Eiger, G., Shamir, U., and Ben-Tal, A. (1994). Optimal design of water distribution networks. *Water Resources Research*, 30(9), 2637–2646, doi:10.1029/94WR00623.

Ekinci, O. and Konak, H. (2009). An optimization strategy for water distribution networks. *Water Resources Management*, 23(1), 169–185, doi:10.1007/s11269-008-9270-8.

Fonseca, C.M. and Fleming, P.J. (1996). On the performance assessment and comparison of stochastic multiobjective optimizers. In: *Parallel Problem Solving from Nature-PPSN IV*, Springer-Verlag Lecture Notes on Computer Science No. 1141, ed. H.-M. Voigt. Springer-Verlag, London, U.K., pp. 584–593.

Germanopoulos, G., Jowitt, P.W., and Lumbers, J.P. (1986). Assessing the reliability of supply and level of service for water distribution systems. *Proceedings of the Institute of Civil Engineers, Part I*, 80(2), 413–428, doi:10.1680/iicep.1986.741.

Giustolisi, O., Laucelli, D., and Colombo, A.F. (2009). Deterministic versus stochastic design of water distribution networks. *Journal of Water Resources Planning and Management*, 135(2), 117–127, doi:10.1061/(ASCE)0733-9496(2009)135:2(117).

Goulter, I.C. (1987). Current and future use of systems analysis in water distribution network design. *Civil Engineering Systems*, 4(4), 175–184, doi:10.1080/02630258708970484.

Goulter, I.C. (1988). Measures of inherent redundancy in water distribution network layouts. *Journal of Information Optimization Science*, 9(3), 363–390.

Goulter, I.C. (1992). Assessing the reliability of water distribution networks using entropy based measures of network redundancy. In: *Entropy and Energy Dissipation in Water Resources*, eds. V.P. Singh and M. Fiorentino. Kluwer Academic Publishers, Dordrecht, the Netherlands, pp. 217–238.

Goulter, I.C. and Bouchart, F. (1990). Reliability constrained pipe network model. *Journal of Hydraulic Engineering*, 116(2), 211–229, doi:10.1061/(ASCE)0733-9429(1990)116:2(211).

Goulter, I. and Coals, A.V. (1986). Quantitative approaches to reliability assessment in pipe networks. *Journal of Transportation Engineering*, ASCE, 112(3), 287–301, doi:10.1061/(ASCE)0733-947X(1986)112:3(287).

Hobbs, B. and Beim, G.K. (1988). Analytical simulation of water supply capacity reliability: 1. Modified frequency duration analysis. *Water Resources Research*, 24(9), 1431–1444, doi:10.1029/WR024i009p01431.

Kullback, S. and Leibler, R.A. (1951). On information and sufficiency. *Annals of Mathematical Statistics*, 22(1), 79–86.

Kwon, H.J. and Lee, C.E. (2008). Reliability analysis of pipe network regarding transient flow. *KSCE Journal of Civil Engineering*, 12(6), 409–416, doi:10.1007/s12205-008-0409-1.

Lansey, K.E., Duan, N., Mays, L.W., and Tung, Y.-K. (1989). Water distribution system design under uncertainties. *Journal of Water Resources Planning and Management*, ASCE, 115(5), 630–645, doi:10.1061/(ASCE)0733-9496(1989)115:5(630).

Niven, R.K. (2004). The constrained entropy and cross-entropy functions. *Physica* A, 334, 444–458.

Quimpo, R.G. and Shamsi, U. (1991). Reliability-based distribution system maintenance. *Journal of Water Resources Planning and Management*, ASCE, 117(3), 321–339, doi:10.1061/(ASCE)0733-9496(1991)117:3(321).

Perelman, L. and Ostfeld, A. (2007). An adaptive heuristic cross entropy algorithm for optimal design of water distribution systems. *Engineering Optimization*, 39(4), 413–428, doi:10.1080/03052150601154671.

Perelman, L., Ostfeld, A. and Salomons, E. (2008). Cross entropy multiobjective optimization for water distribution systems design. *Water Resources Research*, 44, W09413, doi:10.1029/2007WR006248.

Rubinstein, R.Y. (1997). Optimization of computer simulation models with rare events. *European Journal of Operations Research*, 99(1), 89–112, doi:10.1016/S0377-2217(96)00385-2.

Shannon, C.E. (1948). The mathematical theory of communication, I and II. *Bell System Technical Journal*, 27, 379–423.

Shibu, A. and Janga Reddy, M. (2013). Cross entropy optimization for optimal design of water distribution networks. *International Journal of Computer Information Systems and Industrial Management Applications*, 5, 308–316.

Singh, V.P. and Oh, J. (2014). A Tsallis entropy-based redundancy measure for water distribution networks. *Physica A*, 421, 360–376.

Su, Y.C., Mays, L.W., Duan, N., and Lansey, K.E. (1987). Reliability-based optimization model for water distribution systems. *Journal of Hydraulic Engineering*, 114(12), 1539–1556, doi:10.1061/(ASCE)0733-9429(1987)113:12(1539).

Tanyimboh, T.T. and Templeman, A.B. (1993a). Maximum entropy flows for single-source networks. *Engineering Optimization*, 22(1), 49–63, doi:10.1080/03052159308941325.

Tanyimboh, T.T. and Templeman, A.B. (1993b). Calculating maximum entropy flows in networks. *Journal of Operations Research*, 44(4), 383–396.

Templeman, A.B. (1992). Entropy and civil engineering optimization. In: *Optimization and Artificial Intelligence in Civil and Structural Engineering*, Kluwer Academic Publishers, Dordrecht, the Netherlands, 1, 87–105.

Todini, E. (2000). Looped water distribution networks using a resilience index based heuristic approach. *Urban Water*, 2(2), 115–122, doi:10.1016/S1462-0758(00)00049-2.

Tsallis, C. (1988). Possible generalization of Boltzmann-Gibbs statistics. *Journal of Statistical Physics*, 52(1/2), 479–487, doi:10.1007/BF01016429.

Wagner, J.M., Shamir, U., and Marks, D.H. (1988a). Water distribution reliability: Analytical methods. *Journal of Water Resources Planning and Management*, 114(3), 253–275.

Wagner, J.M., Shamir, U., and Marks, D.H. (1988b). Water distribution reliability: Simulation methods. *Journal of Water Resources Planning and Management*, 114(3), 276–294.

Walters, G. (1988). Optimal design of pipe networks: A review. *Proceedings of First International Conferences on Computational Methods in Water Resources*, Marrakesh, Morocco, Vol. 2, Computational Mechanics Publications, Southampton, U.K., pp. 21–31.

Xu, C.-C. and Jowitt, P.W. (1992). Discussion of "Entropy-based redundancy measures in water distribution networks," by K. Awumah, I. Goulter, and S.K. Bhatt. *Journal of Hydraulic Engineering*, 118(7), 1064–1066, doi:10.1061/(A)0733-9429(1922)118:7(1064).

# Index